D0945589

Foundations for Microstrip Circuit Design

SECOND EDITION

Foundations for Microstrip Circuit Design

SECOND EDITION

TERRY EDWARDS
ENGALCO, Knaresborough, UK

JOHN WILEY & SONS
Chichester • New York • Brisbane • Toronto • Singapore

Other Wiley Editorial Offices

John Wiley & Sons, Inc., 605 Third Avenue,
New York, NY 10158-0012, USA

Jacaranda Wiley Ltd, G.P.O. Box 859, Brisbane,
Queensland 4001, Australia

John Wiley & Sons (Canada) Ltd, 5353 Dundas Road West,
Fourth Floor, Etobicoke, Ontario M9B 6H8, Canada

John Wiley & Sons (SEA) Pte Ltd, 37 Jalan Pemimpin 05-04,
Block B. Union Industrial Building, Singapore 2057

Library of Congress Cataloging-in-Publication Data:

Edwards, T. C. (Terence Charles)
 Foundations for microstrip circuit design / Terry Edwards. — 2nd ed.
 p. cm.
 Includes bibliographical references and index.
 ISBN 0 471 93062 8
 1. Microwave integrated circuits. 2. Strip transmission lines.
 I. Title
 TK7876.E38 1991
 621.381'32—dc20 91-13815
 CIP

A catalogue record for this book is available from the British Library

Typeset by Photo·graphics, Honiton, Devon
Printed in Great Britain by Biddles Ltd, Guildford

Contents

Preface

For a variety of techno-economic reasons microwave circuit integration has evolved mainly at the hybrid level, with microstrip transmission lines fulfilling a very wide range of planar circuit requirements. Although monolithic microwave integrated circuits (MMICs) are now commercially available these require microstrip implementation on chip and also for interconnections. MMICs cannot be expected to provide all (or even most) desired microwave and millimeter-wave functions and meet all capabilities. Thus it seems inevitable that the demand for microstrip circuits will continue to expand.

This design text is both a sequel and an update to the original well-received first edition of the same title originally published in 1981. The expanded text provides foundations for the accurate design of microstrip components and circuits applicable to microwave, millimeter-wave or high-speed digital subsystems.

The text is primarily intended for design engineers and research and development specialists who are active in these areas. It is also likely to prove useful to lecturers and students on advanced undergraduate and graduate electronics and communications engineering courses.

Whilst most of the techniques lead to expressions which may be evaluated using only a moderately sophisticated calculator, the main emphasis throughout is on techniques suitable for fast CAD/CAE using reasonably powerful desktop personal computers. The direction is strongly toward useful design formulae and approaches—a repeat coverage of well-documented theoretical analyses of the microstrip structure has been considered unnecessary and out of place here.

This work is partly based upon the following activities of the writer during the past decade: teaching new, graduate design-oriented microwave courses and conducting research involving microstrip transmission lines. The microwave courses were presented at La Trobe University (Melbourne, Australia) and the University of Bradford (UK) as part of Masters and related degrees, whilst most of the research was performed at the Royal Military College of Science (Shrivenham, UK).

The text is organized into ten chapters, leading through fundamental aspects of microstrip design (principally Chapters 3 and 4) on to circuit applications in Chapter 8 and later. Four Appendices are provided covering fundamental theory on parallel-coupled TEM-lines, Q-factor analysis, scattering parameters and an overview of a range of appropriate commercially available CAD/CAE software packages.

A basic review of TEM-mode transmission line theory is presented in Chapter 1. This is intended mainly for reference in connection with concepts and/or expressions required in many later chapters. It also serves as an appropriate introductory chapter.

Several structures which can conceivably be incorporated in microwave integrated circuits are described and critically compared in Chapter 2, which also deals with manufacturing processes. This chapter may be used as a source of initial information to assist in the choice of a particular technology other than microstrip.

A large number of significant items are discussed for what is believed to be the first time within a book. These include the following:

Proven-accurate expressions for microstrip analysis or synthesis, under static-TEM or dispersive conditions, and on an isotropic or sapphire substrate (Chapters 3 and 4)

A detailed appraisal of the frequency dependence of characteristic impedance and operating-frequency restrictions (Chapter 4)

Design information on many discontinuities, including short circuits and the transverse slit (Chapter 5)

Accurate design of parallel-coupled lines, special types of coupler, and crosstalk calculations (Chapter 6)

Details regarding several types of transitions and many measurement procedures (Chapter 7)

Most of the material and the design information in Chapters 8, 9 and 10.

High-speed digital applications are given in addition to microwave applications. Summaries of design recommendations are provided at the conclusion of appropriate chapters, viz. 3, 4, 5 and 6.

Acknowledgements

During the preparation of both editions of this text there were a number of people internationally who made significant contributions and thanks to whom the quality of the material was decidedly enhanced. Most of these were located at La Trobe University (Melbourne, Australia), the Royal Military College of Science (RMCS, Shrivenham, England) and the University of Bradford (England). I especially wish to thank several of my previous colleagues at RMCS. Particularly, at what is now known as the School of Electrical Engineering and Science, Past (Honorary) Professors Michael Potok, Chris Harris and (currently holding the Wolfson chair) Professor Jim James for their advice and stimulation during the preparation of the original manuscript. Dr Ann Henderson also read certain parts of the text for technical content and her comments and suggestions proved extremely helpful.

At La Trobe University my thanks are extended to the original holder of the Tad Szental Chair of Communication Engineering—the late Professor Daryl Hooper who exhibited such infectious drive and encouragement. Also to Dr Laurie Cahill for many deep and helpful ideas and discussions and for first indicating to me the microstrip field diagram concepts resulting in Figure 3.4 here. Both La Trobe and Bradford Universities were fortunate with the quality of their advanced electronics undergraduate and post-graduate student intakes who were responsible for many stimulating ideas which found their place in this text.

Dr Roger Owens, now with Thorn EMI Electronics (Computer Systems Division, Wells, England), was of great assistance in his critical reading of much of the original manuscript and also with helpful suggestions for this second edition. Roger first brought to my attention most of the theory presented in Appendix B.

It is beyond doubt that the final text is greatly enhanced by the supportive and most positive feedback provided by various referees and plan reviewers for both the original and second editions.

I wish especially to thank Mrs Julie Swann who meticulously typed (in that daunting era prior to the word-processor) the manuscript for the first

edition and Mrs Norma Chubb who word-processed parts of the second edition, notably the reference lists. Precise and faithful diagram production by several engineering drawing experts (particularly Mrs Sheila Greener at RMCS and, more recently, Steve Webb of York, England) helped immensely with this project.

The attention of the reader is drawn to those persons and organisations specifically acknowledged beneath appropriate Figures reprinted here. Finally I should like to thank the publishers, John Wiley & Sons Ltd (notably Ann-Marie Franks and Ian McIntosh at Chichester, England) for their strong support and highly professional backing leading to the production of this text.

Finally I wish to acknowledge here the wonderful patience of my wife Patricia (of less than one year) during the preparation of this text.

Terry Edwards
ENGALCO
Knaresborough
England

1991

1 A Basic Review of TEM-mode Transmission Line Theory

1.1 THE CONCEPT OF RADIO-FREQUENCY TRANSMISSION LINES

A review of important TEM-mode radio-frequency transmission line theory is essential here because microstrip lines can be treated, at least to a first approximation, by assuming a static-TEM situation. We shall see that many significant microstrip design calculations require TEM-mode results from transmission line theory.

When transmission line structures have dimensions which are substantially smaller than the wavelengths of the signals being transmitted, then the structures may be satisfactorily analysed on the basis of line voltages and currents. As the frequency goes higher, and therefore the wavelength becomes smaller, it becomes necessary to set up a complete electromagnetic field solution in order to analyse the line structures. Here we can often assume that small line dimensions prevail and therefore a number of useful results are obtained on a voltage and current basis. The theory thus developed is called distributed circuit theory. Some common TEM-mode line cross-sections which have been fully analysed and are well understood are shown in Figure 1.1.

The parallel line structure of Figure 1.1(a) is rarely used for carrying radio-frequency signals because, being unshielded, it radiates energy and is easily electrically disturbed by any nearby objects. Notice that both the parallel and the coaxial lines need only have dielectric supports at intervals along their lengths to maintain correct separation of the conductors. Furthermore, these supports may be filamentary, so that the 'dielectric filling' can consist almost entirely of air and the dielectric losses are low. On the other hand, the stripline inherently requires a solid, low-loss, dielectric filling.

Electric and magnetic field patterns associated with the TEM mode of propagation are well known, and the electric field is shown here in Figure 1.2 for a high-Q triplate stripline.

(a) Parallel, two-wire
line

(b) Coaxial
line

(c) 'Stripline'
(triplate line)

Fig. 1.1. Cross-sections of some common TEM-mode transmission lines

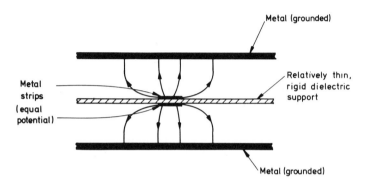

Fig. 1.2. High-Q triplate stripline, showing electric field

The remainder of this chapter is concerned with definitions and results from transmission line theory.

1.2 PRIMARY TRANSMISSION LINE CONSTANTS

In this chapter we are concerned only with uniform transmission lines. Regardless of the actual structure, such lines are shown schematically as in Figure 1.3(a). *Primary* constants are indicated in Figure 1.3(b) and can be defined as follows:

Resistance along the line = R
Inductance along the line = L all specified
Conductance shunting the line = G 'per loop metre'
Capacitance shunting the line = C

The 'per loop metre' specification means that each quantity is determined on a 'go and return' basis. For example, to find the total value of R per loop metre for a coaxial line we must add the resistance of one metre of the inner conductor to that of the outer conductor. In most radio-frequency transmission lines the effects due to L and C tend to dominate, because of the relatively high inductive reactance and capacitive susceptance respectively. In

(a) UNIFORM TRANSMISSION LINE (b) PRIMARY CONSTANTS

Fig. 1.3. The uniform transmission line: primary constants

such cases we refer to 'loss-free' lines, although in practice some information about R or G may be necessary to determine actual power losses. The 'loss-free' concept is just a useful and good approximation.

The propagation of a wave along the line is characterized by the (complex) *propagation coefficient* γ:

$$\gamma - \sqrt{(R + j\omega L)(G + j\omega C)} \tag{1.1}$$

or

$$\gamma = \alpha + j\beta \tag{1.2}$$

where

$\alpha = $ *attenuation coefficient*, Nepers per metre

$\beta = $ *phase-change coefficient*, degrees, or radians, per metre

At sufficiently high radio frequencies, eqns (1.1) and (1.2) yield the important result:

$$\beta = \omega\sqrt{LC} \tag{1.3}$$

1.3 SECONDARY CONSTANTS FOR TRANSMISSION LINES

In one complete wavelength λ_g along the line the travelling wave must experience 2π radians phase shift. Thus

$$\beta\lambda_g = 2\pi$$

or

$$\beta = \frac{2\pi}{\lambda_g} \tag{1.4}$$

(Hence β is also called the 'wave number'.)

From the relationship for the velocity $v_p = f\lambda_g$ we can also write

$$v_p = \frac{\omega}{\beta} \quad \text{m/s} \tag{1.5}$$

By using eqn (1.3) we also obtain

$$v_p = \frac{1}{\sqrt{LC}} \tag{1.6}$$

The velocity of propagation v_p is also given in terms of the absolute permeability μ and permittivity ϵ of the medium through which the wave passes:

$$v_p = \frac{1}{\sqrt{\mu\epsilon}} = \frac{c}{\sqrt{\mu_r\epsilon_r}} \tag{1.7}$$

where $c = 2.99793 \times 10^8$ m/s, the velocity of light in free space—often approximated to 3×10^8 m/s. In coaxial lines designed for microwave applications the velocity is very close to c because there is so little dielectric filling, only well-separated spacers. (This is *not* true for coaxial cables, whose foam filling reduces the velocity.)

Most lines do not possess any ferromagnetic materials and thus have $\mu_r = 1$. For stripline (Figure 1.1c) the filling consists of a uniform dielectric and eqn (1.7) then gives the velocity as

$$v_p = \frac{c}{\sqrt{\epsilon_r}} \tag{1.8}$$

We note that the wave is slowed by the dielectric medium.

By noting that $c = f\lambda_0$ and $v = f\lambda_g$ we can use eqn (1.8) to show that

$$\lambda_g = \frac{\lambda_0}{\sqrt{\epsilon_r}} \tag{1.9}$$

(The subscript 'g' means 'guide'; we can look upon the transmission line as a 'guiding' structure.) This expression, eqn (1.9), is particularly significant. It explicitly states that the wavelength is reduced according to the square root of the relative permittivity of the material in which the line is embedded. For example, a coaxial transmission line is filled with alumina ($\epsilon_r \simeq 9$) and excited at 10 GHz. The *free-space* wavelength would be 3 cm, but the wavelength in the coaxial line is only $3/\sqrt{9} = 1$ cm. Thus, distributed components become shorter and occupy less space, if high-permittivity materials can be used. This has important implications for microstrip.

The final secondary constant to be discussed is the characteristic impedance Z_0. This is given generally by

$$Z_0 = \sqrt{\frac{R + j\omega L}{G + j\omega C}}$$

At high radio frequencies this simplifies to

$$Z_0 = \sqrt{\frac{L}{C}} \tag{1.10}$$

which is a *purely resistive* result.

For coaxial lines two values are commonly encountered:

At v.h.f.: $Z_0 = 75 \ \Omega$ usually

At u.h.f. and s.h.f. (microwave frequencies):

$$Z_0 = 50 \ \Omega \text{ usually}$$

Changing Z_0 requires a change in the dimensions of the line. Specifically, in the case of coaxial lines, it requires changing the ratio of the radii applicable to the inner and outer conductors. Dimensional alterations are much more easily carried out with stripline or its derivatives and therefore such structures can be fabricated for a wide range of characteristic impedances (typically 20 to 150 Ω for stripline).

It is not often appreciated that eqn (1.10) can be re-expressed, with the aid of eqn (1.6), in the following alternative forms:

$$Z_0 = v_\text{p}L \tag{1.11a}$$

or

$$Z_0 = \frac{1}{v_\text{p}C} \tag{1.11b}$$

Equation (1.11b) is particularly useful in establishing some fundamental microstrip parameters.

1.4 TRANSMISSION LINE IMPEDANCES

The input impedance to a transmission line varies with the distance progressed along the line. Therefore a suitable distance notation is required, which is shown in Figure 1.4. Many standard textbooks derive the expressions given as eqns (1.12) to (1.17) here.

By setting down the distance limits into expressions for the voltage and current at *any* point along the line we obtain, for the load impedance Z_L, the following expression for the input impedance at $z = 0$:

$$Z_\text{in} = Z_0 \left(\frac{Z_\text{L} \cosh \gamma l + Z_0 \sinh \gamma l}{Z_0 \cosh \gamma l + Z_\text{L} \sinh \gamma l} \right) \tag{1.12}$$

Fig. 1.4. Notation applicable to a general, uniform line

This expression is completely general. For loss-free lines the result simplifies to

$$Z_{in} = Z_0 \left(\frac{Z_L + jZ_0 \tan \beta l}{Z_0 + jZ_L \tan \beta l} \right) \quad (1.13)$$

The input impedances for lines having short-circuit or open-circuit terminations (see Figure 1.5) follow from eqn (1.13), and these impedances can now be easily written down.

For the short-circuit case $Z_L = 0$ and eqn (1.13) gives

$$Z_{sc} = jZ_0 \tan \beta l \quad (1.14)$$

For the open-circuit case $Z_L \to \infty$ and eqn (1.13) gives

$$Z_{oc} = -jZ_0 \cot \beta l \quad (1.15)$$

Short-circuit or open-circuit terminated lines are very useful structures in a wide variety of circuits.

By choosing the correct lengths (l) inductive or capacitive circuit elements are automatically realized in transmission line form, as can be seen from eqns (1.14) and (1.15). There are many applications in resonators, filters, and matching and coupling networks.

When the line is terminated in a load impedance exactly equal to the characteristic impedance (Z_0) of the line itself it is perfectly broad-band

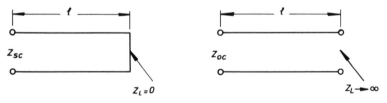

(a) SHORT-CIRCUIT TERMINATED LINE *(b) OPEN-CIRCUIT TERMINATED LINE*

Fig. 1.5. Short-circuit and open-circuit terminated lines

matched. The denominator is equal to the numerator on the right-hand parentheses of eqn (1.12) and hence

$$Z_{in} = Z_0$$

This completely matched condition is often used in test and measurement procedures.

1.5 REFLECTION AND VOLTAGE STANDING-WAVE RATIO

In all cases, except the completely matched condition, the load termination reflects some of the energy originally sent down the line. Interference between the incident and reflected waves, travelling at the same velocity but in opposite directions, causes a 'standing-wave' field pattern to be set up.

The voltage reflection coefficient Γ (the ratio of reflected to incident voltage at the load) is given by

$$\Gamma = \frac{Z_L - Z_0}{Z_L + Z_0} \qquad (1.16)$$

The ratio of maximum to minimum amplitude of the standing wave is called the voltage standing-wave ratio r (or 'vswr'), given by

$$r = \frac{1 + |\Gamma|}{1 - |\Gamma|} \qquad (1.17)$$

Notice that, in general, Γ may be complex but r is necessarily always real. For the matched condition, $\Gamma = 0$ and $r = 1$. The reflection coefficients and standing-wave ratios for short-circuit and open-circuit terminated conditions are left as an exercise for the reader.

1.6 HALF-, QUARTER-, AND EIGHTH-WAVELENGTH LINES

Transmission lines which are half of a wavelength long at some desired frequency of operation are often encountered. Various combinations of short-circuit or open-circuit terminations (both ends) may be used. Such lines then form 'half-wave resonators'. When both ends are short-circuit terminated, as in Figure 1.6, coupling to the line would have to be provided by some separate probe. The actual length of such a resonator may be extended to $n\lambda_g/2$, where n is any integer. It is again left as an exercise to examine the behaviour of open-circuit terminated resonators.

Lines which are a quarter of a wavelength long can provide useful properties. Clearly, for a short-circuit terminated quarter-wave line we obtain half, longitudinally, of the standing-wave pattern shown in Figure 1.6. At the input plane location P, the standing wave has a voltage maximum which corresponds to an *open*-circuit condition. Thus we effectively transform from the short circuit to an open circuit. This is consistent with an input impedance calculation using eqn (1.14), and substituting $\beta = 2\pi/\lambda_g$ yields

$$Z_{sc}\bigg|_{\lambda_g/4} = jZ_0 \tan (\pi/2)$$

$$= \infty$$

which is certainly the impedance of an open circuit. This feature can be used in many circuit designs to deliberately ensure that an open circuit appears in a desired location—at one frequency.

Eighth-wavelength lines, when short-circuit terminated, provide an input impedance which is just jZ_0 Ω. This can be convenient in some matching circuit designs, to give just one example.

1.7 SIMPLE (NARROW-BAND) MATCHING

In most practical situations a transmission line ultimately feeds some load which has an impedance differing considerably from the characteristic impedance, i.e. it is mismatched—possibly quite severely. Examples include feeds to solid-state amplifying devices such as transistors, and antenna feeders. By connecting a suitably designed network between the feed line and the load we can adjust impedances in order to obtain a good overall match. If this match is required to hold over a specified range of frequencies the network design may become quite complicated.

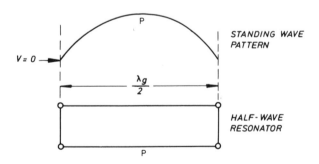

Fig. 1.6. Half-wavelength resonator

This section briefly reviews two simple methods for narrow-band matching: the single-shunt stub and the quarter-wave transformer.

A schematic representation of the single-shunt stub technique is shown in Figure 1.7. Although a short-circuit termination is indicated here an open-circuit termination may alternatively be used, provided that the termination quality is sufficiently good. In this arrangement all lines are shown with identical characteristic impedances. This may not always be so and some extra flexibility is afforded if the stub and main line impedances are of different values. The best method for 'hand calculating' the required lengths l_1 and l_2 to obtain the matched condition $Z_{in} = Z_0$ is to use a Smith chart. This is fully described in a number of standard texts and will not be outlined here.

For computer-aided design (CAD) suitable expressions are required, which are now stated for l_1 and l_2. The results are (for a *resistive* load):

$$l_1 = \frac{\lambda_g}{2\pi} \tan^{-1} \sqrt{\frac{Z_L}{Z_0}} \qquad (1.18)$$

and

$$l_2 = \frac{\lambda_g}{2\pi} \tan^{-1} \left(\frac{\sqrt{Z_L Z_0}}{Z_L - Z_0} \right) \qquad (1.19)$$

It is important to note that these results are satisfactory only when Z_L is real (resistive). If Z_L is complex the basic expressions must be reexamined in order to proceed with the analysis, which becomes quite lengthy. With the use of two, three, or sometimes more stubs the distance l_1 can be kept fixed and broader-band matching is achievable. Naturally, full analyses yielding lengths and separations become very unwieldy.

There is a simpler technique available for matching resistive loads: the quarter-wave transformer. In the single-section quarter-wavelength basic method this represents, again, a narrow-band approach. The basic arrangement is shown in Figure 1.8.

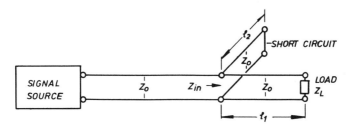

Fig. 1.7. Single, short-circuit terminated, shunt stub matching

Fig. 1.8. Quarter-wavelength transformer matching. (Note that the junction between the two lines is assumed to be perfect—zero length and zero discontinuity.)

We set $Z_L = R_L$ (resistive) and $l = n\lambda_g/4$ in eqn (1.13) to obtain

$$Z_{in} = \frac{Z_{0T}^2}{R_L}$$

whence the required characteristic impedance of the quarter-wave line is

$$Z_{0T}^2 = Z_0 R_L \qquad (1.20)$$

For example, to match a 50 Ω line into a 10 Ω resistive load the impedance required is

$$Z_{0T}^2 = 500$$

or

$$Z_{0T} = 22.4 \ \Omega$$

In stripline or microstrip circuits this would be relatively easy to implement.

A broader-band match is obtained if several quarter-wave sections are connected in cascade. The drawbacks are that space may be wasted on the circuit layout, the match may remain poor at the band edges, and only resistive loads are accommodated.

1.8 EQUIVALENT TWO-PORT NETWORKS

There are applications where it is useful to employ a lumped-parameter, equivalent, two-port network for a particular length of transmission line. One example is in the realization of filters consisting of cascaded lengths of transmission lines.

Any equivalent network must be constrained to have the same terminal characteristics as those of the length of transmission line it represents, and this will only hold at one single frequency.

Having said this, the behaviour may well be satisfactorily represented over at least a narrow band of frequencies. The propagation constant γ will usually be the most frequency-sensitive parameter. The general equivalent *T*-network is given in Figure 1.9(a) and the equivalent π-network is given in Figure 1.9(b).

Fig. 1.9. Equivalent two-port networks

Expressions for the various impedances, given next, are readily obtained by equating transmission line terminal voltages and currents with network terminal voltages and currents:

$$Z_a = Z_0 \tanh \frac{\gamma l}{2} \tag{1.21a}$$

$$Z_b = \frac{Z_0}{\sinh \gamma l} \tag{1.21b}$$

and

$$Z_q = Z_0 \sinh \gamma l \tag{1.22a}$$

$$Z_p = Z_0 \coth \frac{\gamma l}{2} \tag{1.22b}$$

For loss-free lines these expressions become reactance terms, as follows:

$$Z_a = j Z_0 \tan \frac{\beta l}{2} \tag{1.23a}$$

$$Z_b = \frac{-j Z_0}{\sin \beta l} \tag{1.23b}$$

and

$$Z_q = j Z_0 \sin \beta l \tag{1.24a}$$

$$Z_p = -jZ_0 \cot \frac{\beta l}{2} \tag{1.24b}$$

This set of formulae is useful, for example, in some microstrip filter designs.

For very short lengths of loss-free transmission line, such that $l \ll \lambda_g/4$, a very simple approximate equivalent network, shown in Figure 1.10, is adequate.

1.9 CHAIN (*ABCD*) PARAMETERS FOR A UNIFORM LENGTH OF LOSS-FREE TRANSMISSION LINE

The notation for the equivalent two-port network, to be characterized in terms of *ABCD* parameters, is shown in Figure 1.11. With this choice of quantities, including *directions*, the independent variables are now assumed to be the output variables (V_2 and I_2). The ABCD parameters then directly provide the related input (dependent) variables, assuming the network is linear:

$$V_1 = AV_2 + BI_2 \tag{1.25a}$$

$$I_1 = CV_2 + DI_2 \tag{1.25b}$$

where A, B, C, and D are constants that characterize the network.

In matrix form (1.25) becomes

$$\begin{bmatrix} V_1 \\ I_1 \end{bmatrix} = \begin{bmatrix} A & B \\ C & D \end{bmatrix} \begin{bmatrix} V_2 \\ I_2 \end{bmatrix} \tag{1.26}$$

Fig. 1.10. Simple *T*-equivalent network for a line having length $l \ll \lambda_g/4$

Fig. 1.11. Chain (*ABCD*) parameters: notation for a two-port network

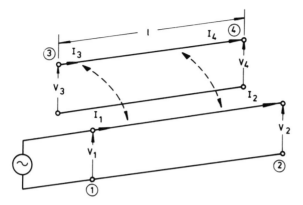

Fig. 1.12. Two parallel-coupled transmission lines

This is often called the *chain* matrix because it is so useful in analysing cascaded (or 'chained') networks.

For a length l of uniform loss-free transmission line the voltages and currents are interrelated thus:

$$V_1 = V_2 \cos \beta l + j I_2 Z_0 \sin \beta l \qquad (1.27a)$$

and

$$I_1 = j \frac{V_2}{Z_0} \sin \beta l + I_2 \cos \beta l \qquad (1.27b)$$

These expressions can be written in the following matrix form:

$$\begin{bmatrix} V_1 \\ I_1 \end{bmatrix} = \begin{bmatrix} \cos \beta l & j Z_0 \sin \beta l \\ j Y_0 \sin \beta l & \cos \beta l \end{bmatrix} \begin{bmatrix} V_2 \\ I_2 \end{bmatrix} \qquad (1.28)$$

Comparing eqn (1.26) with eqn (1.28) gives the final result for the transmission line chain matrix:

$$\begin{bmatrix} A & B \\ C & D \end{bmatrix} = \begin{bmatrix} \cos \beta l & j Z_0 \sin \beta l \\ j Y_0 \sin \beta l & \cos \beta l \end{bmatrix} \qquad (1.29)$$

This result is useful in analysing many single and multiple transmission line structures. One area where it will be required in this work is in coupler considerations, leading to coupler design, as described in detail in Chapter 6.

1.10 PARALLEL-COUPLED TRANSMISSION LINES: BASIC PROPERTIES

1.10.1 Even and odd modes

Any parallel-coupled pair of transmission lines, regardless of their practical realization, may be described by the four-port configuration indicated schematically in Figure 1.12. The chain lines indicate mutual coupling; input and output connections and terminations have been omitted for simplicity.

To examine the coupling modes which are established we now concentrate on a pure TEM-mode type of structure. The simplest structure to consider is the dual, coaxial, coupled-line system, in which electric and magnetic fields couple through a narrow slot cut into the adjacent outer conductor along a certain length. Such a structure is shown in Figure 1.13. Notice that the consistent field patterns correspond to waves travelling in opposite directions along the respective lines, and it may be stated that any 'parallel'-coupled pair of transmission lines yield contra-directional travelling waves.

The coupled coaxial arrangement constitutes a pure TEM-mode system, assuming either a uniform dielectric filling or an air-filled structure. At any instant the relative polarities of the voltages, taken at any specific plane along the structure, will either be alike or opposite. We refer to the different field configurations set up by such polarities as the even mode and the odd mode respectively. The diagram of Figure 1.14 illustrates these effects. The complete behaviour of the coupled structure can be obtained by superposition of the effects due to these two modes.

Because of the assumed uniform dielectric filling, any signal travelling in such a pure TEM-mode system will always travel at the same velocity, $v = c/\sqrt{\epsilon_r}$ (eqn. 1.8). Thus the velocity associated with the even mode is identical to that associated with the odd mode here.

The equivalent primary constants for the coupled lines having even or odd modes, taken separately, must differ because of the different field distributions. As a result of this two distinct characteristic impedances may

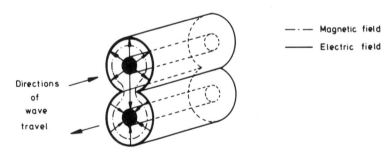

Fig. 1.13. Parallel-coupled coaxial lines

(a) EVEN MODE (b) ODD MODE

Fig. 1.14. Even mode and odd mode for parallel-coupled coaxial lines (for simplicity only the electric field is indicated)

be defined, one for each mode, expressed in terms of the different primary constants by equations of the form of eqn (1.10). We term these impedances the even-mode and odd-mode characteristic impedances, denoted by Z_{0e} and Z_{0o}. This applies to any TEM-mode or quasi-TEM mode parallel-coupled structure, and the characteristic impedances are major parameters in design procedures.

1.10.2. Overall parameters for couplers

A coupled-line circuit element may be specified in terms of the following overall parameters:

Coupling factor (or coupling parameter)	$= C$
Transmission factor	$= T$
Directivity factor	$= D$

The meaning of these parameters is clarified by Figure 1.15.
 We can now define the quantities more closely as follows:

$$C = V_3/V_1 = \text{voltage fraction 'transferred'}$$
$$\text{(coupled) across to the opposite arm}$$

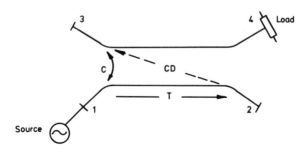

Fig. 1.15. Schematic diagram of a directional coupler arrangement, indicating the major parameters

$T = V_2/V_1 =$ transmission directly through the 'primary' arm of the structure

$D = V_4/V_3 =$ measure of the undesirable coupling which can occur to port 4, which we call the directivity

$I = V_4/V_1 =$ degree of isolation between port 4 and port 1

It is usual to quote all these quantities in decibels, so that we talk in terms of 'a 10 dB coupler', for example. This is a somewhat loose description, since the voltage of port 3 is -10 dB relative to that on port 1. In ideal couplers $D = 0$ and the power is split entirely between ports 2 and 3. This applies quite accurately to coaxial and triplate stripline couplers for which TEM modes exist, but it is not accurate for microstrip couplers of simple parallel-coupled construction, and the considerably reduced directivity can be a serious problem, as will be seen in Chapter 6.

In practice the specification for a coupled-line circuit element would include information on the following:

degree of coupling at the centre frequency (coupling C, in decibels) type of transmission structure
characteristic impedance of the 'external' environment (usually 50 Ω) bandwidth and centre frequency
lowest acceptable 'directivity', D (in decibels)

From data in this form the circuit designer must eventually calculate and decide upon the structural dimensions of the coupler.

BIBLIOGRAPHY

A substantial number of textbooks are available which deal in detail with the material covered, in the main, only briefly in this chapter. Only a short list of particularly useful and well-presented texts is given here.

Collins, R. E., *Foundations for Microwave Engineering*, McGraw-Hill Book Co., New York, 1966.
Ramo, S., Whinnery, J. R., and Van Duzer, T., *Fields and Waves in Communication Electronics*, John Wiley and Sons, Inc., New York, 1965.
Sander, K. F., and Reed, G. A. L., *Transmission and Propagation of Electromagnetic Waves*, Cambridge University Press Cambridge, 1978.

2 Transmission Line Options for Microwave Integrated Circuits (MICs)

2.1 INTRODUCTORY REMARKS

Microstrip, and circuits using it, forms the principal subject of this book. However, it is important to establish the general setting with which we are concerned and also to briefly examine some alternative transmission line structures which sometimes exhibit special advantages over other lines. In this chapter we shall therefore cover the following topics:

(a) microwave frequency and wavelength ranges; application
(b) a selection of alternative transmission line structures
(c) choice of substrate for hybrid microwave microcircuits
(d) thin-film manufacture
(e) thick-film manufacture.

2.2 MICROWAVE FREQUENCIES AND APPLICATIONS

The use of the term 'microwave' is actually rather ambiguous—particularly at the lower end of the range. Sometimes signals at frequencies of only a few hundred megahertz are referred to as 'microwaves'. For present purposes we shall assume that appropriate frequencies extend from around 1 GHz to several hundred gigahertz, as indicated in Figure 2.1.

The bulk of microwave technology occupies the spectrum including higher u.h.f. through s.h.f., i.e. from roughly 1 to 30 GHz. At lower frequencies more conventional radio techniques take over. In the e.h.f. bands the wavelengths are extremely short, being less than 10 mm, and hence signals at such wavelengths are often called 'millimetric'.

Fig. 2.1. Microwave frequency and wavelength ranges

A strong and significant core of applications exists for signals at microwave frequencies, among them:

radars (a very large family with a wide variety of demands and techniques)
point-to-point multichannel radio links ('microwave radio')
satellite communications systems (both fixed and mobile)
'troposcatter' systems
cellular (mobile) communications systems
microwave/millimeter-wave video distribution systems (MVDS)
microwave transceivers for optical fibre transmission
microwave instrumentation
'wireless' local area networks

The appropriate type of technology and its impact on systems costing will vary greatly depending upon the application. At one extreme we can consider a simple vehicle proximity radar, based upon a milliwatt-level Doppler or clutter-reference system, attracting a unit price of only a few dollars. On the other hand, an earth-sation for a satellite communications system represents a considerable capital investment costing up to hundreds of thousands of dollars.

Circuits using microstrip can be economically implemented in many low-to-medium power radars, associated electronic countermeasures, some segments of point-to-point microwave radio links, and certain portions of satellite communications systems. Microstrip transmission lines are widely used in microwave (hybrid) integrated circuits (MICs) for these systems.

Most of the structures are also suitable for various high-speed digital applications and some design techniques for these applications are also discussed later in this book.

Before proceeding further, let us momentarily pause to consider the position of coaxial and waveguide design. The evolution of MIC technologies does *not* mean that coaxial line and waveguide-based design is to be abandoned; on the contrary, such propagating structures assume continuing and significant roles.

Carefully constructed coaxial lines and waveguides are used in precision instrumentation equipment—the network analyser, for example—and this situation is not likely to change radically in the foreseeble future.

Waveguides also remain useful in at least three areas:

(a) high power transmission (e.g. kilowatt-to-megawatt transmitters)
(b) circuits where a high-Q component is deemed necessary
(c) millimetric-wavelength systems.

Developments in MIC technology are already tending to overcome problems in areas (b) and (c), but not (a).

The dramatic evolution of microwave semiconductor devices has added extra impetus to the development of MICs and, more recently, MMICs.

For several years both bipolar and field-effect transistors have been readily available. These devices, silicon up to almost 10 GHz and gallium arsenide above this, are capable of providing useful power gain and/or low-noise performance. Also diode devices, such as Schottky, Gunn and IMPATT, are available to detect or generate the microwave power. Furthermore, it has become apparent that hybrid microcircuit technology can readily be adapted to provide MICs. This combination of thick- and thin-film hybrid microcircuit technology and 'chip' components such as microwave transistors has formed a powerful design vehicle for modern integrated microwave systems.

2.3 MIC TRANSMISSION LINE STRUCTURES

Many structures have been conceived; Figure 2.2 depicts a collection of eight. Every structure comprises a combination of metal, shown as dense black in Figure 2.2, and dielectric, shown shaded in Figure 2.2. In most cases the dielectric principally supports the metal pattern, acting as a *substrate*, and clearly influences the wave propagation, but the image guide is an exception in that propagation is actually via the dielectric slab. The actual choice of structure depends upon several factors including the type of circuit or subsystem and its operating frequency. For example, the image line is only suitable for consideration in millimetric wavelength applications whereas microstrip is capable of propagating d.c., or a.c. at frequencies up to some tens of gigahertz at least. Each type of line structure has potential advantages for various applications and therefore each is briefly discussed here.

2.3.1 Microstrip (Figure 2.2ii)

Although microstrip has a very simple geometric structure the electromagnetic fields involved are actually complex. Accurate and thorough analysis requires

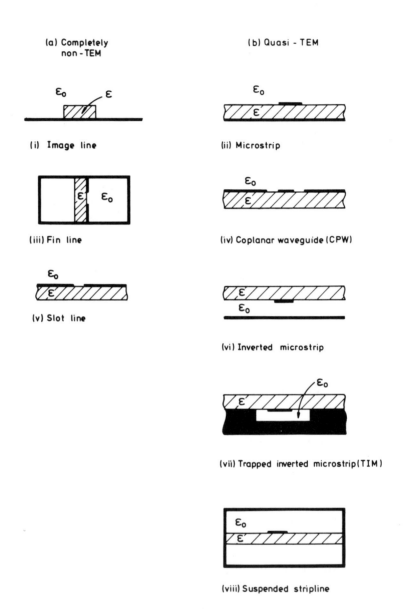

Fig. 2.2. Some transmission line structures suited to microwave circuit integration. (Reproduced by permission of B. E. Spielman, 'Integrated-circuit media for millimeter-wave applications', which was originally published in AGARD Conference Proceedings 245 by the Advisory Group for Aerospace Research and Development, North Atlantic Treaty Organization, AGARD/OTAN, February 1979.)

quite elaborate mathematical treatment, and this is briefly described in Chapter 4. However, simple approaches to the quasi-TEM mode calculations combined with frequency-dependent expressions yield quite acceptable design accuracy for many applications. Accuracies within 1 per cent are generally achievable.

MICs using microstrip can be designed for frequencies ranging from a few gigahertz, or even lower, up to many tens of gigahertz.

At higher frequencies, particularly into the millimetre wavelength ranges, losses (including radiation) increase greatly, higher-order modes become a considerable problem, and fabrication tolerances become exceedingly difficult to meet. It is thought that the frequency limit for the extensive use of microstrip is probably around 60 GHz.[1]

Since design with microstrip is the principal subject of this book no more need be said in this section.

2.3.2 Inverted microstrip (Figure 2.2vi)

The strip conductor pattern for this type of structure is manufactured in the same manner as for microstrip. The main difference is the absence of any conductor on the opposite side of the substrate—a ground plane is air-space separated instead. Thus the substrate mainly supports the microstrip lines and most of the field lies within the air between the strip and the ground plane. This ensures that the wavelength is long relative to that in conventional microstrip operated at similar frequencies. (It also infers that inverted microstrip can be operated satisfactorily at substantially higher frequencies; this has been achieved.)[2] Compared with conventional microstrip a wider line is obtained for a prescribed characteristic impedance, and this both reduces conductor dissipation and relaxes fabrication tolerances.

Substrate/ground-plane spacers have to be well separated from the circuit to avoid interference. Some design information is available.[3]

2.3.3 Trapped inverted microstrip (TIM) (Figure 2.2vii)

This is rather similar to inverted microstrip except that now the ground plane is made into a slot or 'channel'-shaped section effectively grounding and shielding three planes, away from the strip conductor. This provides for inherent suppression of some higher-order modes which could otherwise propagate.

Both inverted microstrip and TIM structures are potential candidates for applications at frequencies extending up to and through 95 GHz, i.e. for millimetric wave applications. Manufacturing difficulties are, however, considerable—due to the tight fabrication tolerances which must be imposed on conductor/ground-plane spacing and channel dimensions.

A limited amount of design information is available[4, 5] for TIM and it is recommended that considerable further work should be carried out on this structure.

2.3.4 Suspended stripline (Figure 2.2vii)

If the ground 'plane' of inverted microstrip is made to totally enclose the structure, with air gaps at top and bottom, we have suspended stripline. Although the losses are all relatively low, resulting in a fairly high Q-factor, waveguide modes can easily be excited at higher frequencies. This could possibly be prevented by bringing the top walls very close together and/or by introducing various structures to act as mode suppressors. In the first place the separations became so small that there is insufficient space for the incorporation of semiconductor devices, etc., and the introduction of mode suppressors represents a new design problem. In any case the basic structure is not compatible with shunt-mounted semiconductor devices.

Some work has also been performed on 'modified' suspended striplines.[6]

2.3.5 Coplanar waveguide (CPW) (Figure 2.2iv)

This is the fifth and last structure briefly considered here which supports a form of quasi-TEM mode of transmission. While this is true, the actual shape of the structure is distinctly different from any of those considered so far because the metallization is formed on one side of the substrate alone. Each 'side-plane' conductor is grounded and the centre strip carries the signal; thus much less field enters the substrate when compared with microstrip. Considerable design information is available and the structure has been used in special applications such as mixers for frequencies a little lower than those where microstrip is useful.[7, 8] Semiconductor diodes may be strapped across the intermetallic gaps (no need to drill the substrate). The open nature of the structure represents one limitation since this results in considerable radiation at higher microwave frequencies. The book by Gupta et al.[8] is recommended for a detailed treatment of coplanar waveguide.

2.3.6 Slotline (Figure 2.2v)

This transmission line structure consists of a dielectric substrate metallized on one side only. The metallization has a completely separating narrow slot etched into it to form the slotline. The structure was first proposed by Cohn in 1968 and an extensive account has been given by Gupta et al.[8] (see Chapters 5 and 6). Gupta's material includes a wide range of basic design details as well as information on discontinuities, transitions, and applications.

Some clever circuit applications have been devised but the following disadvantages should be noted for slotline:

(a) Characteristic impedances below about 60 Ω are difficult to realize.
(b) Circuit structures often involve difficult registration problems (especially with metallization on the opposite side to the slot).
(c) The Q-factor is significantly lower than the other structures considered here.

2.3.7 Finline ('E-plane' circuits) (Figure 2.2iii)

A totally shielding rectangular conducting box (like rectangular waveguide but avoiding waveguide modes) has a dielectric substrate fixed, usually centrally, across two of its faces.[9, 10] A metal circuit is deposited on one side of the substrate and a slot pattern in this metal forms the finline circuit. It is essential to maintain an excellent continuous microwave short circuit to the upper and lower waveguide walls, but this is fairly straightforward to achieve in practice. Since the structure is connected in the E-plane, circuits made using it are also called 'E-plane' integrated circuits. Because of this the transmission mode is also far removed from TEM. Some considerable advantages over other structures are obtained, mainly

(a) low losses (typically a factor about three better than microstrip)
(b) simpler fabrication.

The second advantage, simpler fabrication, is very significant because many applications have been studied for finline at frequencies up to 100 GHz and thus finline has to compete with structures such as inverted and trapped-inverted microstrip.

Despite some biasing problems, compatibility with solid-state devices is fairly good, especially in the case of beam-lead devices. Satisfactory band-pass filters, with approximately 10 per cent bandwidth, have been built; quadrature hybrids, transitions to waveguide, and balanced mixer circuits have also been designed and manufactured.[11, 12] Discontinuities in finline have been analysed on a modal basis and the results have been presented in a form appropriate to the design of impedance transforming and filter networks.[13]

2.3.8 Imageline (Figure 2.2i)

At very high microwave frequencies, into hundreds of GHz, the wavelengths become comparable with practically realistic dielectric 'slab' cross-sectional dimensions. If such a dielectric 'slab', in the form of a continuous strip of dielectric, is placed on a backing sheet of conductor it forms what is known as an imageline.

This structure then behaves like a dielectric waveguide which propagates a range of TE and TM modes trapped within it, due principally to the dielectric–air interface. A substantial amount of work on this and related

structures has been reported in the literature.[14, 15] Imageline is formed by adhering a dielectric slab (or thick strip) on to a metal ground plane, and typical dielectrics which may be used include boron nitride, alumina, and high-resistivity silicon. The dielectric slab is usually machined to a regular, approximately rectangular, shape.

Although unloaded Q-factors of several thousand are common and operation above 100 GHz is best, many problems remain. There is poor compatibility with active devices, mutual coupling, and radiation from discontinuities and bends. It is also apparent that the dielectric/metal adhesives are very lossy and this reduces the practical, loaded, Q-factor.

It is therefore clear that much more work is necessary before image guide, or some derivative thereof, becomes a really attractive practical proposition.

Due to these difficulties some further low-cost contenders have been considered for operation at frequencies around and exceeding 100 GHz. One interesting possibility is afforded by dielectric-filled waveguides of approximately semicircular cross-section.[16] Although measurements have been made available at somewhat lower frequencies (28–40 GHz), unloaded Q-factors of about 10^3 are obtained. The following advantages are also evident:

(a) No radiation losses, especially at bends (several decibels are typical for 90° bends in imagelines).
(b) Line-packing density is potentially high. Again, this is due to the inherent shielding.
(c) Tolerance requirements in the manufacturing process will be somewhat less restrictive than for imagelines since the flowing process involved in the manufacture of these waveguides equalizes irregularities in the longitudinal direction.

Many features, especially parameters determining predictable manufacture, need to be examined before this structure becomes useful in engineering terms.

2.3.9 Summary: impedances and Q-factors of the various structures

For applications in any circuits the characteristic impedances and Q-factors achievable with these various structures need to be known. Table 2.1 lists the ranges of characteristic impedance which may be expected.

Each of the quasi-TEM structures is capable of being designed at lower characteristic impedances than those indicated in Table 2.1. The general technique is simply to increase the width of the strip conductor, but this results in a risk of transverse resonance modes (dealt with in detail for microstrip in Chapter 4). At low-to-moderate frequencies microstrip impedances down to just below 10 Ω are thus achievable.

In most cases the upper limit to characteristic impedance is set by manufacturing tolerances and the unreliability associated with a very narrow

Table 2.1. Characteristic impedance (Z_0) ranges for the various structures[1]

Structure	$Z_0(\Omega)$
Microstrip	20–125
Inverted microstrip	25–130
Trapped inverted microstrip (TIM)	30–140
Suspended stripline	40–150
Coplanar waveguide (CPW)	40–150
Slotline	60–200
Finline	10–400
Imageline	≈ 26

strip. For the quasi-TEM lines, especially microstrip, this is an embarrassment since some circuit applications can demand $Z_0 > 150 \ \Omega$. Unless slotline or finline can then be used alternative clever circuit/system design is necessary. The characteristic impedance range of finline is remarkably large— 10–400 Ω—which is more than adequate for nearly all conceivable filter or matching network applications. On the other hand, imageline is restricted to around one value ($\sim 26 \ \Omega$) which is a further drawback with this structure. However, several useful circuits have been developed employing various directional couplers, ring resonators, etc., which do not depend upon changes in characteristic impedance for their design and operation.[15]

Unloaded Q-factors are also of importance, and these are indicated in Figure 2.3. Most of the results are typical practical values but imageline suffers dramatic reductions in unloaded Q-factors when any bending or other shape discontinuity is introduced. Many practical circuits or subsystems are likely to demand a range of characteristic impedances, different operating frequencies, and different bandwidths, etc. Spielman[1] has commented that, especially for millimetric wave applications, the integration of multifunction circuits is likely to require the use of more than one type of transmission structure to achieve the desired level of performance. Transitions would also be required to efficiently couple energy from one structure into another, e.g. waveguide to imageline, imageline to finline, TIM to finline, etc. Extensive advancements will almost certainly be seen in these areas.

2.4 CHOICE OF SUBSTRATE FOR HYBRID MICROCIRCUITS

Many facets, mechanical and thermal as well as electronic and economic, influence the decision process leading to the correct choice of a particular substrate for a specific type of MIC and application. The kinds of questions which should be asked must include:

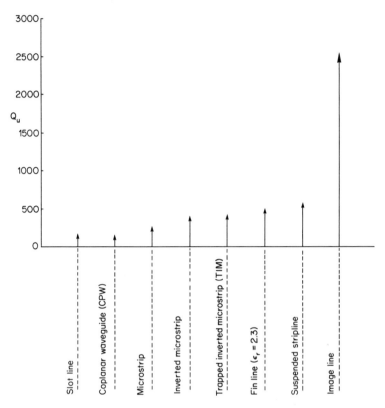

Fig. 2.3. Comparison of typical unloaded Q-factors (Q_u) at a frequency of 30 GHz for the structures shown in Figure 2.2. The substrate relative permittivity is 10 for all the structures except finline, where it is 2.3. (Reproduced by permission of Dr U. H. Gysel, and of B. E. Spielman, 'Integrated-circuit media for millimeter-wave applications', which was originally published in AGARD Conference Proceedings 245 by the Advisory Group for Aerospace Research and Development, North Atlantic Treaty Organization, AGARD/OTAN, February 1979.)

(a) Can the *cost* of the substrate under consideration be justified in the light of the application and attendant circuit/system costs?

(b) Is the technology to be *thin* or *thick* film?

(c) What frequency range or ranges are involved? (This will influence thickness and permittivity.)

(d) Will the surface finish be sufficiently good to keep conductor losses tolerably low—yet maintain metal-film adhesion?

(b) Are the mechanical strength and thermal conductivity sufficient for the application in mind?

(c) Are the substrates readily available with sufficient surface area, considering the circuit complexity and the operating frequencies?

This is not an exhaustive list and other considerations may well arise in certain situations. Before comparing materials it is useful to list a variety of possible substrate materials; these are shown in Table 2.2.

As a rough and ready general indication, not to be taken as a definite guide, the following substrate materials are often employed in the applications indicated:

(a) Plastic substrates: *experimental* circuits at frequencies below a few gigahertz as well as array antennas for frequencies up to and beyond 20 GHz. Irradiated polyolefin, RT-duroid 5880, Cu-flon, and other materials are used.

(b) Alumina: production circuits for frequencies up to about 20 GHz. The grade to be used depends on a number of factors including the choice of basic fabrication technique: thick film or thin film. Comparatively high conductor and dielectric losses and poor reproducibility usually exclude the choice of 85 per cent purity (or worse) material.

(c) Production circuits for millimetric wave applications from some tens of gigahertz up to perhaps 300 GHz: quartz, or an alternative structure such as finline or imageline. The lower permittivity of quartz allows larger distributed circuit elements to be incorporated.

Since alumina substrates have become very popular for manufacturing circuits to operate at any frequency up to about 20 GHz, a further table is now presented (Table 2.3) which summarizes some important features regarding alumina substrates.

A wide variety of commercially available substrates exists, one example being MRC Superstrate.* Extensive microstrip measurements indicate that the maximum permittivity of such high-purity substrates must lie somewhere between the limits 10.1 and 10.2. This is interesting because the weighted average permittivity of sapphire is also within this range. Using the information given in Table 2.2 this average can be calculated quite accurately:

$$\epsilon_{r(\text{sapphire} \rightarrow \text{average})} = \frac{1}{3}(2\epsilon_\perp + \epsilon_\parallel)$$

Substituting the given values (Table 2.2) of $\epsilon_\perp = 9.4$ and $\epsilon_\parallel = 11.6$,

$$\epsilon_{r(\text{sapphire} \rightarrow \text{average})} = \frac{1}{3}(18.8 + 11.6)$$

$$= 10.13$$

MRC's Superstrate material has a permittivity of about 9.8 at microwave frequencies. The surface finish is generally good and the material finds much application in MICs because film adhesion is so good.

*Company tradename.

Table 2.2. Properties of some typical substrate materials

Material	Surface roughness (μm)	$10^4 \tan \delta$ (10 GHz)	Approximate ϵ_r	Thermal conductivity K (W/cm²/°C)	Dielectric strength (kV/cm)
Air (dry)	—	$\simeq 0$	1	0.00024	30
Alumina					
99.5%	0.05–0.25	1–2	10	0.37	4×10^3
96%	5–20	6	9	0.28	4×10^3
85%	30–50	15	8	0.20	4×10^3
Sapphire	0.005–0.025	0.4–0.7	(9.4 and 11.6)† (single crystal)	0.4	4×10^3
Glass ('typical')	0.025	20	5	0.01	—
Polyolefin (irradiated)	1	1	2.3	0.001	$\simeq 300$
Quartz (fused)	0.006–0.025	1	3.8	0.01	10×10^3

Beryllia* (BeO)	0.05–1.25	1	6.6	2.5	—
Rutile	0.25–2.5	4	100	0.02	—
Ferrite/garnet	0.25	2	13–16	0.03	4×10^3
RT-duroid™ 5880	0.75–1.0(i) / 4.25–8.75(ii)	5–15	2.16–2.24	0.0026	—
RT-duroid™ 6010	0.75–1.0(i) / 4.25–8.75(ii)	10–60	10.2–10.7	0.0041	—
Epsilam-10™	—	20	10.0–13.0	0.0037	—
Cu-flon	—	4.5	2.1	—	—
GaAs (high resistivity)	0.025	6	13	0.3	350
Si (high resistivity)	0.025	10–100	12	0.9	300

* Beryllia is, however, a very dangerous ceramic to handle. Its dust is toxic for humans and it *must not* be machined at all. It is best avoided altogether if possible.

† The properties of sapphire are discussed in more detail in Chapter 3.

(i) Average peak-to-valley height difference for rolled copper construction.

(ii) Average peak-to-valley height difference for electro-deposited copper construction.

Table 2.3. Some specific observations regarding alumina substrates

Permittivity	(a) Average values	Within the approx. range $8.2 \leq \epsilon_r \leq 10.2$ Depends on purity and other factors Considerable
	(b) Batch-to batch variation	Therefore has to be taken into account in design (Worsens with deteriorating purity)
	(c) Material dispersion	Significant $\dfrac{1}{\epsilon_r}\left(\dfrac{\partial \epsilon_r}{\partial f}\right)$ through microwave frequencies (e.g. 8 to 12 GHz for some aluminas). Some dipolar behaviour, amount depends on impurities
Dielectric loss (tan δ)		Generally $< 1 \times 10^{-3}$ (1×10^{-4} for $> 99.5\%$ purity material)
Anisotropy		Present to varying extents Depends upon method of manufacture (caused by preferential orientation of crystallites)
Surface roughness and flatness		Surface roughness, see Table 2.2 Flatness: bowing can often occur to a significant extent with cheaper substrates (*not* necessarily just low-purity substrates)

Although sapphire is usually the most expensive substrate likely to be entertained for MICs it does offer some unique advantages:

(a) It is optically transparent. Hence chip devices can be accurately registered for attachment since the underside may be viewed.

(b) It has a fairly high permittivity ($\simeq 11$ typically). This is almost compatible with high-resistivity silicon. It also means that microstrip components are fairly small due to the reduced wavelength.

(c) Sapphire substrates are highly reproducible; all pieces are essentially identical in dielectric properties and they are not subject to bowing.

(d) Surface finish is optical grade which results in low power loss from surface roughness.

(e) Thermal conductivity is quite high, being about 30 per cent higher than the best alumina.

Against these advantages, and apart from the relatively high cost, we must bear in mind that sapphire substrate areas are limited (usually little more than 25 mm square) and the dielectric anisotropy poses some additional circuit design problems.

2.5 THIN-FILM MANUFACTURE

Microwave-grade hybrid circuits place stringent demands on the accuracy and repeatability of the circuit technology. These demands can be met by carefully controlled thin-film manufacturing processes which are therefore used for the great majority of MICs. 'Clean-room' areas are used for the purpose.

Various metals form the bulk material, gold, copper or aluminium being in common use. Because of its superior conductivity and also its lower price, copper is preferred—but a final gold film must be evaporated over the copper to provide an environmental protective layer. The necessary thickness of the metal film may be determined by considering the skin depth in the metal at the lowest frequency of interest. Taking this lowest frequency to be about 4 GHz the skin depth in copper is then approximately 1 μm. At least four skin depths are recommended and therefore the required thickness should be 4–5 μm.[20] The processes actually used for making the circuit may be visualized, rather crudely, as precision (small-scale) printed circuit processes. Practical techniques involve three main steps:

(a) Evaporation or sputtering a very thin 'seed' layer of a suitable metal on to the surface of the substrate
(b) Evaporating or sputtering a very thin layer of conductor metal on to this 'seed' layer
(c) Electroplating the bulk conductor metal on to layer (b)

Steps (a) and (b) provide, respectively, mechanical and electrical foundation layers on which to electroplate good-quality bulk conductor metal. The techniques differ chiefly in the choice of evaporation or sputtering, and in the manner in which (using photolithographic methods) the circuits are defined. In one well-proven technique[21] step (a) comprises \leqslant 200 Å of chromium, step (b) comprises a layer of copper of similar thickness, and approximtely 5 μm of copper is plated in step (c). The very thin layers are produced by magnetron sputtering,[22] where the combined electric and magnetic fields produce efficient ionization of the metal (copper) charge. The ions are attracted towards the substrate target which behaves as an anode.

Circuit definition may be accomplished either by a plate-through technique or by an etch-back technique, which are now described.

2.5.1 Plate-through technique

Substrates which are cleaned to a very high specification have their surfaces sputter coated as explained above. Next a photoresist layer, of similar thickness to the final metal film required, is 'spun' on to the surfaces. Exposure with ultraviolet light through a precision contact photomask (with

the required circuit pattern) is followed by subsequent processing to yield a pattern of slots in the photoresist layer. The copper is plated through these slots to form the circuit pattern and then the photoresist layer is washed away and the excess sputter coating is etched off. There are further details in the cited reference.[21]

2.5.2 Etch-back technique

Again the process starts by coating the substrates with thin films of metal by sputtering, exactly as explained before, and a photoresist layer is also spun on to the surfaces. In this case, however, the photomask exposure step is arranged so that those larger areas of the surface where metal is not required are left unprotected by the photoresist layer. Thus, at this stage, the circuit pattern is effectively defined by a fine pattern of the photoresist layer across the very thin sputtered layers. 'Etching-back' then removes the unwanted areas of metal which were originally sputtered on. Electroplating yields the desired final thickness. This technique has the following advantages:

(a) A thinner photoresist layer may be used—typically 0.5 μm.

(b) Plating is more uniform.

(c) Microstrip edges are relatively square.

However, for the thinner photoresist layer to work satisfactorily the substrate surface must be extremely smooth—and free from dust or grit (otherwise the layer becomes perforated). Sapphire substrates, with their optical surface finish, are very good in this respect.

2.5.3 Equipment required

For precise and repeatable circuit production the following items of equipment are recommended:

(a) Class 10 000 clean room, achieving class 500 in the main working area

(b) Coordinatograph for preparing 25 × size of artwork (in the 'cut'n strip' method)

(c) Precision 25 × reduction camera

(d) Magnetron sputtering operating in high vacuum, typically of the order of 10^{-6} Torr

(e) Vacuum spinner, having adjustable speed in the range 1000–7000 rev/min, and timer

(f) High-intensity collimated ultraviolet light source, for exposing photoresist layers

(g) Ultrapure water (UPW) system, delivering water of about 18 MΩ resistivity

Many other details should be considered by anyone contemplating the establishment of a thin-film circuit manufacturing facility, and the paper by Ladbrooke et al.[21] treats the topic very well.

2.5.4 Thin resistive films

Resistive films are required for many types of MICs including circuit elements such as terminations, bias networks and attenuators.

These resistive films must have a low temperature coefficient of resistivity and be of good stability. Commonly used resistive materials include nichrome and tantalum. Resistance values within the range 50–5000 Ω typify MIC requirements and such values can readily be realized by the deposition of films using these materials. With abrasive trimming or, alternatively, laser trimming it is possible to achieve resistance tolerances to an accuracy exceeding 1 per cent. The creation of these resistive films demands additional processes of deposition and etching beyond those of the thin-film metallization described above, and this complexity may be obviated by bonding directly chip resistors onto the conducting pattern (i.e. using surface mount components).

2.6 THICK-FILM MANUFACTURE

Thick-film technology has become well established in the low-cost manufacture of hybrid microcircuits at lower-than-microwave frequencies and for densely packaged digital subsystems.[23] Large advantages in cost reductions are obtainable compared with thin-film technology, and even more so with monolithic integrated circuit technology, especially where small 'batch production' runs are involved.

Thick-film materials and techniques have been developed for MICs operating at frequencies up to at least 20 GHz.[24] The losses associated with the thick-film conductors have been shown to be (typically) about 20 per cent worse than those associated with comparable thin-film conductors and this limitation, combined with tolerancing difficulties, tends to restrict production circuits to a maximum frequency of around 10 GHz. Circuits such as parallel-coupled and open-circuit stub filters, compensated couplers, attenuators, and power levellers have all been satisfactorily realized.

Two methods are commonly encountered for the manufacture of 'thick-film' cheap MICs:

(a) Thick-film patterns are printed and fired on to the ceramic substrate—usually alumina but occasionally quartz.[25, 26]
(b) A printed circuit technique is used to etch the desired pattern in the copper cladding of what is usually a plastic substrate (usually polyolefin for microwave applications; see Table 2.2).

Either method is substantially simpler and less demanding on both equipment and environment than thin-film technology. Method (b) is very well known and need not be described here. The first method, (a), is used for MIC manufacture to an increasing extent and therefore it will be briefly described.

2.6.1 Pastes, printing, and processing for thick-film MICs

The metal, usually gold, which eventually forms the microcircuit conductors is initially available to the circuit manufacturer as a paste or 'ink' contained in a jar. A few millilitres of this paste are placed on a fine-mesh 'screen', with areas open for the circuit pattern, and some of the paste is squeezed through these open areas and on to the surface of a substrate held rigidly just beneath the screen. Settling, drying, and firing sequences complete the 'thick' deposit, a very high proportion of which is pure metal. Although the process has earned the generic name 'thick film' the actual film thicknesses are only usually about twice that of thin films; i.e. 10 μm or so. The best way to explain the process is to briefly describe each step, and so we start with the pastes.

(a) The conductor pastes have to be special 'fritless' materials, and manufacturers such as Du Pont or Electro-Science Labs provide such pastes. For substantial shelf life pastes should be kept in a refrigerator.

(b) The artwork is prepared using similar procedures to those described for thin-film technology and photographic processing is then used to obtain a *positive* transparency.

(c) Suitable screens are available, often with dimensions about 15 \times 20 cm and made of at least 305 mesh stainless steel or polyester. These are stretched to 'drum tightness' over a rigid frame which will fit into the screen printer. They are also coated with a suitable photoresist layer.

(d) The positive transparency, obtained in step (b), is held in intimate contact (image side) with the coated surface of a screen, and then standard ultraviolet exposure and wash-and-bake processes leave a screen with apertures for the required circuit. All other areas of the screen are then opaque with a durable photoresist layer.

(e) This screen is placed in a printing jig (preferably a machine with well-defined adjustments and controls). The substrate which is to receive the circuit is placed beneath the aperture region of the screen—with accurate registration. This substrate is held firm by vacuum suction and is usually about one-substrate-thickness clear of the base of the screen.

(f) A few millilitres of conductor paste are placed on the screen, between the squeegee and the aperture region.

(g) When the parameters have all been optimized, the aperture region of the screen is wet, and the action smooth, a satisfactory wet deposit of paste will be transferred on to the substrate.

(h) This 'wet-circuit' deposit must next be left to settle by placing it horizontally in a clean area for about 15 min. Following this it is dried at approximately 100°C for about 15 to 20 min under an infrared drying unit.

(i) The firing process returns the material of the deposit to a predominantly metallic substance of high electrical conductivity. For the gold-suspension pastes this generally entails about 10 min at a temperature set somewhere in the range 900–1000°C, depending upon the specification of the paste.

(j) Immediately after firing the circuit is, in principle at least, ready for use. In many instances, however, the precise definition required *cannot* be achieved with the as-fired circuit, and either laser trimming or an etch-back process is necessary. This is especially a problem in proximity-coupled microstrips which are often required in MICs.[24, 27]

As noted in Section 2.6.4 above, there exists a requirement for resistive films and these can also be provided in thick film form. The deposition and processing mainly follows the steps given above the conductive films except that the firing profile is more critical since, in conjunction with the resistive paste formulation, it controls the final resistivity of the film.

2.7 THIN-FILM TRANSFER COMPONENTS

The one main drawback of thick-film MICs is the problem of occasional imprecise definition, outlined in step (j) above. England[28] has developed a novel and inexpensive technique to overcome this problem by introducing *thin*-film components as 'add-on' elements for a thick-film circuit. In this method the thick-film circuit is first made as described in the foregoing, but without those portions which demand precise definition. Instead, gaps are appropriately left in the circuit ready to receive bonded-in thin-film elements. These elements are produced on a polystyrene base or 'carrier' which merely supports the circuit element. Several successful circuits have been made and full details are given in the cited reference.[28]

2.8 MID-FILM TECHNOLOGY

It has long been appreciated that thin-film technology, on the one hand, and thick-film technology, on the other, provide extremes of resolution capability for MICs, with opposite extremes of capital and running costs. A manufacturing technology which provides a compromise between these extremes is now available and is termed mid-film technology.[29] In common with both thin- and thick-film technologies high purity alumina substrates are particularly suitable for the realization of mid-film circuits. Whilst this

is true several other types of substrates have been used for this purpose including fused quartz and glass. The mid-film manufacturing process begins by coating the substrate with a thin layer of an organic photosensitive resin which is then exposed to UV through a positive photomask. The exposed sections of the mid-film resin then cross-link under this process and become non-receptive to particles. Next, a powder composed of solid metallic conductors, most usually gold or silver, combined with glass frit is then applied to this surface—a controlled 'blow-on' technique can be used for this process. The powder particles become embedded in the unexposed sections of this light sensitive film. This sequence of steps is followed by firing the entire assembly in a belt conveyor furnace held at a centre temperature of appoximately 920°C, in a similar fashion to that previously described here for thick-film technology. This step removes the organic resin as a result of oxidation and it also bonds the particles of powder onto the substrate surface.

By comparison with the thick-film technology approach which is limited to line widths rather greater than around 100 μm, this mid-film process enables the fabrication of significantly narrower line widths—actually down to approximately 50 μm. Thick-film resistors and dielectric materials may be applied to create appropriate resistive or capacitive components using similar processes to those described above. Alternatively discrete (surface-mount) chip components may be added to the mid-film conductor pattern. This mid-film technology can be used to provide MICs operating over various bands within the approximate frequency range 3–12 GHz.

2.9 MONOLITHIC TECHNOLOGY

Conceptually, true electronic integrated circuits (ICs) comprise the simultaneous, or near simultaneous, processing of all electronic functions within a semiconductor chip. At lower frequencies and in by far the greater number of electronic applications worldwide, the basic semiconductor material used is silicon. At microwave frequencies, up to several GHz at least, silicon can still be used and indeed this is sometimes the case for production monolithic microwave integrated circuits (MMICs).

Semiconductor substrates of suitable form for MMICs must have high resistivities, in the order of 2000 Ω-cm or more in order to maintain the transmission losses at sufficiently low levels and also to provide enough isolation between components on the chip. High-resistivity silicon does meet these requirements and so also does semi-insulating GaAs. In fact, semi-insulating GaAs provides significantly higher resistivity than the high resistivity silicon (by a factor of at least 10^4).

Various lithographic processes are used to delineate device geometries and circuit patterns in MMICs. For example, with photolithography, diffraction effects limit the resolution of microstrip and other lines to dimensions in the order of a wavelength of light. For substantially higher resolution electron beam photolithography is used, particularly for the realization of sub-micron gates for devices such as GaAsFETs to operate at the higher microwave and into millimeter-wave frequencies. Combined processing techniques, taking the best advantages of photolithography and electron beam lithography are often used in the fabrication of commercial MMICs. Since the passive circuit patterns use precisely the same lithographic techniques that are employed to define the sub-micron transistor gates it follows that dimensional tolerances in the circuit patterns using this technology are controlled far more closely than they are in hybrid MIC technology.

In this text we are principally concerned with the design of MICs, both hybrid and monolithic, but with a specific emphasis on the circuit design itself, i.e. not the active devices. Therefore in our brief discussion of monolithic technology here we shall exclude any details regarding the processes required for the realization of active devices in MMICs (i.e. including the various transistor types like GaAsFETs, HEMTs and HBTs). Some remarks are, however, in order regarding the substrate requirements. Selected substrates should be capable of retaining their intrinsic properties during the required annealing cycle in the production process. Chrome-compensated semi-insulating GaAs offers a stable resistivity over the temperature range of 850–900°C necessary in the fabrication of the microwave devices within the MMIC. In these terms the implementation of MMICs in silicon presents a drawback due to the fact that silicon's intrinsic insulating qualities become degraded at temperatures exceeding 800°C. It is, however, possible to process silicon wafers, using a combination of ion implantation and selective laser annealing and this avoids increasing the temperature of the entire substrate which would otherwise have resulted in the degradation.

In spite of the relatively very high capital costs associated with wafer fabrication facilities MMICs provide several distinct advantages for an increasing variety of microwave and millimeter-wave applications. To introduce yet another abbreviation, millimeter-wave MMICs are sometimes termed MMMICs. The main application outlets for MMICs or MMMICs, as distinct from hybrid MICs, are those with relatively high volume requirements. Some microstrip-orientated MMIC design approaches are described in Chapter 9.

An outline comparison of hybrid MIC *vis-à-vis* MMIC technologies, focusing on their respective features and attributes, is provided in Table 2.4.

An excellent detailed coverage of economics, costs and other comparative aspects of MIC versus MMIC technology is provided by Sweet.[30]

Table 2.4. MIC/MMIC technology: manufacturing costs and operational capability comparisons

Technology	Capital cost ($)* (Fab facility)	MM-wave capabilities	Volume applications
Thin-film hybrid MICs	10^4–10^5	Yes—to over 100 GHz	Not usually†
Thick-film hybrid MICs	10^3–10^4	No—max about 10 GHz	Not usually†
Mid-film hybrid MICs	about 10^4	No—max about 12 GHz	Not usually†
Monolithic microwave ICs (MMICs)	10^6–10^8	Yes	Yes—well suited to high volume§

* Ball-park indications of the start-up costs for an MIC or MMIC fabrication ('fab') facility.
† Hybrid MICs are best restricted to low-to-moderate volume applications, on economic grounds.
‡ MMICs (MMMICs) have been designed to operate at frequencies exceeding 100 GHz, but for relatively simple circuits only so far.
§ There are many other considerations apart from volume that may influence the decision to realize a circuit in MIC or MMIC form.

REFERENCES

1. Spielman, B. E., 'Integrated-circuit media for millimetre-wave applications', in *Millimeter and Submillimeter Wave Propagation* (Eds E. Spitz and G. Cachier) AGARD Conference Proceedings No. 245. Papers and discussions presented at the Twenty-fifth Meeting of the Electromagnetic Wave Propagation Panel, Munich, 4–8 Sept. 1978, pp. 20–1 to 20–7. (The basis of some of the material was originally presented by Dr U. Gysel, at 'mm Workshop', San Diego, CA, 24 June 1977, no proceedings published).
2. Schneider, M. V., 'Micostrip lines for microwave integrated circuits', *The Bell System Technical Journal*, May–June **1969**, 1421–1444.
3. Gunston, M. A. R., *Microwave Transmission-line Impedance Data*, Van Nostrand Reinhold Company, London, 1972.
4. Spielman, B.E., 'Computer-aided analysis of dissipation losses in isolated and coupled transmission lines for microwave and millimetre-wave applications', *NRL Formal Report*, No. 8009, Washington, D.C., 1976.
5. Ganguly, A. K., and Spielman, B. E., 'Dispersion characteristics for arbitrarily configured transmission media', *IEEE Trans.*, **MTT-25**, No. 12, December 1977, 1138–1141.
6. Villotte, J. P., Aubourg, M., and Garault, Y., Modified suspended striplines for microwave integrated circuits', *Electron. Lett.*, **14**, No. 18, 31 Aug. 1978, 602–603.
7. Holder, P. A. R., 'X-band microwave integrated circuits using slotline and coplanar waveguide', *The Radio and Electronic Engineer*, **48**, No. 1/2, January/February 1978, 38–42.
8. Gupta, K. C., Garg, Ramesh and Bahl, I. J., *Microstrip Lines and Slotlines*, Artech House Inc., Dedham, MA, 1979.

9. Meier, P. J., 'Integrated finline millimeter components', *IEEE Trans.* **MTT-22**, No. 12, December 1974, 1209–1216.
10. Meier, P. J., 'Millimeter integrated circuits suspended in the E-plane of rectangular waveguide', *IEEE Trans.* **MTT-26**, No. 10, October 1978, 726–733.
11. Cohen, L. D., and Meier, P. J., 'E-plane mm-wave circuits', *Microwave Journal*, August **1978**, 63–66.
12. Bates, R. N., and Coleman, M. D., 'Millimetre wave finline balanced mixers', *Proceedings of Ninth European Microwave Conference*, 17–21 September 1979, Brighton, pp. 721–725.
13. El Hennawy, H., and Schunemann, K., 'Analysis of fin-line discontinuities', *Proceedings of Ninth European Microwave Conference*, 17–21 September 1979, Brighton, pp. 448–452.
14. Barnoski, M. K. (Ed.), *Introduction to Fibre Optics*, Plenum Publishing Corporation, New York, 1974.
15. Knox, R. M., 'Dielectric waveguide microwave integrated circuits—an overview', *IEEE Trans.*, **MTT-24**, No. 11, November 1976, 806–814.
16. Hinken, J. H., 'Waveguides for microwave integrated circuits', *Proceedings of Ninth European Microwave Conference*, 17–21 September 1979, Brighton, pp. 656–660.
17. Abbott, D. A., *et al.*, 'Monolithic gallium arsenide circuits show great promise', *Microwave Systems News*, August **1979**, 73–96.
18. Greiling, P. T., *et al.*, 'Why design logic with GaAs—and how?', *Microwave Systems News*, January **1980**, 48–60.
19. Upadhyayula, L. C., 'Monolithic logic circuits with transferred-electron logic devices (TELDS)', *First Speciality Conference on Gigabit Logic for Microwave Systems*, Florida, 3–4 May 1979, pp. 136–142.
20. Caulton, M., and Sobol, H., 'Microwave integrated-circuit technology—a survey', *IEEE J. Solid-State Circuits*, **SC-5**, December 1970, 292–303.
21. Ladbrooke, P. H., England, E. H., and Potok, M. H. N., 'A specific thin-film technology for microwave integrated circuits', *IEEE Trans. on Manufacturing Technology*, **MFT-2**, No. 2, December 1973, 53–59.
22. England, E. H., Private communication.
23. Harper, C. A. (Ed.), *Handbook of Thick Film Hybrid Microelectronics*, McGraw-Hill Book Co., New York, 1974.
24. Rickard, D. C., 'Thick film MIC components in the range 10–20 GHz', *Proceedings of the Sixth European Microwave Conference*, Rome, 1976, pp. 687–691.
25. Ramy, J.-P, *et al.*, 'Optimization of the thick- and thin-film technologies for microwave circuits on alumina and fused silica substrates', *IEEE Trans.*, **MTT-26**, No. 10, October 1978, 814–820.
26. Goloubkoff, M., and Thebault, C., 'M.I.C. technology of fused silica substrate: Ku band transponder application', *Proceedings of the Ninth European Microwave Conference*, 17–21 September, 1979, Brighton, pp. 651–655.
27. Bates, R. N., 'Design of microstrip spur-line bandstop filters', *IEE Journal on Microwaves, Optics and Acoustics*, **1**, No. 6, November 1977, 209–214.
28. England, E. H., 'Thin film transfer components for cheap MICs', *Proceedings of the Ninth European Microwave Conference*, 17–21 September 1979, Brighton, pp. 646–650.
29. Conn, D. R., Naguib, H. M., and Anderson, C. M., 'Mid-film for microwave integrated circuits', *IEEE Trans. Components, Hybrids, Manuf. Technol.*, **CHMT-5**, No. 1, March 1982, 185–191.
30. Sweet, A., *MIC and MMIC Amplifier and oscillator circuit design*, (Particularly Chapter 9), Artech House, Inc., 1990.

3 Static-TEM Parameters and Design at Lower Frequencies

3.1 THE MICROSTRIP MIC DESIGN PROBLEM

As remarked in Chapter 2, microstrip is a particularly useful transmission line medium for implementation in distributed circuit designs at frequencies from a few gigahertz through some tens of gigahertz. In this chapter we are concerned with some fundamental aspects of the microstrip structure and with the realization of dimensional design parameters chiefly for applications at the lower microwave frequencies.

3.1.1 A transistor amplifier input network

One of the many areas where a microstrip design approach is really attractive is that of microwave transistor amplifiers. Two examples are considered in more detail in Chapter 8, but the case of an input network for a bipolar transistor is examined briefly here. Such transistors are used at frequencies of only a few gigahertz and the design techniques described in this chapter are usually of sufficient accuracy. The general input arrangement for this type of amplifier is shown in Figure 3.1(a), where the low-noise optimizing network is sometimes referred to as a 'noise-matching' network. This is typical of a microwave receiver 'front-end'. In other cases power matching may be required and a different network would therefore be developed.

A fairly simple microstrip realization of the input network is illustrated schematically in Figure 3.1(b). (Detailed design is given in Chapter 8.) The signal source is assumed to be connected without any discontinuity through the 50 Ω microstrip to the 'L-shaped' microstrip network. This consists of two lines having characteristic impedances and electrical lengths determined by the network design. The main design problem is then: *to evaluate the physical widths and lengths of the microstrip lines* (*this applies to all microstrip designs*).

a) Block diagram

b) Circuit requirements (schematic plan view)

Fig. 3.1. Input network arrangement (a) and microstrip circuit (b) for a microwave bipolar transistor amplifier operating at a frequency of 2 GHz. (Reproduced by permission of Hewlett-Packard Company, Palo Alto, Calif.)

We shall see that the width of a microstrip line is principally a function of its characteristic impedance and the thickness of the substrate. The physical length depends upon the wavelength, which is a function of the width, the substrate permittivity, and, of course, the signal frequency. Explicit techniques and formulae will be given here for the calculation of width and length to accuracies within about 1 per cent.

From Figure 3.1(b) it should be clear that various *discontinuities* are inherent in the structures as follows: where the 50 Ω and 31.3 Ω lines join, at the right-angled junction of the 31.3 Ω and 65.8 Ω lines, and (less obviously) at the short circuit on the end of the 65.8 Ω line. These, and other, discontinuities are described in more detail in Chapter 5, where the details of the chamfer to reduce the discontinuity effect of the right-angled bend are also considered.

3.1.2 The geometry of microstrip

The general geometry of microstrip is shown in Figure 3.2. Materials and techniques for its manufacture were discussed in Chapter 2, where the basic advantages were also described in comparison with other types of MIC transmission lines. The most important dimensional parameters are the microstrip width w and the height h (equal to the thickness of the substrate). Also of main importance is the relative permittivity of the substrate (ϵ_r). The thickness t of the metallic, top-conducting strip is generally of much lesser importance and may quite often be neglected. Effects of this thickness

Fig. 3.2. The general geometry of a microstrip line, including choice of coordinates

are dealt with in Section 3.10. We shall consistently refer to the *xyz* coordinate description as shown in Figure 3.2.

Some of the particularly useful characteristics of microstrip include the following:

(a) D.c. as well as a.c. signals may be transmitted.
(b) Active devices, diodes, and transistors may readily be incorporated (shunt connections are also quite easily made).
(c) In-circuit characterization of devices is straightforward to implement.
(d) Line wavelength is reduced considerably (typically one-third) from its free-space value, because of the substrate fields. Hence distributed component dimensions are relatively small.
(e) The structure is quite rugged and can withstand moderately high voltages and power levels.

In the next six main sections of this chapter we are concerned with basic considerations and microstrip *synthesis* where the substrate is a homogeneous, isotropic dielectric. Materials such as alumina, which is widely used, approximate to this description. We shall then consider synthesis using a dielectrically anisotropic substrate and a ferrite substrate.

The chapter concludes with discussions on the effects of finite thickness and manufacturing tolerances and pulse propagation.

3.2 THE QUASI-TEM MODE OF PROPAGATION

It is clear (Figure 3.2) that the microstrip involves an abrupt dielectric interface between the substrate and the air above it. Any transmission line which is filled with a *uniform* dielectric can support a single, well-defined *mode* of propagation, at least over a specified range of frequencies (TEM

for coaxial lines, TE for waveguides, etc.). Transmission lines which do not have such a uniform dielectric filling cannot support a single mode of propagation; microstrip is within this category. Although this is true the bulk of the energy is transmitted along microstrip with a field distribution which quite closely resembles TEM; it is usually referred to as 'quasi-TEM'. The detailed field distribution is quite complicated, but the main transverse electric field can be visualized as shown in Figure 3.3.

Gupta *et al.*[1] have used Maxwell's equations to convincingly demonstrate the necessity for longitudinal components of electric and magnetic fields. This is clearly inconsistent with a pure TEM or a TE propagating mode.

Representative views of the magnetic and electric field distributions are given in Figure 3.4. (These are *not* precisely determined field contours, and they must only be regarded as diagrammatic illustrations of the situation.) The longitudinal components can clearly be seen, and these become increasingly significant as the frequency is raised.

At most transverse cross-sectional planes, taken across the shielded microstrip, the electric (E) and magnetic (H) fields follow the distributions indicated in Figure 3.5. Note the abrupt change in direction of the electric field line as it passes through the air–substrate interface. These fields have been analysed by a number of workers using various *static* techniques.[2-7] We shall not study any of these techniques in detail here, but the results are powerful and significant for the circuit designer wishing to use microstrip. There are two reasons for this:

(a) For the majority of microstrip lines suitable for MICs, the statically derived results are quite accurate where the frequency is below a few gigahertz.

(b) At higher frequencies, up to the limits for the useful operation of microstrip, these 'static' results can still be used in conjunction with frequency-dependent functions in closed formulae. This is developed in detail in Chapter 4.

3.3 STATIC-TEM PARAMETERS

The microstrip synthesis problem consists of finding the values of width w and length l corresponding to the characteristic impedance Z_0 and electrical

Fig. 3.3. Transverse cross-section of microstrip, showing electric field

(a) Magnetic field distribution

(b) Electric field distribution (partial view)

Fig. 3.4. Three-dimensional views of approximate magnetic and electric fields surrounding a shielded microstrip line (for simplicity here, only the electric field in the air is shown in b)

length θ defined at the network design stage. Initially a suitable substrate, of thickness h and relative permittivity ϵ_r, will have to be chosen. A basis for this choice is given in Section 2.13 of Chapter 2, and the choice also depends upon certain frequency limitations which are discussed in Section 4.9. The synthesis actually yields the normalized width-to-height ratio w/h initially, as well as a quantity called the *effective microstrip permittivity* ϵ_{eff}. This quantity is unique to mixed-dielectric transmission line systems and it provides a useful link between various wavelengths, impedances, and propagating velocities. We shall shortly define the static-TEM effective microstrip permittivity (ϵ_{eff}) precisely.

Fig. 3.5. A typical transverse cross-section of a shielded microstrip line showing representative magnetic (H) and electric (E) fields

3.3.1 The characteristic impedance Z_0

For any TEM-type transmission line the characteristic impedance at high frequencies may be expressed in any one of three alternate forms (see Section 1.4):

$$Z_0 = \sqrt{\frac{L}{C}} \tag{3.1}$$

or

$$Z_0 = v_p L \tag{3.2a}$$

or

$$Z_0 = \frac{1}{v_p C} \tag{3.2b}$$

both of which involve the phase velocity v_p of the wave travelling along the line. It is also recalled (Chapter 1) that this phase velocity is given by

$$v_p = \frac{1}{\sqrt{LC}} \tag{3.3}$$

When the substrate of the microstrip line is (effectively) removed we have an air-filled line along which the wave will travel at c, the velocity of light in free space ($c = 2.99793 \times 10^8$ m/s). The characteristic impedance of this air-filled 'microstrip', Z_{01}, is given by

$$Z_{01} = \sqrt{\frac{L}{C_1}} \tag{3.4}$$

or

$$Z_{01} = cL \qquad (3.5a)$$

or

$$Z_{01} = \frac{1}{cC_1} \qquad (3.5b)$$

where L remains unaltered and C_1 is the capacitance per unit length for this structure.

Combining eqns (3.1), (3.5a), and (3.5b) yields the very significant result

$$Z_0 = \frac{1}{c\sqrt{CC_1}} \qquad (3.6)$$

This means that we have the required characteristic impedance if only we can evaluate the capacitances per unit length of the structure, with and without the presence of the dielectric substrate. One way in which this has been achieved will be outlined shortly (Section 3.6).

3.3.2 The effective microstrip permittivity ϵ_{eff}

For the air-spaced microstrip line the propagation velocity is given by

$$c = \frac{1}{\sqrt{LC_1}} \qquad (3.7)$$

and by dividing eqn (3.7) by eqn (3.3) and squaring, we obtain

$$\frac{C}{C_1} = \left(\frac{c}{v_p}\right)^2 \qquad (3.8)$$

The capacitance ratio C/C_1 is termed the *effective microstrip permittivity* ϵ_{eff}, an important microstrip parameter. From eqn (3.8), ϵ_{eff} is given by

$$\epsilon_{eff} = \left(\frac{c}{v_p}\right)^2 \qquad (3.9)$$

Although this result has been derived on a *static* basis here, it is fundamentally important for microstrip and will be used again where v_p is taken to be frequency dependent (Chapter 4).

A useful relationship between Z_0, Z_{01}, and ϵ_{eff} can be obtained by combining eqns (3.2b), (3.5b), (3.8), and (3.9). The result is

$$Z_0 = \frac{Z_{01}}{\sqrt{\epsilon_{eff}}}$$

that is

$$Z_{01} = Z_0\sqrt{\epsilon_{\text{eff}}} \tag{3.10}$$

This will be useful in several respects, including the operation of a graphical design technique which is shortly described.

Upper and lower bounds can readily be found for ϵ_{eff}, in the static low-frequency limit, by considering the effects of very wide and very narrow lines as indicated in Figure 3.6.

For the very wide lines nearly all of the electric field is confined to the substrate dielectric, the structure resembles a parallel-plate capacitor, and therefore, at this extreme,

$$\epsilon_{\text{eff}} \rightarrow \epsilon_r$$

In the case of very narrow lines the field is almost equally shared by the air ($\epsilon_r = 1$) and the substrate so that, at this extreme,

$$\epsilon_{\text{eff}} \simeq \frac{1}{2}(\epsilon_r + 1)$$

The range of ϵ_{eff} is therefore

$$\frac{1}{2}(\epsilon_r + 1) \leq \epsilon_{\text{eff}} \leq \epsilon_r \tag{3.11}$$

It can be convenient to express the effective microstrip permittivity as follows

$$\epsilon_{\text{eff}} = 1 + q(\epsilon_r - 1) \tag{3.12}$$

where the new quantity, the *filling factor q*, has the bounds

$$\frac{1}{2} \leq q \leq 1 \tag{3.13}$$

Wheeler[6, 7] has evaluated this filling factor, and we shall employ it in the approximate graphical design technique.

3.3.3 Synthesis: the width-to-height ratio w/h

The width-to-height ratio (w/h) is a strong function of Z_0 and of the substrate permittivity ϵ_r. Wheeler's results[7] are particularly useful in this respect,

Fig. 3.6. Extremely wide ($w \gg h$) and extremely narrow ($w \ll h$) microstrip lines

although it has more recently been found that some modifications are necessary for high accuracy (within 1 per cent) to be achieved.[8] Closed formulae for w/h will be given in Section 3.5.

3.3.4 Wavelength λ_g and physical length l

For any propagating wave the velocity is given by the appropriate frequency–wavelength product. In free space we have $c = f\lambda_0$ and in the microstrip the velocity is $v_p = f\lambda_g$. Substituting these products into eqn (3.9) we obtain

$$\epsilon_{\text{eff}} = \left(\frac{\lambda_0}{\lambda_g}\right)^2$$

or

$$\lambda_g = \frac{\lambda_0}{\sqrt{\epsilon_{\text{eff}}}} \tag{3.14}$$

where λ_0 is the free-space wavelength.

More conveniently, where the frequency is given in gigahertz and denoted by F, the wavelength can be expressed directly in millimetres as follows:

$$\lambda_g = \frac{300}{F\sqrt{\epsilon_{\text{eff}}}} \text{ mm} \tag{3.15}$$

The physical length l of a microstrip line to yield a specified electrical length θ (in degrees) is easily determined. In Chapter 1 (Section 1.3) we derived the phase coefficient β, and βl is equal to this electrical length in radians, i.e.

$$\beta l = \theta$$

and hence

$$\frac{2\pi l}{\lambda_g} = \theta$$

With θ in degree this gives

$$l = \frac{\theta\lambda_g}{360} \tag{3.16}$$

Thus, with λ_g evaluated using eqn (3.15), we can simply find l.

3.4 APPROXIMATE GRAPHICAL SYNTHESIS

Presser[9] has devised a graphical technique for the analysis or synthesis of microstrip lines. The method is quick and useful where errors of a few per

cent can be tolerated and where a calculator is not readily available. This technique is based upon the work of Wheeler[6, 7] and thus it uses the static-TEM approach which is applicable for frequencies up to a few gigahertz. The 'design curves' given by Presser are reproduced in Figure 3.7. Notice that the width-to-height or 'shape' ratio w/h and filling factor q are both plotted to a base of the *air-spaced* characteristic impedance Z_{01}. This helps to make the graph fairly universal, as the following design synthesis steps will indicate:

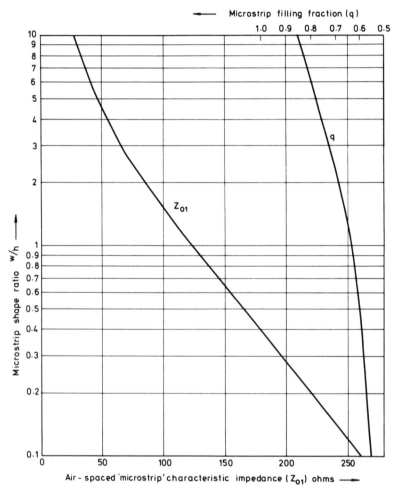

Fig. 3.7. 'General' curves for the analysis or synthesis of microstrip. (Reproduced by permission of *Microwaves* from Presser,[9] Figure 2.)

(a) Make the initial, very rough, assumption—$\epsilon_{eff} \simeq \epsilon_r$—to obtain starting values.
(b) Calculate the air-spaced characteristic impedance approximately, from eqn (3.10), with $\epsilon_{eff} \simeq \epsilon_r$:

$$Z_{01} \simeq \sqrt{\epsilon_r} Z_0$$

(c) Use the general two-curve graph (Figure 3.7) to find w/h applicable to this Z_{01}, and also note the corresponding value of the microstrip filling factor q.
(d) Hence calculate the up-dated value of ϵ_{eff} using eqn (3.12):

$$\epsilon_{eff} = 1 + q(\epsilon_r - 1)$$

This now ends one iteration. Steps (b) to (d) must be repeated, using the progressively improved values of ϵ_{eff} (instead of ϵ_r in steps i and ii). Fast convergence is generally obtained here and only two or three iterations are usually necessary before ϵ_{eff} is finally within about 1 per cent of its previous value; i.e. convergence has occurred.

The final value of w/h, and hence the width w, is of course the value appropriate to the final Z_{01}. Lastly, the wavelength must be calculated using eqn (3.15). The *static-TEM* part of the microstrip design procedure is then complete using this graphical technique.

However, *any* graphical approach will inevitably involve errors of several per cent and also clearly cannot form part of a computer-aided design (CAD) algorithm. This form of graphical technique is mainly suitable for approximate guidance purposes. Most microstrip MIC design work necessarily proceeds on a CAD basis and demands relatively high accuracy (typically better than 1 per cent). Therefore closed formulae are very useful for synthesis; in the next section suitable formulae are presented.

3.5 FORMULAE FOR ACCURATE STATIC-TEM CALCULATIONS

Closed formulae are highly desirable for use in microstrip calculations for the following reasons:

(a) A fairly high accuracy is achievable.
(b) Fast CAD algorithms can be implemented, including the following.
(c) Incorporation of static-TEM formulae in frequency-dependent calculations for higher-frequency design (Chapter 4).

Various workers have reported formulae for microstrip calculations.[7,8,10,11] Owens[8] carefully investigated the ranges of applicability of many of the expressions given by Wheeler,[7] comparing calculated results with numerical

computations requiring elaborate algorithms—but known to be very accurate.
New changeover values for the shape ratio (w/h) were advocated for greatest
accuracy by using alternate expressions dependent upon the w/h range. The
formulae are applicable to alumina-type substrates ($8 \leq \epsilon_r \leq 12$), the most
useful expressions being given in the following sections.

3.5.1 Synthesis formulae (Z_0 and ϵ_r given)

For narrow strips (i.e. when $Z_0 > (44 - 2\epsilon_r) \ \Omega$):

$$\frac{w}{h} = \left(\frac{\exp H'}{8} - \frac{1}{4 \exp H'} \right)^{-1} \qquad (3.17)$$

where

$$H' = \frac{Z_0 \sqrt{2(\epsilon_r + 1)}}{119.9} + \frac{1}{2} \left(\frac{\epsilon_r - 1}{\epsilon_r + 1} \right) \left(\ln \frac{\pi}{2} + \frac{1}{\epsilon_r} \ln \frac{4}{\pi} \right) \qquad (3.18)$$

We may also use, with slight but significant shift of changeover value to
$w/h < 1.3$ (i.e. when $Z_0 > (63 - 2\epsilon_r) \ \Omega$):

$$\epsilon_{\text{eff}} = \frac{\epsilon_r + 1}{2} \left\{ 1 - \frac{1}{2H'} \left(\frac{\epsilon_r - 1}{\epsilon_r + 1} \right) \left(\ln \frac{\pi}{2} + \frac{1}{\epsilon_r} \ln \frac{4}{\pi} \right) \right\}^{-2} \qquad (3.19)$$

where H' is given by eqn (3.18) (as a function of Z_0) or, alternatively, as
a function of w/h, from eqn (3.17):

$$H' = \ln \left\{ 4\frac{h}{w} + \sqrt{16 \left(\frac{h}{w} \right)^2 + 2} \right\} \qquad (3.20)$$

Another, somewhat simpler, expression for $\epsilon_{\text{eff}}(Z_0)$ is furnished by eqn (15)
of Owens,[8] which is

$$\epsilon_{\text{eff}} = \frac{\epsilon_r + 1}{2} \left\{ 1 + \frac{29.98}{Z_0} \left(\frac{2}{\epsilon_r + 1} \right)^{1/2} \left(\frac{\epsilon_r - 1}{\epsilon_r + 1} \right) \left(\ln \frac{\pi}{2} + \frac{1}{\epsilon_r} \ln \frac{4}{\pi} \right) \right\}^2 \qquad (3.21)$$

We now consider the ranges and formulae for 'wide' strips.
 For wide strips (i.e. when $Z_0 < (44 - 2\epsilon_r) \ \Omega$):

$$\frac{w}{h} = \frac{2}{\pi} \left\{ (d_\epsilon - 1) - \ln (2d_\epsilon - 1) \right\}$$

$$+ \frac{\epsilon_r - 1}{\pi \epsilon_r} \left\{ \ln (d_\epsilon - 1) + 0.293 - \frac{0.517}{\epsilon_r} \right\} \qquad (3.22)$$

where

$$d = \frac{59.95 \pi^2}{Z_0 \sqrt{\epsilon_r}} \qquad (3.23)$$

With the same slight shift of changeover value as before (i.e. where $w/h > 1.3$ and $Z_0 > (63 - 2\epsilon_r)\ \Omega$) Owens found that a modified form of an earlier expression due to Schneider[12] gave very accurate results. Owens' formula is

$$\epsilon_{\text{eff}} = \frac{\epsilon_r + 1}{2} + \frac{\epsilon_r - 1}{2}\left(1 + 10\frac{h}{w}\right)^{-0.555} \tag{3.24}$$

Alternatively, where Z_0 is known at first

$$\epsilon_{\text{eff}} = \frac{\epsilon_r}{0.96 + \epsilon_r(0.109 - 0.004\epsilon_r)\{\log(10 + Z_0) - 1\}} \tag{3.25}$$

For microstrip lines on alumina ($\epsilon_r = 10$) this expression appears to be accurate to ± 0.2 per cent over the impedance range

$$8 \le Z_0 \le 45\ \Omega$$

Finally, some accurate expressions for analysis (calculating Z_0) are also given.

3.5.2 Analysis formulae (w/h and ϵ_r given)

For 'narrow' strips ($w/h < 3.3$):

$$Z_0 = \frac{119.9}{\sqrt{2(\epsilon_r + 1)}}\left[\ln\left\{4\frac{h}{w} + \sqrt{16\left(\frac{h}{w}\right)^2 + 2}\right\}\right]$$

For 'wide' strips ($w/h > 3.3$):

$$Z_0 = \frac{119.9\pi}{2\sqrt{\epsilon_r}}\left[\frac{w}{2h} + \frac{\ln 4}{\pi} + \frac{\ln(e\pi^2/16)}{2\pi}\left(\frac{\epsilon_r - 1}{\epsilon_r^2}\right)\right.$$

$$\left. + \frac{\epsilon_r + 1}{2\pi\epsilon_r}\left\{\ln\frac{\pi e}{2} + \ln\left(\frac{w}{2h} + 0.94\right)\right\}\right]^{-1} \tag{3.27}$$

where 'e' is the exponential base: e = 2.7182818

These expressions may readily be handled using a pocket calculator. They can also be incorporated in fairly extensive CAD routines which involve frequency-dependent expressions for effective microstrip permittivity $\epsilon_{\text{eff}}(f)$ (Chapter 4).

3.5.3 Overall accuracies to be expected from the foregoing expressions

In all cases the shape ratio w/h will be accurate to ± 1 per cent.

For narrow lines ($w/h < 1.3$), ϵ_{eff} has the error range $+ 0.5$–0.0 per cent. When calculated using eqn (3.24), ϵ_{eff} is accurate to ± 0.25 per cent. The expressions for Z_0 result in accuracies to ± 1 per cent.

3.6 COMPUTER-BASED ANALYSIS TECHNIQUES

It would be very time consuming and expensive to employ *fundamental* approaches, requiring substantial computer 'power', for microstrip synthesis (or analysis) leading to computer-aided circuit design. Instead, expressions such as those given in the preceding section are much more useful and economical. However, as pointed out early in Section 3.5, the formulae only apply to a restricted range of substrates, and for design work using different substrates (e.g. quartz, plastics, etc.) a fundamental computation is necessary. Eventually a family of closed formulae would again be derived, with accuracies determined by comparison with computed results using a technique such as that outlined here.

Since an electrostatic field analysis is performed it is useful to be able to express the characteristic impedance entirely in capacitive terms.[13] This is already available in the form of eqn (3.6), which is repeated here for convenience:

$$Z_0 = \frac{1}{c\sqrt{C_1 C}} \tag{3.28}$$

Thus, in order to compute Z_0, only the capacitances per unit length, C_1 and C, are required.

Now the capacitance may, in general, be expressed in terms of net charge Q and voltage V_T as

$$C_x = \frac{Q}{V_T} \tag{3.29}$$

The charge is obtainable by applying Gauss' theorem

$$Q = \oint \mathbf{D} \cdot \mathbf{ds} \tag{3.30}$$

in which \mathbf{D} is the displacement and s is a surface enclosing the strip conductor.

Where v is a point voltage, Laplace's equation may be written

$$\nabla^2 v = 0 \tag{3.31}$$

and for every voltage v appropriate values of electric field **E** may be determined. The constitutive relation

$$\mathbf{D} = \epsilon\mathbf{E} \qquad (3.32)$$

then enables **D** to be evaluated and substituted in eqn (3.30).

In practice the computations are performed using some numerical method to solve Laplace's equation. Also, a shielded microstrip is generally considered (the shield can be progressively moved away from the strip to approach the unshielded case, if desired).

Of course, two complete calculations of capacitance, using eqn (3.29), are required: one with the substrate present, yielding C, and a second with the substrate removed, yielding C_1. Then eqn (3.28) gives the characteristic impedance. The effective microstrip permittivity ϵ_{eff} is also known, via eqns (3.8) and (3.9):

$$\epsilon_{\mathrm{eff}} = \frac{C}{C_1} \qquad (3.33)$$

Great precision is readily achievable if a sufficiently fine grading of points is set up for the computation.

3.7 A WORKED EXAMPLE OF STATIC-TEM SYNTHESIS

Let us consider the 31.3 Ω series element of the microstrip matching circuit shown for a transistor amplifier in Figure 3.1 which was given in Section 3.1. We are required to calculate the width and approximate length of this element, and for the moment we shall neglect the effects of the discontinuities present at each end, i.e. the step to 50 Ω microstrip and the bend. These and other discontinuities are dealt with in Chapter 5.

The initial specification is, from Figure 3.1:

$$Z_0 = 31.3\,\Omega, \qquad l = \lambda_g/4, \qquad f = 2\,\mathrm{GHz}$$

(We could alternatively have specified $\theta = 90°$, which is identical to $l = \lambda_g/4$.)

Also assume that a substrate is chosen having the following specification:

Relative permittivity $\epsilon_r = 9.8$

Thickness $h = 0.6\,\mathrm{mm}$

Since the frequency is only 2 GHz the static-TEM methods described here will definitely be of 'sufficient' accuracy (i.e. within ± 1 per cent when the accurate expressions are used). At first we carry out a rapid, approximate determination using Presser's graphical technique.

3.7.1 Graphical determination (see Section 3.4)

Referring to step (a) given in Section 3.4 we have

$$Z'_{01} \approx \sqrt{9.8 \times 31.3} \approx 98 \, \Omega$$

From the graph of Figure 3.7 we obtain, for this value of Z_{01}:

$$q' \approx 0.66$$

From the remaining steps in Section 3.4,

$$\epsilon'_{\text{eff}} \approx 6.8$$

and hence the improved value of Z_{01} is

$$Z_{01} \approx \sqrt{6.8 \times 31.3} \approx 81.7 \, \Omega$$

Using the graph again, with this new value of Z_{01}:

$$\frac{w}{h} = 2.2 \quad \text{and} \quad q = 0.7$$

whence

$$\epsilon_{\text{eff}} = 7.16$$

The reader is advised to check these figures independently.
Attention is now given to the use of accurate formulae.

3.7.2 Accurately calculated results

An algorithm has been devised, using BASIC language on a minicomputer, to evaluate any desired design quantities using the expressions given in Section 3.5. The correct choice of expression is made automatically in the program.
For this 31.3 Ω line the results are

$$\frac{w}{h} = 2.225 \quad \text{and} \quad \epsilon_{\text{eff}} = 7.095$$

Notice that the approximate graphical technique gives a fairly accurate result for w/h, but involves substantial error in ϵ_{eff}. This is because q cannot be accurately determined from the graph, e.g. a change of value to $q = 0.69$ yields $\epsilon_{\text{eff}} = 7.07$, which is actually slightly lower than the computed value.

3.7.3 Final dimensions of the microstrip element

The accurately calculated results are employed. Since the substrate thickness $h = 0.6$ mm, with $w/h = 2.225$, we find the microstrip width:

$$w = 1.335 \text{ mm}$$

The length is now required. An expression for the microstrip wavelength eqn (3.15) was given earlier:

$$\lambda_g = \frac{300}{F\sqrt{\epsilon_{eff}}} \text{ mm}$$

where F is in gigahertz.

Since the physical length l is required to be $\lambda_g/4$ we obtain

$$l = \frac{\lambda_g}{4} = \frac{75}{2\sqrt{7.095}}$$

so that

$$l = 14.08 \text{ mm}$$

The length calculation was *not* automatically included in the computer program because this embodies a frequency-dependent calculation (dealt with in Chapter 4) *before* microstrip wavelength is determined. Due to the effects of the discontinuities the actual length will be slightly less than the calculated value of 14.08 mm.

3.8 MICROSTRIP ON A DIELECTRICALLY ANISOTROPIC SUBSTRATE

Many substrates have been found to possess a degree of dielectric anisotropy. One particularly useful substrate material has turned out to be sapphire, which possesses well-defined dielectric anisotropy and has been quite intensively investigated.[14, 15] Because of this we only consider sapphire here and before proceeding further it is worth listing the advantages of sapphire, as follows:

(a) The material is extremely pure and the permittivity values are now accurately known (and very stable).
(b) There is negligible batch-to-batch variation. That is to say, one piece of sapphire, taken from anywhere, will be almost exactly like any other piece of sapphire.
(c) An optical surface finish is readily achievable, resulting in very uniform and precisely defined thin metal films for microstrip conductors.
(d) The material is completely optically transparent. This means that the process of mounting chip devices can be accurately registered because the operator can see the underside.
(e) Since the thermal conductivity of sapphire is about 30 per cent higher than that of alumina it may be preferred to the latter for mounting power devices.

Some other advantages were discussed in Chapter 2.

In this section we show that microstrip circuit design on sapphire may proceed in the manner so far described for isotropic substrates, provided a preferred orientation is chosen and a suitable function for an equivalent permittivity is used.

With the nomenclature shown in Figure 3.8 the permittivity tensor $\overline{\epsilon}$ may be written

$$\overline{\epsilon} = \begin{bmatrix} \epsilon_x & 0 & 0 \\ 0 & \epsilon_y & 0 \\ 0 & 0 & \epsilon_z \end{bmatrix} \tag{3.34}$$

and this interrelates the electric displacement and electric field vectors thus

$$\overline{D} = \overline{\epsilon}\epsilon_0\overline{E} \tag{3.35}$$

In all practical cases the principal axis of the sapphire is aligned to be very nearly perpendicular to the substrate–air interface, i.e. along the y direction in Figure 3.8. Therefore eqn (3.34) may be written

$$\overline{\epsilon} = \begin{bmatrix} \epsilon_\perp & 0 & 0 \\ 0 & \epsilon_\parallel & 0 \\ 0 & 0 & \epsilon_\perp \end{bmatrix} \tag{3.36}$$

Furthermore, the values are known accurately for sapphire, yielding

$$\overline{\epsilon}_{\text{sapphire}} = \begin{bmatrix} 9.40 & 0 & 0 \\ 0 & 11.60 & 0 \\ 0 & 0 & 9.40 \end{bmatrix} \tag{3.37}$$

Thus, with this particular orientation, and the sapphire substrate forming the dielectric between *two completely parallel plates with no fringing*, only the 11.60 permittivity value affects the TEM propagation which would then apply. Since microstrip is certainly not a parallel plate and does involve considerable fringing fields the effects of this anisotropy are significant. The problem was studied fundamentally by Owens et al.[14] who set up special finite difference equations and solved for the potentials around the structure. The effective microstrip permittivity ϵ_{eff} is finally found as a ratio of capacitances, as described in Section 3.6. However, because of the anisotropy, ϵ_{eff} depends upon w/h in a diferent manner to the isotropic dependence. To appreciate this more completely, the electric field distribution is shown in Figure 3.8 for three different w/h values. In case (c) the fields are almost entirely affected by ϵ_y (= 11.60). In the other extreme shown, case (a), for relatively small w/h, ϵ_x (= 9.40) significantly affects the field because there is a considerable fraction of electric field in the x direction.

To simplify the synthesis of microstrip on a sapphire substrate a new permittivity concept has been suggested to account for the gross anisotropic

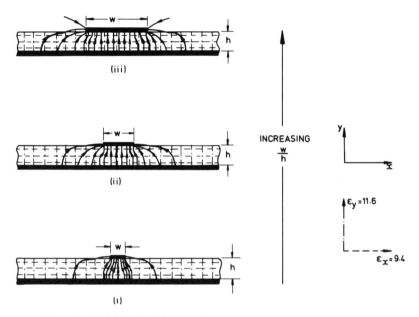

Fig. 3.8. Electric field distributions, in the transverse plane, applicable to microstrip on an anisotropic substrate (sapphire) for three shape (w/h) ratios: (i) 'small', (ii) 'moderate', and (iii) 'large'

effects. This is termed the *equivalent isotropic substrate relative permittivity* ϵ_{req}, and it is, of course, dependent upon the width-to-height ratio w/h. A strict definition of ϵ_{req} is as follows:

> The equivalent isotropic substrate relative permittivity ϵ_{req} is the permittivity of some fictitious uniform and isotropic substrate, supporting a microstrip transmission line, which would yield microstrip static-TEM design parameters having values identical to those for the same line geometry (w/h) on sapphire.

The situation is illustrated in Figure 3.9.

For a known line geometry, w/h, an appropriate value of ϵ_{req} may be calculated and then used as 'ϵ_r' in any desired microstrip design formulae. Three design curves for microstrip on sapphire, originally published in Owens *et al.*,[14] are reproduced here as Figure 3.10.

Quite accurate values of ϵ_{req} may be obtained by a combination of simple considerations and conventional (isotropic substrate) calculation. A cross-sectional model for microstrip on an anisotropic substrate assists this calculation of ϵ_{req} and is shown in Figure 3.11.

The electric field immediately beneath the strip and down to the ground plane is almost entirely affected by ϵ_y alone. Elsewhere in the substrate and

(a) SAPPHIRE SUBSTRATE (b) EQUIVALENT FICTITIOUS
 ISOTROPIC SUBSTRATE

Fig. 3.9. Illustrating the concept of equivalent isotropic substrate relative permittivity (ϵ_{req})

in the surrounding air the direction and density of the electric flux is such that ϵ_x and ϵ_y have approximately equal influences on the remaining capacitance associated with the microstrip. Thus, in Figure 3.11, we can treat the regions bounded by the magnetic walls separately. The total capacitance C per unit length of the microstrip line can thus be decomposed into separate quantities:

(a) The simple parallel-plate capacitance directly beneath the strip given by

$$C_y = \epsilon_y \epsilon_o \left(\frac{w}{h}\right) \tag{3.38}$$

(b) A fringing capacitance C_f accounting for the remaining flux regions.

As shown in Figure 3.11, there is a hypothetical permittivity quantity ϵ_i ascribed to the remaining substrate regions. If we calculate the total capacitance C_i per unit length for the microstrip, *but on some uniform isotropic substrate of permittivity* ϵ_i, the fringing capacitance is then approximately the difference:

$$C_f \simeq C_i - \epsilon_i \epsilon_o \left(\frac{w}{h}\right) \tag{3.39}$$

where C_i may be evaluated using any appropriate conventional technique as described earlier—the method outlined in Section 3.6, for example. At present ϵ_i is unknown, but it is quite a straightforward task to find a value for this quantity, as indicated shortly.

Finally, the equivalent relative permittivity ϵ_{req} is given by a combination of ϵ_i and ϵ_y, weighted by the respective capacitance ratios C_f/C and C_y/C:

$$\epsilon_{req} = \epsilon_i \left(\frac{C_f}{C}\right) + \epsilon_y \left(\frac{C_y}{C}\right)$$

However, the total capacitance $C = C_y + C_f$; therefore,

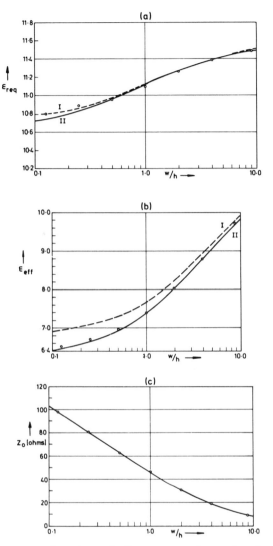

Fig. 3.10. Design curves for microstrip lines on an anisotropic substrate of sapphire (C axis perpendicular to the ground plane). (© 1976 IEEE. Reprinted, with permission, from Owens *et al.*[14])

$$\epsilon_{\text{req}} = \frac{\epsilon_i C_f + \epsilon_y C_y}{C_f + C_y} \qquad (3.40)$$

An optimized value $\epsilon_i = 10.60$ was obtained by curve fitting to the finite difference theory values making up the discrete points shown alongside curve I of Figure 3.10(a)—where curve II is the actual plot of eqn (3.40)

MAGNETIC WALLS

Fig. 3.11. A model for microstrip on an anisotropic substrate with C axis perpendicular to the ground plane. (© 1976 IEEE. Reprinted, with permission, from Owens et al.[14])

using this value of ϵ_i. This value can be used as a constant in determining ϵ_{req} for particular microstrip-on-sapphire designs. According to Figure 3.10(a) the worst discrepancy, amounting to 0.65 per cent, occurs at the $w/h = 0.1$ extreme.

As a purely empirical alternative, the following expression gives ϵ_{req} and the *only* independent variable involved is w/h:

$$\epsilon_{req} = 12.0 - \frac{1.21}{1 + 0.39\{\log(10w/h)\}^2} \tag{3.41}$$

The accuracy of eqn (3.41) when tested against other theoretical calculations and measurements—with results extrapolated to zero frequency—appears to be ± 0.5 per cent throughout the range

$$0.1 \le \frac{w}{h} \le 10$$

which, on the sapphire substrate, approximates to the characteristic impedance range

$$10 \le Z_0 \le 100 \ \Omega$$

and this is quite sufficient for a wide variety of circuit design purposes.

When static-TEM analysis is applied to microstrip-on-sapphire designs the effective permittivity evaluated, ϵ_{eff}, will depend upon the shape ratio w/h. This is clearly shown by the curves of Figure 3.10(b) where curve I has been generated using ϵ_y as 'ϵ_r' in the calculations, but curve II is generated using eqn (3.40) for ϵ_{req}. The importance of employing the correct permittivity (ϵ_{req}) is evident since only then are the results close to the theoretical predictions (discrete points). For Figure 3.10(c), ϵ_{req} was used in determining the continuous curve. Thus, curve II of Figure 3.10(b) and the curve of Figure 3.10(c) represent design curves for microstrip on sapphire.

There are several other dielectrically anisotropic substrate materials, notably Epsilam-10 and single-crystal quartz. In any design using these materials the principles outlined above may be developed in the same manner to provide a design basis.

Mariki and Yeh[16] have reported a three-dimensional TLM analysis of microstrip on an anisotropic substrate. Their results are entirely concerned with *dispersion* for this case and are considered in rather more detail in Chapter 4.

3.9 MICROSTRIP ON A FERRITE SUBSTRATE

A number of circuits have been designed and built on various ferrite substrates and it appears that the earliest significant papers appeared as long ago as 1957–58.[17, 18] However, the realization of production circuits using ferrite substrates is expensive, on account of the relatively high cost of the ferrite and its finishing. Further, many circuit functions certainly do not require ferrite for their operation—indeed, greater quality is generally achievable on purely dielectric substrates such as sapphire, alumina, quartz, etc. Where non-reciprocal properties are demanded ferrite devices may be inserted into the substrate. Therefore only a brief treatment of the design problem is presented here.

For magnetically saturated ferrite, magnetized in the z direction and with an r.f. field propagating in this direction, the Polder tensor is

$$\bar{\mu} = \mu_0 \begin{bmatrix} \mu & -j\kappa & 0 \\ j\kappa & \mu & 0 \\ 0 & 0 & \mu_z \end{bmatrix} \qquad (3.42)$$

where $\mu_z = 1$ and the permeability elements μ and κ are exactly related to precession frequency, microwave frequency, saturation magnetization, d.c. magnetic field, macroscopic relaxation time, and resonance line width.

In the limit of high applied field or high frequency the ferrite behaves as a dielectric medium with a moderately low loss (tan $\delta \simeq 0.001$ typically) and permittivity range

$$9 \le \epsilon_r \le 16$$

In the partially magnetized state (common in practice) μ is somewhat less than unity and

$$\kappa = \frac{\gamma(4\pi M)}{\omega}$$

where γ and $4\pi M$ are respectively the gyromagnetic ratio and the average magnetizations for the ferrite.

Effective permeability μ_{eff} is an important quantity which is analogous to a bulk substrate relative permeability and is thus very useful in design. The defining formula is

$$\mu_{\text{eff}} = \frac{\mu^2 - \kappa^2}{\mu} \tag{3.43}$$

At sufficiently high frequency $\omega \gg \gamma\,(4\pi M)$ and $\kappa \to 0$, so that $\mu_{\text{eff}} \to \mu$. Under these conditions μ_{eff} is not much less than unity:

$$0.7 \leq \mu_{\text{eff}} \leq 1.0 \qquad (\text{as } \omega \to \infty)$$

The quantity κ/μ is usually termed the (ferrite) anisotropy. In the case of microstrip an empirical correction has been given for μ_{eff}.[19] But the above comments remain approximately true.

It is useful to define an *effective microstrip permeability* μ_e as being analogous to the effective microstrip permittivity defined earlier in this chapter by eqn (3.12), repeated here,

$$\epsilon_{\text{eff}} = 1 + q(\epsilon_r - 1) \tag{3.44}$$

where q, and hence ϵ_{eff} are functions of w/h and ϵ_r.

Provided some basic and reasonable assumptions are made regarding a microstrip line (perfect conductors; isotropic, homogeneous, non-gyromagnetic substrate)[19] then its dielectric properties will not affect the magnetic field distribution and its magnetic properties will not affect its electric field distribution. We can then write down a magnetic–electric duality statement:

$$\mu_e \left\{ \frac{w}{h}, \mu_{\text{eff}} \right\} = \left(\epsilon_{\text{eff}} \left\{ \frac{w}{h}, \epsilon_r \right\} \right)^{-1} \tag{3.45}$$

By analogy with eqn (3.44) we can also write

$$\mu_e = \left(1 + q_m \left\{ \frac{1}{\mu_{\text{eff}}} - 1 \right\} \right)^{-1} \tag{3.46}$$

where q_m is the *magnetic* filling factor. This completely defines μ_e. However, although not explicitly indicated here, it must be appreciated that both q_m and μ_{eff} are functions of w/h for the microstrip. For computer-aided design purposes Pucel and Masse[19] gave analytical expressions for μ_e and μ_{eff}. These expressions will not be repeated here, but a graph is given (Figure 3.12) as a design aid.

We also require expressions for the characteristic impedance and the guide wavelength. Using eqn (3.1), $Z_0 = (L/C)^{1/2}$, we obtain

$$Z_0 = \left(\frac{\mu_e L_1}{\epsilon_{\text{eff}} C_1} \right)^{1/2} \tag{3.47}$$

where L_1 and C_1 are for the air-spaced case. Hence

$$Z_0 = Z_{01} \sqrt{\frac{\mu_e}{\epsilon_{\text{eff}}}} \tag{3.48}$$

Fig. 3.12. Effective microstrip *permeability* μ_e as a function of w/h. The effective relative permeability of the substrate material μ_{eff} is the parameter. (© 1972 IEEE. Reprinted, with permission, from Pucel and Masse[19])

also, using eqn (3.3), $v_p = (LC)^{-1/2}$, we obtain

$$v_p = \frac{c}{\sqrt{\mu_e \epsilon_{eff}}} \tag{3.49}$$

which leads to

$$\lambda_g = \frac{\lambda_0}{\sqrt{\mu_e \epsilon_{eff}}} \tag{3.50}$$

or

$$\lambda_g = \frac{300}{F\sqrt{\mu_e \epsilon_{eff}}} \text{ mm} \tag{3.51}$$

(compare eqn 3.15).

For synthesis, where w and λ_g are desired, it is recommended that the design is first carried out assuming a *non-ferrite* substrate (but, of course, with the appropriate permittivity). One or two stages of iteration, using the above techniques and information, should then suffice to obtain quite accurate final values for w and λ_g. This is assisted by the fact that μ_{eff} and μ_e both vary only slowly over the range

$$0.1 \leq \frac{w}{h} \leq 10.0$$

3.10 EFFECTS OF FINITE STRIP THICKNESS, METALLIC ENCLOSURE, AND MANUFACTURING TOLERANCES

3.10.1 Effects of finite strip thickness

Referring back to Figure 3.2 we see that microstrip, in general, will have a finite thickness t which must influence the field distribution. For *most* single microstrip lines the effect of this thickness on the design parameters is very small and may often be neglected. Even in the case of microstrip circuits manufactured using 'thick-film' technology (Section 2.12) there is usually no need to allow for thickness when calculating impedance or effective microstrip permittivity because such films invariably taper towards the strip edges.

With some etched microstrip circuits, on plastic substrates, and circuits where the microstrips are designed to carry at least moderate power, the thickness may be significant. An indication of the change in electric field distribution is provided by Figure 3.13, and several workers have investigated the effects of finite strip thickness. Some straightforward and accurate expressions have been reported[20] which are now stated here, starting with characteristic impedances:

For $w/h \leq 1$:

$$Z_0 = \frac{60}{\sqrt{\epsilon_{\text{eff}}}} \ln \left(8\frac{h}{w_e} + 0.25\frac{w_e}{h} \right) \tag{3.52a}$$

For $w/h \geq 1$:

$$Z_0 = \frac{120\pi}{\sqrt{\epsilon_{\text{eff}}}} \left\{ \frac{w_e}{h} + 1.393 + 0.667\left(\frac{w_e}{h} + 1.444 \right) \right\}^{-1} \tag{3.52b}$$

where
With $w/h \leq 1/2\pi$:

$$\frac{w_e}{h} = \frac{w}{h} + \frac{1.25t}{\pi h}\left(1 + \ln\frac{4\pi w}{t} \right) \tag{3.53a}$$

With $w/h \geq 1/2\pi$:

$$\frac{w_e}{h} = \frac{w}{h} + \frac{1.25t}{\pi h}\left(1 + \ln\frac{2h}{t} \right) \tag{3.53b}$$

Finally, the effective microstrip permittivity ϵ_{eff} should be evaluated using expressions given previously (Section 3.5) and the following term subtracted:[20]

$$\Delta\epsilon_{\text{eff}}(t) = \frac{(\epsilon_r - 1)t/h}{4.6\sqrt{w/h}} \tag{3.54}$$

(a) Microstrip showing thickness t

E-field

(b) (c)

Increasing
thickness t

Fig. 3.13. Changes in the distribution of electric field (transverse cross-section) as
the thickness of microstrip is altered

so that the final value is given by

$$\epsilon_{\text{eff}}(t) = \epsilon_{\text{eff}} - \Delta\epsilon_{\text{eff}}(t) \tag{3.55}$$

For microstrip having $t/h \le 0.005$, $2 \le \epsilon \le 10$, and $w/h \ge 0.1$, the effects
of this thickness are negligible (within approximately 1 per cent on Z_0 or
ϵ_{eff}). At smaller values of w/h *or* greater values of t/h the significance
increases. These trends are clearly shown in curves published by Schwan,[21]
although he gave no details of their derivation. It is interesting to compare
the trends exhibited in Schwan's results with those predicted by the
expressions given here (eqns 3.52a to 3.54). Rather arbitrarily an alumina
substrate is chosen, with relative permittivity 10 and thickness 0.5 mm, and
we also assume a strip conductor thickness of 0.032 mm, that is $t/h = 0.064$.
With a shape ratio $w/h = 0.1$ Schwan indicates a 20 to 30 per cent reduction
in Z_0 and approximately a 14 per cent reduction in ϵ_{eff}, both compared with
the $t/h = 0$ values. Careful checks with the formulae given here confirm
these trends, but show much smaller changes—approximately 7 per cent
for *both* the permittivity and the impedance. Measurements and other
indications tend to support these smaller changes.

The effect of the thickness of the conducting strip is quite significant when considering conductor losses; this is examined in Chapter 4.

3.10.2 Effects of a metallic enclosure

In many practical applications MICs entail a metallic enclosure which serves the purposes of hermetic sealing, mechanical strength, electromagnetic shielding, connector mounting, and ease of handling of the module. The presence of conducting top and side walls lowers both the characteristic impedance and the effective microstrip permittivity, which is due to the increased proportion of electric flux in air. The effects of a top cover alone, as well as those due to a complete enclosure, have been reported by several workers. Closed formulae[22] are given here which predict the effects of a conducting top cover and show how both the characteristic impedance and effective microstrip permittivity are modified in comparison with the unshielded static-TEM expressions given in Section 3.5. The nomenclature of Figure 3.14 is used throughout.

Characteristic impedance is considered first:

Where $w/h \leq 1.3$:

$$Z_0(\text{shielded}) = Z_0(\text{unshielded}) - \Delta Z_{0s1} \qquad (3.56)$$

Where $w/h \geq 1.3$:

$$Z_0(\text{shielded}) = Z_0(\text{unshielded}) - \Delta Z_{0s2} \qquad (3.57)$$

in which

$$\Delta Z_{0S1} = 270\{1 - \tanh(0.28 + 1.2\sqrt{h'/h})\} \qquad (3.58)$$

and

$$\Delta Z_{0S2} = \Delta Z_{0S1}\left(1 - \tanh\left[1 + \frac{0.48\{(w_e/h) - 1\}^{0.5}}{\{1 + (h'/h)\}^2}\right]\right) \qquad (3.59)$$

Fig. 3.14. Cross-section of shielded microstrip

The effective microstrip permittivity of a shielded (top cover only) microstrip line is given by

$$\epsilon_{\text{eff}} + \frac{\epsilon_r + 1}{2} + \left(\frac{\epsilon_r - 1}{2} R\right) \tanh\left\{0.18 + 0.237(h'/h) - \frac{0.415}{(h'/h)^2}\right\} \quad (3.60)$$

in which

$$R = \left\{ \begin{array}{ll} \{1 + 12(h/w)\}^{-0.5} + 0.04\{1 - (w/h)\}^2 & \text{for } w/h \leq 1 \\ \{1 + 12(h/w)\}^{-0.5} & \text{for } w/h \geq 1 \end{array} \right.$$

The expressions quoted in Section 3.5 might alternatively be used, but the nature of the shielding-modifying term is then uncertain.

In all these equations, (3.56) to (3.60), w_e has already been defined by eqns (3.53a) and (3.53b), Z_0 (unshielded) can be determined using expressions given previously in Section 3.5, and $b' = a - h$, referring to Figure 3.14.

Comparison with the theoretical results due to Kowalski and Pregla[23] shows discrepancies within 1 per cent. However, more extensive experimental investigation is highly desirable and design expressions allowing for the proximity *of at least one side wall* would be particularly useful.

3.10.3 Effects due to manufacturing tolerances

The effects of tolerances on various parameters has been quite intensively investigated and reported in detail by Gupta *et al.*[1] Although often difficult, it is important to control the variations in thickness h and relative permittivity ϵ_r of substrates. Some batch-to-batch variations are, however, inevitable, and the effects of these on electrical quantities such as impedance and VSWR must be determined. In addition to the effects of thickness and permittivity variations, the surface finish of the substrate affects the precision with which the strip width can be manufactured.

The maximum excursion in the characteristic impedance is given by

$$\frac{|\Delta Z_0|_{\max}}{Z_0} = \left|\frac{\Delta w}{w} S_{Z_0(w)}\right| + \left|\frac{\Delta h}{h} S_{Z_0(h)}\right| + \left|\frac{\Delta \epsilon_r}{\epsilon_r} S_{Z_0(\epsilon_r)}\right| \quad (3.61)$$

and the maximum excursion in the effective microstrip permittivity is given by

$$\frac{|\Delta \epsilon_{\text{eff}}|_{\max}}{\epsilon_{\text{eff}}} = \left|\frac{\Delta w}{w} S_{\epsilon_{\text{eff}}(w)}\right| + \left|\frac{\Delta h}{h} S_{\epsilon_{\text{eff}}(h)}\right| + \left|\frac{\Delta \epsilon_r}{\epsilon_r} S_{\epsilon_{\text{eff}}(\epsilon_r)}\right| \quad (3.62)$$

where the sensitivity is, in general, defined by

$$S_{Q(p)} = \frac{p}{Q} \frac{\partial Q}{\partial p} \quad (3.63)$$

in which Q is a quantity whose sensitivity is to be determined (Z_0, ϵ_{eff}) and P is a controlling parameter (w, h, ϵ_r).

It is particularly important that the demanded accuracy in the manufacture of strip width is evaluated. This is the minimum selected from the two values of Δw obtained by rearranging eqns (3.61) and (3.62) for

$$\frac{\Delta w}{w} S_{Z_0(w)}$$

and

$$\frac{\Delta w}{w} S_{\epsilon_{eff}(w)}$$

respectively. Any of the required sensitivities may be obtained by appropriately differentiating expressions given in Section 3.5.

Gupta et al.[1] show that the mismatch (VSWR) due to tolerances in the various quantities (ϵ_r, h, and w) may be directly related to the resultant characteristic impedance change. The following expression is given for VSWR(r):

$$r = \left(1 - \frac{|\Delta Z_0|_{max}}{Z_0}\right)^{-1} \tag{3.64}$$

where the term in Z_0 is determined from eqn (3.61). Detailed graphs of parameter variations are given for an alumina substrate in the cited reference. In broad terms it is found that the VSWR can be held below 1.05 provided the tolerances are less than approximately 5 per cent.

For similar tolerances the worst VSWR is approximately 1.1 (alumina *or* plastic substrates). The maximum change in effective microstrip permittivity (eqn 3.62) is approximately 3.3 per cent, on alumina, when the relative permittivity tolerance is also approximately 3.5 per cent, *or* the substrate thickness tolerance about 4 per cent, *or* the microstrip width tolerance is about 8 per cent.

3.11 PULSE PROPAGATION ALONG MICROSTRIP LINES

There are many instances in which microstrip lines are used for carrying pulses, rather than microwave energy, in high-speed systems. This area of application tends to be increasing in importance, especially with the introduction of digital circuits in gallium arsenide monolithic technology. Examples of such applications include:

very high speed computer logic
high bit-rate digital communications (e.g. direct pulsing of a laser or LED for optical communications)

high-speed samplers for oscilloscopes or time-domain reflectometers
'pico-second' resolution counters
radar

One of the most significant properties of microstrip in these applications is
that of *propagation delay*. The delay time per metre, τ_d, for a microstrip
line is given simply by

$$\tau_d = \frac{1}{v_p} \quad \text{s/m}$$

or, using eqn (3.9),

$$\tau_d = \frac{\sqrt{\epsilon_{eff}}}{c} \quad \text{s/m} \tag{3.65}$$

where ϵ_{eff} and c have the definitions given previously.

The static-TEM effective microstrip permittivity ϵ_{eff} is determined using
any appropriate method as described earlier. As an example, let us consider
a 50 Ω microstrip line on high-purity alumina: $\epsilon_{eff} \approx 6.7$. We obtain, using
eqn (3.65),

$$\tau_d = \frac{\sqrt{6.7}}{3 \times 10^8} \quad \text{s/m}$$

$$\approx 8.6 \text{ ns/m}$$

In practice this means that the delay is 8.6 ps/mm and a length of 20 mm
produces 172 ps of delay for a pulse. High-speed gates may typically have
around 300 ps delay per gate, so that we often require considerably more
than 20 mm of microstrip in systems using such gates. A good example is
the triggered phase-locked oscillator. In this, a 2.5 ns delay is typically
required and it is estimated that a total microstrip run approaching 300 mm
is necessary. A cursory study of Figure 3 in the paper by Chu *et al.*[24] shows
that a very long run of microstrip is indeed implemented.

One possible way in which to increase the effective propagation delay is
to introduce capacitive loading at intervals along the microstrip. The delay
then becomes influenced by these capacitances ($v_p = (LC)^{-1/2}$) and v_p is
reduced, i.e. the pulse is slowed, if C is increased. Further, the effective
inductance per unit length may also be increased and the microstrip takes
on the general appearance shown in Figure 3.15.

This structure is, of course, in the frequency domain, a form of low-pass
filter, and such circuits are discussed in Chapter 8. The delay associated
with such a structure is given by $n\sqrt{LC}$, where n is the (equal) number of
sections representing L (or C). Design techniques for the microstrip sections
are given in Chapter 8.

Fig. 3.15. A microstrip circuit structure yielding substantial delay

Crosstalk between adjacent circuits is a serious problem in pulse systems. Methods for calculating the amount of crosstalk require a study of the coupled microstrip line problem, discussed in Section 6.14.

3.12 RECOMMENDATIONS RELATING TO THE STATIC-TEM APPROACHES

This section provides a summary of the principal design methods and formulae which have been presented in this chapter. Although the static-TEM approaches, when used directly, exhibit significant errors at frequencies beyond a few gigahertz they form a basis for powerful design expressions which are used at higher frequencies. This will be discussed in Chapter 4.

3.12.1 The principal static-TEM synthesis formulae

These were given in Section 3.5. They are of proven accuracy to within 1 per cent for w/h and 0.5 per cent for ϵ_{eff}; some are repeated here for convenience.

(a) For narrow strips (i.e. when $Z_0 > (44 - 2\epsilon_r)\ \Omega$):

$$\frac{w}{h} = \left(\frac{\exp H'}{8} - \frac{1}{4 \exp H'}\right)^{-1}$$

under the slightly different condition, $w/h < 1.3$:

$$\epsilon_{\text{eff}} = \frac{\epsilon_r + 1}{2}\left\{1 - \frac{1}{2H'}\left(\frac{\epsilon_r - 1}{\epsilon_r + 1}\right)\left(\ln\frac{\pi}{2} + \frac{1}{\epsilon_r}\ln\frac{4}{\pi}\right)\right\}^{-2}$$

where, for both expressions,

$$H' = \frac{Z_0\sqrt{2(\epsilon_r + 1)}}{119.9} + \frac{1}{2}\left(\frac{\epsilon_r - 1}{\epsilon_r + 1}\right)\left(\ln\frac{\pi}{2} + \frac{1}{\epsilon_r}\ln\frac{4}{\pi}\right)$$

(b) For wide strips (i.e. when $Z_0 < (44 - 2\epsilon_r)\ \Omega$):

$$\frac{w}{h} = \frac{2}{\pi}\left\{(d_\epsilon - 1) - \ln(2d_\epsilon - 1)\right\}$$

$$+ \frac{\epsilon_r - 1}{\pi \epsilon_r} \left\{ \ln (d_\epsilon - 1) + 0.293 - \frac{0.517}{\epsilon_r} \right\}$$

where

$$d_\epsilon = \frac{59.95 \pi^2}{Z_0 \sqrt{\epsilon_r}}$$

and, again with the slightly different condition, $w/h > 1.3$:

$$\epsilon_{\text{eff}} = \frac{\epsilon_r + 1}{2} + \frac{\epsilon_r - 1}{2} \left(1 + 10 \frac{h}{w} \right)^{-0.555}$$

For the effects of shielding (top cover), finite thickness of strip, and manufacturing tolerances Section 3.10 should be consulted.

The approximate graphical method described in Section 3.4 is useful at least for a 'first run' analysis or synthesis and is applicable to *any substrate*. Apart from lack of accuracy its main drawback is the restricted range encompassed.

3.12.2 Microstrip on a sapphire (anisotropic) substrate

This problem was discussed in detail in Section 3.8. The main design expression is the empirical formula for equivalent relative permittivity ϵ_{req}

$$\epsilon_{\text{req}} = 12.0 - \frac{1.21}{1 + 0.39 \{\log(10w/h)\}^2}$$

Once more, this expression is only known to be accurate within ± 0.5 per cent over the range

$$0.1 \leq \frac{w}{h} \leq 10$$

On the sapphire substrate this approximates the characteristic impedance range

$$10 \leq Z_0 \leq 100 \; \Omega$$

The value of ϵ_{req} obtained using the above expression is directly substituted for 'ϵ_r' in the static-TEM formulae which apply to the more useful isotropic substrate situation. However, w/h is *initially* not accurately known in design and some iterative CAD routine is therefore required. The designer might start with a rough initial value $\epsilon_{\text{req}} \approx 11$ for sapphire, compute w/h from a static-TEM expression, evaluate an up-dated ϵ_{req}, etc. Since $\epsilon_{\text{req}} (w/h)$ is a fairly slowly varying function this would be a good, fast-converging approach.

Table 3.1.

$\sim w/h$	$\sim Z_0$ (Ω)	\sim Percentage reduction in w
0.1	110	3
1.0	50	2.5
8.0	12	1

3.12.3 Design corrections

Once an MIC has been manufactured there is often little that one can do to 'trim' the structure and hence improve upon the performance of the complete circuit. Where the passive (i.e. microstrip) circuitry can be tested in the absence of any active devices the deviations in actual measured performance may be observed and used to calculate the required artwork/processing adjustments for an improved second attempt. Usually the spread of characteristics and the rejection costs preclude this approach when active devices have been incorporated. Sometimes it may also be possible to etch or laser-trim microstrip elements.

Whatever the method, the following guidelines are applicable to microstrip on an alumina-type substrate:

(a) *Always* start with a slightly lower impedance than actually desired, i.e. larger w/h, if trimming is contemplated. Table 3.1 indicates the approximate magnitude of adjustments for a 1 per cent increase in Z_0.

(b) The physical lengths of sections should also be made, initially, slightly longer than finally required if trimming is contemplated. This is because the operating frequency for these lengths will then be slightly lower than required and the physical length (proportional to wavelength) must be reduced by etching or laser-trimming to increase the operating frequency back to the correct value. The adjustment may be made on a *pro rata* basis, i.e. with a 1 per cent reduction in length we can expect approximately a 1 per cent increase in frequency.

(c) As a final suggestion the length of a top-cover shield might be adjusted to 'trim' the performance of MICs. See Section 3.10.2 for suitable expressions which would yield the values of adjustments required.

REFERENCES

1. Gupta, K. C., Garg, Ramesh, and Bahl, I. J., *Microstrip Lines and Slotlines*, Artech House Inc., Dedham, Massachusetts, 1979, Chap. 1.
2. Greig, D. D., and Engelmann, H. F., 'Microstrip—a new transmission technique for the kilomegacycle range', *Proc. IRE*, **40**, No. 12, December 1952, 1644–1650.

3. Fromm, W. E., and Fubini, E. G., 'Characteristics and some applications of stripline components', *Proc. National Electronics Conf.*, **10**, held at Hotel Sherman, Chicago, Ill., 4–6, October 1954, 58–59.
4. Dukes, J. M. C., 'The application of printed circuit techniques to the design of microwave components', *Proc. IEE*, **105**, Part B, March 1958, 155–172.
5. Yamashita, E., and Atsuki, K., 'Design of transmission-line dimensions for a given characteristic impedance', *IEEE Trans.*, **MTT-17**, August 1969, 638–639.
6. Wheeler, H. A., 'Transmission-line properties of parallel wide strips by a conformal mapping approximation', *IEEE Trans.*, **MTT-12**, May 1964, 280–289.
7. Wheeler, H. A., 'Transmission-line properties of parallel strips separated by a dielectric sheet', *IEEE Trans.*, **MTT-13**, No. 3, March 1965, 172–185.
8. Owens, R. P., 'Accurate analytical determination of quasi-static microstrip line parameters', *The Radio and Electronic Engineer*, **46**, No. 7, July 1976, 360–364.
9. Presser, A., 'RF Properties of microstrip lines', *MicroWaves*, **7**, March 1968, 53–55.
10. Hammerstad, E. O., and Bekkadal, F. 'A microstrip handbook', *ELAB Report*, STF44 A74169, University of Trondheim, Norway, February 1975.
11. Cahill, L. W., 'Approximate formulae for microstrip transmission lines', *Proc. IREE, Australia*, **3**, October 1974, 317–321.
12. Schneider, M. V., 'Microstrip lines for microwave integrated circuits', *Bell Syst. Tech. J.*, **48**, May/June 1969, 1421–1444.
13. Bryant, T. G., and Weiss, J. A., 'Parameters of microstrip transmission lines and of coupled pairs of microstrip lines', *IEEE Trans.*, **MTT-16**, No. 12, December 1968, 1021–1027.
14. Owens, R. P., Aitken, J. E., and Edwards, T. C., 'Quasi-static characteristics of microstrip on an anisotropic sapphire substrate', *IEEE Trans.*, **MTT-24**, No. 8, August 1976, 499–505.
15. Hayashi, Y., and Kitazawa, T., 'Analysis of microstrip transmission lines on a sapphire substrate', *Trans. Inst. Electron & Common. Eng. Jpu Sect. E (Japan)*, **E62**. No. 6, June 1979, 426–427.
16. Mariki, G. E., and Yeh, C., 'Dynamic Three-Dimensional TLM Analysis of Microstriplines on Anisotropic Substrate', *IEEE Trans.*, **MTT-33**, No. 9, 1985, 789–799.
17. Lewin, A., 'A resonance absorption isolator in microstrip at 4Gc/s, *Proc. IEE*, **B 104**, Suppl, 6, 1957, 364–365.
18. Brodwin, M. E., 'Propagation in ferrite-filled microstrip', *IRE Trans. Microwave Theory Tech.*, **6**, 1958, 153–155.
19. Pucel, R. A., and Masse, D. J. 'Microstrip propagation on magnetic substrates, Part I. Design theory. Part II. Experiment', *IEEE Trans.*, **MTT-20**, 1972, pp. 304–308 and 309–313.
20. Bahl, I. J., and Garg, Ramesh, 'Simple and accurate formulas for microstrip with finite strip thickness', *Proc. IEEE*, **65**, 1977, 1611–1612.
21. Schwan, K. P., 'Take the guesswork out of thick microstrip K', *MicroWaves*, December **1977**, 184–185.
22. Bahl, I. J., 'Use exact methods for microstrip design', *MicroWaves*, December **1978**, 61–62.
23. Kowalski, G., and Pregla, R., 'Dispersion characteristics of shielded microstrips with finite thickness', *AEU*, **25**, 1971, 193–196.
24. Chu, D. C., Allen, M. S., and Foster, A. S., 'Universal counter resolves picoseconds in time interval measurements', *Hewlett-Packard Journal*, August **1978**, 2–11.

4 Behaviour and Design at High Frequencies

4.1 THE SCOPE OF THIS CHAPTER

In Chapter 3 it was explained that microstrip cannot support a pure TEM, or any other simple electromagnetic field mode, and the necessary longitudinal field components which exist lead to the propagation of hybrid modes. It was further indicated that the 'quasi-TEM' nature of the propagating fields can be approximately calculated by the use of some statically derived expressions. Such expressions have accuracies to within 1 per cent where they are applicable—generally at frequencies below 1 or 2 GHz. They also form an important basis for calculations at higher frequencies.

In this chapter we study the behaviour of microstrip rather more deeply, leading to fairly straightforward design calculations for frequencies above a few gigahertz.

The following list gives the principal topics discussed:

(a) dispersion in microstrip

(b) a model leading to design calculations which allows for dispersion

(c) improved accuracy calculations

(d) an example (a specific design calculation)

(e) a frequency-dependent electromagnetic field solution (using a numerical method and requiring a substantial computer program)

(f) characteristic impedance as functions of frequency

(g) operating frequency limitations

(h) power losses and parasitic coupling.

4.2 DISPERSION IN MICROSTRIP

When the frequency of a signal exciting a microstrip line is (say) doubled, the phase constant or wave number $\beta(= 2\pi/\lambda_g)$ is *not* exactly doubled. Several transmission line structures exhibit this type of behaviour, indicated in Figure 4.1; it is called *dispersion*.

Thus all microstrip lines are dispersive and it follows that the 'exact' relationship between wavelength and frequency is very complicated. Detailed time and space-variant electromagnetic field analyses have been performed requiring numerical methods and substantial computer programs to solve them. Fortunately, as we shall see, useful and accurate (typically within 1 per cent) design calculations may be carried out with suitable closed formulae.

The hybrid mode of propagation along microstrip leads to the dispersive mechanism, and an examination of the types of fields occurring in a *shielded* microstrip line, itself an important structure, helps to explain this dynamic situation (Figure 4.2).

If the centre strip is removed from the shielding waveguide, it reduces to a guide with partial dielectric filling that can support longitudinal section

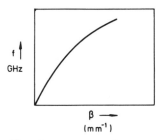

Fig. 4.1. The dispersive effect in *any* general structure or system—non-linearity when frequency f is plotted against wave number or phase coefficient β

Fig. 4.2. Transverse cross-section of shielded microstrip, showing the nomenclature

electric (LSE) or longitudinal section magnetic (LSM) modes.[1] Pure TE or TM modes cannot, however, be supported—unlike the conventional unfilled waveguide. When the centre strip is inserted it causes currents to flow in both x (transverse) and z (longitudinal) directions on this strip; these currents serve to couple the LSE and LSM modes so that the final mode configuration is hybrid. As the frequency is increased, the mode-coupling efficiency increases and the fields become more concentrated in the region beneath the strip—where the substrate permittivity has already resulted in a relatively large electric displacement. Since the fields are forced into the dielectric substrate to an increasing extent as the frequency rises we can define a frequency-dependent effective microstrip permittivity $\epsilon_{\text{eff}}(f)$. This quantity clearly increases with frequency and the wave is progressively slowed down. We may continue to use the more general definition of *effective microstrip permittivity* given by eqn (3.8), except that ϵ_{eff} must now be replaced by $\epsilon_{\text{eff}}(f)$ and v_p by $v_p(f)$ giving

$$\epsilon_{\text{eff}}(f) = \left\{ \frac{c}{v_p(f)} \right\}^2 \tag{4.1}$$

Fundamentally, the dispersion problem then consists of solving the microstrip fields for the velocity $v_p(f)$. The limits of $\epsilon_{\text{eff}}(f)$ are readily established; at the low-frequency extreme it reduces to the static-TEM value, while as frequency is increased indefinitely $\epsilon_{\text{eff}}(f)$ approaches the substrate permittivity itself. This is summarized as follows:

$$\epsilon_{\text{eff}}(f) \rightarrow \begin{cases} \epsilon_{\text{eff}} & \text{as } f \rightarrow 0 \\ \epsilon_r & \text{as } f \rightarrow \infty \end{cases}$$

Between these limits $\epsilon_{\text{eff}}(f)$ changes continuously, as shown by the general curve of Figure 4.3.

There have been several approaches to the detailed hybrid-mode analysis of microstrip,[1–13] one of which is described towards the end of this chapter. A number of closed formulae based upon various models of microstrip have also been developed, and some of these are briefly discussed and compared in the next section.

4.3 APPROXIMATE CALCULATIONS WITH DISPERSION

As explained in Chapter 3 the circuit designer wishes to know the wavelength λ_g of the dominant mode in the microstrip whose characteristic impedance Z_0 and aspect ratio w/h are already known. This implies that $\epsilon_{\text{eff}}(f)$ is required, since

$$\lambda_g = \frac{c}{f \sqrt{\epsilon_{\text{eff}}(f)}} \tag{4.2}$$

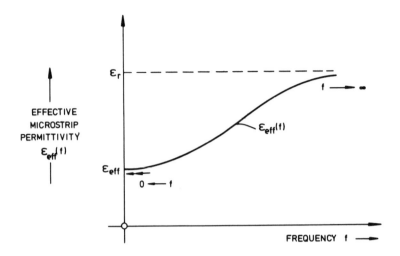

Fig. 4.3. Dispersion in microstrip interpreted as effective microstrip permittivity $\epsilon_{\text{eff}}(f)$ plotted to a base of frequency

It is generally better to find expressions for $\epsilon_{\text{eff}}(f)$ first of all, because changes in $\epsilon_{\text{eff}}(f)$ with frequency are larger and gross errors in the calculations can readily be detected (e.g. it must lie within the limits shown in Figure 4.3).

Fundamentally, this effective microstrip permittivity must be a function of frequency, substrate height, and substrate relative permittivity:

$$\epsilon_{\text{eff}}(f, h, \epsilon_{\text{r}})$$

where it should be noted that the substrate height h enters into the expressions because it influences the surface waves and thus also the complete hybrid field pattern. This can be seen by imagining the effect of increasing the proportion of dielectric substrate in Figure 4.2.

Most of the 'closed-formula' expressions for calculating dispersion also require the static-TEM value of ϵ_{eff}, and sometimes Z_0 as well. These are obtained by the methods given in Chapter 3. Thus, the frequency-dependent effective microstrip permittivity may also be written as a function involving ϵ_{eff} and Z_0 as follows:

$$\epsilon_{\text{eff}}(f, h, \epsilon_{\text{r}}, \epsilon_{\text{eff}}, Z_0)$$

The four methods described here have been superseded by more accurate approaches; therefore they are only dealt with rather briefly.

4.3.1 Earlier approaches

One of the first attempts at developing a closed-formula for dispersion calculations was due to Jain, Makios, and Chudobiak.[14-16] Their first formula

was derived empirically and showed only a first-order relationship with frequency.[14]

This is known to be very inaccurate (clearly it is, at least, incapable of approaching the limit conditions already discussed here) and their more comprehensive method must therefore be examined. Their approach consisted basically of analysing the coupling between the TEM mode and a surface-wave mode TM_0. Rather elaborate algebraic expressions are obtained and either a graphical or a numerical method is required to solve a special effective permittivity which is used. The main expression used takes $\epsilon_{eff}(f) > \epsilon_r$ at high frequencies—which is physically impossible as we have already seen. In fact the full dispersion curves given by Jain, Makios, and Chudobiak clearly show $\epsilon_{eff}(f)$ progressing beyond the limit of 9.9 set by the substrate permittivity. Otherwise the theory shows quite close agreement with experimental results, except that these were obtained using ring resonators which give unreliable results and are discussed later. The actual values of ϵ_{eff} which were used in the work of Jain, Makios, and Chudobiak are all too high due to substantial errors in the static-TEM expressions which were used at that time.

A quite different approach was adopted by Schneider,[17] who took four basic features regarding microstrip dispersion and identified a rational function which satisfied all these features.

They are as follows:

(a) $\epsilon_{eff}(f)$ always *increases* with frequency.

(b) $\epsilon_{eff}(f)_{(f\to 0)} \Rightarrow \epsilon_{eff}$ $\Big\}$ see Figure 4.3.
(c) $\epsilon_{eff}(f)_{(f\to\infty)} \Rightarrow \epsilon_r$

It is also noted that $\epsilon_{eff}(f)$ is asymptotic at both limits. The fourth feature concerns phase velocity $v_p(f)$. This suffers inflection at some frequency close to the cut-off frequency f_c of the TE_1 surface wave, Thus:

(d) $\dfrac{\partial^2 v_p}{\partial f^2} = 0$ when $f = f_c$.

The formulae which Schneider finally obtained are well suited to simple rapid calculations, but Schneider points out that this theory can only be expected to yield an engineering approximation and quoted errors of somewhat less than 3 per cent *in phase velocity* (i.e. about 6 per cent in effective permittivity) calculated by this method compared with other data (for lines having characteristic impedances in the region of 50 Ω). The method relies heavily on satisfying all the discrete conditions, especially (d), and this could be a main reason for the errors which arise. It should be observed that a 3 per cent error in phase velocity, or wavelength, would be unacceptable for the design of many circuits—including filters or matching networks.

Carlin[18] reported a novel and interesting approach which involves setting up and analysing a *circuit model* of microstrip. His circuit model involved coupled lines where one of the lines propagates only a TEM wave while the other line propagates only a TE wave. The details will not be pursued here, but the final expressions involve frequency f, height h, permittivity ϵ_r, static-TEM effective permittivity ϵ_{eff}, and characteristic impedance Z_0— as well as some constants.

A comparison of calculations based upon Carlin's expressions against several other approaches is given shortly in graphical form (see Figure 4.5). Although more accurate methods are known (which are no more complicated than that due to Carlin) it may well be that fairly minor modifications to the expressions could yield improved accuracy.

4.3.2 Getsinger's microstrip model and expressions

An apparently more accurate formula, at least for microstrip on alumina-type substrates, has been developed by Getsinger.[19, 20] The derivation is too lengthy to be given here, but a brief summary of the method will be presented. Getsinger's approach begins with proposing an alternative 'model' for microstrip which is arranged for more straightforward analysis than the microstrip itself. Essentially, the model tends to ovecome the difficulties set up by the 'horizontal' air–dielectric interface in real microstrip and the approach is illustrated in Figure 4.4.

It is important to note that the model is not physically realizable. The model consists of three side-joined transmission lines:

(i) A central parallel-plate line width w' and height h filled with the substrate material of permittivity $\epsilon_{r'}$, and (ii) and (iii) *Two* additional parallel-plate lines, identical in structure and electrically joined to the line (i), having widths a', heights h', and filled with air

Line (i) accounts for the substrate fields in the real microstrip. It is much easier to analyse because it is simply a parallel-plate line. Lines (ii) and (iii) allow for the fields which exist in air with the real microstrip. The assumption

(a) Real microstrip (b) Microstrip model

Fig. 4.4. Getsinger's model for dispersion analysis. (© 1973 IEEE. Reprinted, with permission, from Getsinger.[19])

is made that because the two regions, air filled and dielectric filled of the model and the microstrip are grossly similar, the two structures will have the same dispersive behaviour for the same mode of propagation. This hypothetical 'microstrip model' involves corners with abrupt dielectric discontinuities, and Getsinger makes it clear quite early in his paper that junction capacitance should strictly be included to allow for the corner errors. But this would clearly make the analysis overcomplicated and almost certainly prevent a closed-form expression from being obtained. Therefore no attempt is made to improve the accuracy by including these junction capacitances.

The first step in the analysis consists of mathematically forcing the model to have the same electrical characteristics at the low-frequency limit, as the microstrip. Thus the static-TEM parameters of each structure are made equal.[20] A transverse resonance analysis of the model (Figure 4.4b) is performed when the ultimate aim of obtaining an explicit formula for $\epsilon_{\text{eff}}(f)$ as a function of known parameters. This transverse resonance technique demands that the sum of admittances looking to the left and right from either air–dielectric interface shown in Figure 4.4(b) must be zero. Getsinger shows that this sum may be written

$$\frac{\gamma_a}{h'}\tanh\left(\gamma_a a'\right) + \frac{\gamma_r}{h}\tanh\left(\frac{\gamma_r w'}{2}\right) = 0 \qquad (4.3)$$

where the subscripts a and r refer to air and substrate dielectic respectively.

Also, the propagation coefficients are related by

$$\gamma^2 + \gamma_i^2 = -\omega^2 \mu_0 \epsilon_0 \epsilon_r \qquad (4.4)$$

where $i = a$ or r as appropriate. These expressions are combined, together with appropriate static-TEM relations, and solved to yield an expression for $\epsilon_{\text{eff}}(f)$ as a function of ϵ_{eff}, ϵ_r, Z_0, f, and h'/h. The reader is referred to the details in Getsinger's publications;[19, 20] only the result is given here:

$$\epsilon_{\text{eff}}(f) =$$

$$\epsilon_r - \frac{\epsilon_r - \epsilon_{\text{eff}}}{1 + f^2(2\mu_0 h/Z_0)^2 \left[\frac{\pi^2}{12}\left\{\frac{(\epsilon_{\text{eff}}-1) + (h'/h)^2(\epsilon_r - \epsilon_{\text{eff}})}{\epsilon_r - 1}\right\}\left\{\frac{(\epsilon_{\text{eff}}-1)(\epsilon_r - \epsilon_{\text{eff}})}{\epsilon_{\text{eff}}(\epsilon_r - 1)}\right\}\right]}$$

$$(4.5)$$

There is one really unknown parameter in this expression and that is h'/h. It turns out that, rather fortuitously, this parameter varies little as a function of Z_0 over quite a wide range. For $Z_0 \gtrsim 35\ \Omega$, $h'/h \simeq 3.0$, and as Z_0 falls below 35 Ω the parameter slowly increases—reaching approximately 4.2 at $Z_0 = 15\ \Omega$. Therefore h'/h can be introduced quite readily into eqn (4.5), although this procedure lacks some precision and is uncertain for characteristic impedances of less than 15 Ω or substantially greater than 90 Ω. The formula also remains rather complicated.

A somewhat simpler formula, based on the same foundations, may be written in the form:[19]

$$\epsilon_{eff}(f) = \epsilon_r - \frac{\epsilon_r - \epsilon_{eff}}{1 + G(f/f_p)^2} \tag{4.6}$$

where

$$f_p = \frac{Z_0}{2\mu_0 h} \tag{4.7}$$

and μ_0 is the free-space permeability (of universally constant value $= 4\pi \times 10^{-7}$ H/m). The parameter G is purely empirical, thereby giving some flexibility to the formula. G is dependent mainly upon Z_0 but also to a lesser extent upon h, and Getsinger deduced from measurements of microstrip ring resonators on alumina that

$$G = 0.6 + 0.009Z_0 \tag{4.8}$$

when

$$h = 0.635 \text{ mm}$$

Fortunately, eqn (4.6) is not particularly sensitive to G, and $G = 1.0$ for the most common characteristic impedance of microstrip line on alumina, that is 50 Ω. For a wide range of lines on alumina-type substrates G has been found to vary over the range

$$0.3 \leq G \leq 1.7$$

Substrates where this applies include monocrystalline sapphire, which provides advantages for many MICs, as explained in Chapter 3.

It has been found that eqns (4.6), (4.7), and (4.8) give rise to discrepancies which increase as Z_0 decreases, and can exceed 3 per cent, as Z_0 is reduced below 20 Ω.[21]

The work of Getsinger, as outlined above, provides an excellent basis for the development of expressions having an improved accuracy and which may be used in design. These are described in the next section.

4.4 ACCURATE DESIGN FORMULAE

4.4.1 Edwards and Owens' expressions

When careful measurements are made on microstrip lines over a wide range of frequencies and characteristic impedances, it is found that Getsinger's approach, while more accurate than any of the other methods, requires at least optimization of the G-factor for any given 'family' of microstrip lines on a particular substrate. Edwards and Owens,[21] for example, carried out

extensive measurements on microstrip lines on monocrystalline sapphire substrates and established (by curve fitting) the following expression for G:

$$G = \left(\frac{Z_0 - 5}{60}\right)^{1/2} + 0.004Z_0 \tag{4.9}$$

This works well in conjunction with eqns (4.6) and (4.7) for lines having characteristic impedances with the range

$$10 \leq Z_0 \leq 100\,\Omega$$

and over the frequency range

$$2 \leq f \leq 18\,\text{GHz}$$

It should be noted that the sapphire substrates were nominally 0.5 mm thick and the equivalent isotropic relative permittivities (see Section 3.8) were 10.7 to 11.6.

Some careful measurements have also been carried out on microstrips on an alumina substrate.[21] In this case a suitable expression for the G-factor is

$$G = \left(\frac{Z_0 - 3}{60}\right)^{1/2} + 0.001Z_0 \tag{4.10}$$

which works very well for lines on a 0.65-mm thick alumina substrate (Alsimag 805, $\epsilon_r = 10.15$) where the characteristic impedance range is

$$30 \leq Z_0 \leq 70\,\Omega$$

over the frequency range

$$2 \leq f \leq 18\,\text{GHz}$$

Equation (4.10) may give slightly high results where the characteristic impedance falls significantly below 30 Ω.

Alternatively and applicable accurately at least to both alumina and sapphire, Edwards and Owens[21] suggest the following expression (combining their equations 9 and 11) for dispersion:

$$\epsilon_{\text{eff}}(f) = \epsilon_r - \frac{\epsilon_r - \epsilon_{\text{eff}}}{1 + (h/Z_0)^{1.33}(0.43f^2 - 0.009f^3)} \tag{4.11}$$

where h is in millimetres and f is in gigahertz.

Based upon extensive measurements carried out with microstrip lines on alumina and sapphire over the impedance and frequency ranges indicated above, and also comparisons with computed results using the analysis due to Itoh and Mittra (to be described shortly), the estimated accuracy of eqn (4.11) is better than

$$\pm\,0.8\,\text{per cent}$$

Even this accuracy, which was the best achieved up to 1976, is still insufficient for many filters and some matching networks. Tolerances and substrate repeatability uncertainties reduce the prediction capability further still, and post-production trimming is then mandatory. The use of monocrystalline substrates with controlled thickness removes some repeatability difficulties and tends to place the onus back on to the accuracy of a suitable dispersion formula again.

Since only alumina-type substrates have been tested, eqn. (4.11) is only *guaranteed* to predict with the quoted accuracy, provided the substrate permittivity is within the approximate limits:

$$10 \leq \epsilon_r \leq 12$$

Measurements reported by York and Compton,[22] however, suggest that eqn (4.11) may work well for plastic substrates at frequencies up to 18 GHz (their results indicate discrepancies consistently below 0.9 per cent).

For consistently accurate calculations in cases where the substrates differ markedly from alumina, such as plastics, quartz, ferrites, etc., it is recommended that new experiments should be carried out. Hence a new curve-fitted formula for G is determined for each case, allowing the use of eqns (4.6) and (4.7) to calculate $\epsilon_{eff}(f)$. Figure 4.5 shows a comparison of the five dispersion calculation methods just described. Two characteristic impedances are quite sufficient to indicate the trends: 25 Ω was chosen because, at lower impedances, dispersion is greater; and 50 Ω was chosen because it is a commonly used value. Considerable overall discrepancies are evident although, on these scales, the Getsinger and Carlin curves are fairly close.

A combination of comprehensive measurements, preferably backed by an independent set of theoretical calculations, is needed to decide upon the validity and accuracy of particular formulae. Frequency-dependent analysis, based upon a microstrip field development, is dealt with in outline in Section 4.7.

A summary of results obtained by various methods is indicated by the curves of Figure 4.6. The 'polynomial fit' refers to eqn (4.11), Itoh and Mittra's approach will shortly be outlined, and the experimental results were obtained using a technique which is explained in Section 7.7. It is clear that consistent results are achieved over this wide frequency range and over the moderately wide range of characteristic impedances (31–71 Ω).

4.4.2 Expressions suitable for millimeter-wave design

The design expressions presented so far in this chapter have been well tested and checked against experimental measurements for frequencies up to approximately 18 GHz. This frequency is, however, far below the limits at which microstrip can be used—up to the vicinity of 100 GHz at least—

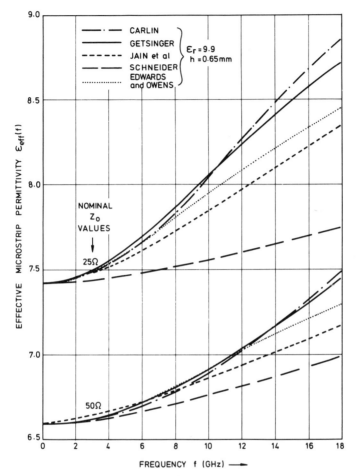

Fig. 4.5. A comparison of the results derived from five different dispersion calculation techniques

and, as outlined in Chapter 2, there are increasing applications available for MICs and MMICs at millimeter-wave frequencies. It has therefore long been recognized that a requirement exists for closed formulae that would enable calculations to be conducted including dispersion in microstrip at these high frequencies.

In 1979 Yamashita et al.[23] published an approximate dispersion formula covering the frequency range 1 to 100 GHz. Their expression can be written:

$$\epsilon_{\text{eff}}(f) = \left(\frac{\sqrt{\epsilon_{\text{r}}} - \sqrt{\epsilon_{\text{eff}}}}{1 + 4F^{-1.5}} + \sqrt{\epsilon_{\text{eff}}} \right)^2 \qquad (4.12)$$

Fig. 4.6. Dispersion curves applicable to three microstrip lines on a microwave-grade high-purity alumina substrate

where

$$F = \frac{4hf\sqrt{\epsilon_r - 1}}{c}\left[0.5 + \left\{1 + 2\log\left(1 + \frac{w}{h}\right)\right\}^2\right] \tag{4.13}$$

However, this expression is not particularly accurate—especially in the lower frequency range up to 18 GHz where the expression due to Edwards and Owens is recommended. Comparison of Yamashita's design formulae with analytical results determined using electromagnetic solutions shows considerable discrepancies over the full 1 to 100 GHz range.

An improved design formula has been developed by Kirschning and Jansen.[24] Their approach begins with a function that bears a close

resemblance to Getsinger's, but they have a new and reportedly more accurate frequency-dependent denominator term. Also, their formula covers a much wider range of permittivities, aspect ratios and frequencies than those due to Getsinger or Edwards and Owens.

Kirschning and Jansen's basic expression is (with frequency f in GHz and thickness h in cm):

$$\epsilon_{\text{eff}}(f) = \epsilon_r - \frac{\epsilon_r - \epsilon_{\text{eff}}}{1 + P(f)} \qquad (4.14)$$

and the form of the denominator frequency function is

$$P(f) = P_1 P_2 \{(0.1844 + P_3 P_4) 10 fh\}^{1.5763} \qquad (4.15)$$

where

$$
\left.
\begin{aligned}
P_1 &= 0.27488 + \{0.6315 + 0.525/(1 + 0.157 fh)^{20}\}(w/h) \\
&\quad - 0.065683 \exp(-8.7513 \, w/h) \\
P_2 &= 0.33622 \{1 - \exp(-0.03442 \, \epsilon_r)\} \\
P_3 &= 0.0363 \exp(-4.6 \, w/h)[1 - \exp\{-(fh/3.87)^{4.97}\}] \\
P_4 &= 1 + 2.751 [1 - \exp\{-(\epsilon_r/15.916)^8\}]
\end{aligned}
\right\} \qquad (4.16)
$$

These workers used computer-based matching in conjunction with the available hybrid mode results (derived from full electromagnetic numerical computations) to force the correct asymptotic behaviour of the function $P(f)$. An accuracy of better than 0.6 per cent is claimed for all frequencies up to 60 GHz (although a full check only appears to have been conducted up to 30 GHz). The restrictions applying are very wide, namely:

$$1 \leqslant \epsilon_r \leqslant 20$$

$$0.1 \leqslant w/h \leqslant 100$$

$$0 \leqslant h/\lambda_0 \leqslant 0.13$$

There are further formulations enabling the effects of dispersion in microstrip to be calculated at frequencies through millimeter-wave. Although the last word has almost certainly not yet been stated in the literature on this subject the position has been reached where reported accuracies are within 1 per cent (of measurements and independent theoretical results over a wide range). Having said this it must be recognised that care is always required in selecting an appropriate accurate dispersion expression.

The final formulation presented here for dispersion calculation is that due to Kobayashi.[25] This work is based upon the concept of a significant 50 per cent dispersion point at which the effective microstrip permittivity is the arithmetic mean of the substrate relative permittivity and the low-frequency

limit value. Kobayashi calculates the frequency for this condition and develops the following dispersion expressions:

$$\epsilon_{\text{eff}}(f) = \epsilon_{\text{r}} - \frac{\epsilon_{\text{r}} - \epsilon_{\text{eff}}}{1 + (f/f_{50})^m} \tag{4.17}$$

where

$$f_{50} = \frac{f_{k,\text{TM}_0}}{0.75 + \{0.75 - (0.332/\epsilon_{\text{r}}^{1.73})\}\,w/h} \tag{4.18}$$

$$f_{k,\text{TM}_0} = \frac{c \tan^{-1}\left(\epsilon_{\text{r}}\sqrt{\dfrac{\epsilon_{\text{eff}} - 1}{\epsilon_{\text{r}} - \epsilon_{\text{eff}}}}\right)}{2\pi h \sqrt{\epsilon_{\text{r}} - \epsilon_{\text{eff}}}} \tag{4.19}$$

$$m = m_0\, m_{\text{c}} \tag{4.20}$$

$$m_0 = 1 + \frac{1}{1 + \sqrt{w/h}} + 0.32\left(\frac{1}{1 + \sqrt{w/h}}\right)^3 \tag{4.21}$$

$$m_{\text{c}} = \begin{cases} 1 + \dfrac{1.4}{1 + w/h}\left\{0.15 - 0.235\exp\left(\dfrac{-0.45f}{f_{50}}\right)\right\} & \text{where } w/h \le 0.7 \\ 1 & \text{where } w/h > 0.7 \end{cases} \tag{4.22}$$

Atwater[26] has reported results of comparative tests on various microstrip dispersion formulae, but it should be noted that Atwater's paper predates the later important Kobayashi work referred to above (Atwater's reference to Kobayashi relates to the 1982 paper). It is, however, clear that Kirschning and Jansen's approach is comparatively accurate within its frequency restrictions.

York and Compton[22] (referred to earlier) report measurements on microstrips ranging in impedance from 35 to 70 Ω, fabricated on both alumina and plastic substrates having thicknesses in the range 0.635–1.605 mm. Although York and Compton's results are limited to frequencies up to 18 GHz it is clear that, out of a total of 9 models tested, only those due to Edwards and Owens, Kirschning and Jansen, and Kobayashi show deviations well within 1 per cent of measurements.

4.5 EFFECTS DUE TO FERRITE AND TO DIELECTRICALLY ANISOTROPIC SUBSTRATES

4.5.1 Effects of ferrite substrates

The static-TEM problem of microstrip on a ferrite substrate was discussed in Section 3.9 of Chapter 3. The frequency dependence of the effective

microstrip *permeability* is of little practical importance so will not be covered here. However, since ferrite also behaves as a dielectric the frequency-dependent effective microstrip *permittivity* is important and should be evaluated as recommended in the foregoing discussion of this section.

Kurushin *et al.*[27] have shown that as the frequency increases in the case of microstrip on a ferrite substrate the conventional dispersion results approach those applicable to the case of a non-ferrite dielectric substrate. Therefore all the design expressions given previously here continue to apply at high frequencies.

Since ferrite devices are often placed on non-ferrite substrates (alumina, semiconductor), and microstrip is then fabricated on the ferrite device surface itself, it is important to gain some design appreciation of the behaviour of this type of structure. The basic properties of such a structure have been evaluated by Hsia *et al.*[28] who used a spectral domain exponential matrix method. These workers generalised their analysis to accommodate isotropic/anisotropic structures—in terms of both permeability and permittivity. Most of the results presented by Hsia *et al.* refer to a ferrite with permittivity of 12.6 providing the top layer whilst GaAs having a permittivity of 12.9 formed the lower substrate. The magnetic bias field was conventionally disposed normally to the plane of this assembly. Hsia *et al.* give curves showing normalized wave number (corresponding to effective permittivity) and characteristic impedance as functions of frequency over the 5–30 GHz range. Their curves for normalized wave number are reproduced as Figure 4.7, in which $z' = b$ refers to the case of microstrip deposited on the lower, non-ferrite, substrate whereas $z' = d$ refers to the case where the microstrip is deposited on the upper ferrite superstrate.

4.5.2 Effects of a dielectrically anisotropic substrate

Static-TEM calculations associated with microstrip on a dielectrically anisotropic substrate were dealt with in Section 3.8. In the case of sapphire, probably the most common anisotropic substrate, the measurements of Edwards and Owens[21] seem to indicate that the influence of the sapphire anisotropy on the detail of the dispersion characteristics is negligibly small in practice (up to 18 GHz).

As indicated in his Table II, and also his reported dispersion results, the data derived by Kobayashi[29] confirm excellent agreement with the predictions due to Edwards and Owens, presented in Section 4.4.1. The use of slightly modified versions of Yamashita's dispersion expression also provides very good agreement and it would be useful to study the effects of suitably adjusting Kirschning and Jansen's formulae in an effort to predict dispersion at high frequencies on dielectrically anisotropic substrates.

A more extensive study of the behaviour of microstrip on a dielectrically anisotropic substrate has been performed by Mariki and Yeh,[30] as mentioned

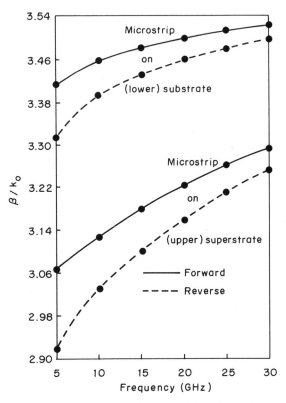

Fig. 4.7. Microstrip propagation coefficient as a function of frequency (© 1990 IEEE. Reprinted, with permission, from Hsia *et al.*[28])

in Chapter 3. These workers use a three-dimensional transmission-line-matrix (TLM) analysis to determine the dispersive effects with this type of substrate. In particular the analysis derives the frequency-dependent effective microstrip permittivity and characteristic impedance. Considering a sapphire substrate Mariki and Yeh's results for microstrip permittivity at frequencies up to 21 GHz are very close to those predicted by Getsinger's original model. Characteristic impedance is also shown to increase with frequency—for example from 37.5 to 54 Ω when the aspect ratio (w/h) is 1.5.

4.6 DESIGN REQUIRING DISPERSION CALCULATION: A WORKED EXAMPLE

As an example we will recall the problem of calculating one element of a transistor matching network, which was calculated for 2 GHz in Section 3.7.

However, we now design the same element for a frequency of 12 GHz, where dispersion will certainly be significant.

The initial data, as previously, are

$$Z_0 = 31.3 \, \Omega, \quad \epsilon_r = 9.8, \quad h = 0.6 \, \text{mm}$$

and in this case the frequency f is 12 GHz. We shall also require the effective microstrip permittivity from the static-TEM design completed in Section 3.7. This is

$$\epsilon_{\text{eff}} = 7.09$$

For a frequency of 12 GHz, eqn (4.11) is used to calculate $\epsilon_{\text{eff}}(f)$:

$$\epsilon_{\text{eff}}(f) = \epsilon_r - \frac{\epsilon_r - \epsilon_{\text{eff}}}{1 + (h/Z_0)^{1.33}(0.43f^2 - 0.009f^3)}$$

The thickness h is directly substituted in millimetres and the frequency f in gigahertz, giving

$$\epsilon_{\text{eff}}(f) = 7.621$$

and the wavelength

$$\lambda_g = \frac{300}{f\sqrt{\epsilon_{\text{eff}}(f)}} = 9.056 \, \text{mm}$$

Both values are obtained from a program on a desk-top computer which aids microstrip design.

Thus we obtain the physical length of the microstrip element, approximately $\lambda_g/4$, i.e.

$$l \approx \frac{9.056}{4} \approx 2.264 \, \text{mm}$$

In the next chapter we deal with the discontinuities. Notice that our new result, for 12 GHz, is significantly different from the previous (2 GHz) result *linearly* scaled for frequency. This yields the erroneous result of 2.346 mm—nearly 4 per cent in error.

Provided we assume here that we will neglect any change in Z_0 with frequency (dealt with in Section 4.8), the microstrip width will remain at the value previously determined: $w = 1.34$ mm.

At higher frequencies the results calculated including dispersion differ more markedly from those calculated using the static-TEM formulae.

For example, let us consider a microstrip line to be designed for use in an MMIC, i.e. fabricated on a semi-insulating GaAs substrate. The thickness is 0.127 mm and the relative permittivity is 13.0.

Let us assume that the operating frequency is 40 GHz and that the line to be designed has an aspect ratio (w/h) of 2.0.

Calculations using the static-TEM expressions yield the low-frequency limit permittivity as 9.136, whereas calculations using the formulae due to Kobayashi yield a value of 9.74 at 40 GHz. The discrepancy is around 6.4 per cent.

Similar ball-park magnitudes of discrepancy occur where plastic (e.g. RT-duroid), quartz or many other substrates are considered.

A comparison of calculated results for this case, using the methods of Yamashita, Kirschning and Jansen, and Kobayashi, is provided by the curves shown in Figure 4.8. Discrepancies between the first two methods continually increase, reaching 1.07 per cent at 40 GHz, whereas the differences between Kirschning and Jansen and Kobayashi remain less than 0.25 per cent over the entire frequency range.

4.7 FIELD SOLUTIONS

4.7.1 One example of a 'classic' frequency-dependent computer-based field solution

In the methods for calculating effective microstrip permittivity outlined so far the emphasis has been on *approximate* approaches using closed formulae. For any of the available approximate formulae it may be said that:

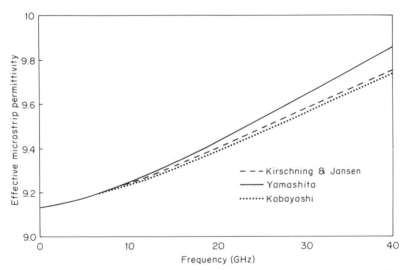

Fig. 4.8. Comparative dispersion curves for microstrip on a GaAs substrate ($\epsilon_r = 13$, $w = 0.254$ mm, $h = 0.127$ mm, $f = 40$ GHz). (From calculations performed independently by Dr Roger P. Owens, to whom the author is indebted.)

(a) The methods are amenable to calculator-based design.
(b) If used for CAD, the methods are relatively inexpensive because they are very fast.

On the other hand, these approximate approaches have a number of drawbacks. For example:

(a) They often suit only a specified microstrip 'family', e.g. those based on alumina substrates.
(b) They require accurate static-TEM data to start with, that is ϵ_{eff} and Z_0.
(c) Such methods are not fundamentally elegant (some empirical element usually has to be introduced).

Frequency-dependent field solutions requiring 'number crunching' on relatively powerful computers have been published by a number of workers.[31–35] While these are comparatively expensive to run they provide the positive benefits of answering drawbacks (a), (b), and (c) above. They generally take up too much computer time and storage to be considered as candidates for inclusion in a larger microwave integrated circuit computer-aided optimization routine. However, such solutions provide an excellent theoretical comparison with measured data and therefore can enable the accuracy of the approximate design formulae to be checked. They can also be readily extended to handle more complex problems such as coupled lines, even involving multiple conductors, etc.

When satisfactory agreement is reached regarding measurements and fundamental theoretical results then 'curve-fitted' closed formulae can be deduced for $\epsilon_{eff}(f)$. It should be appreciated that these frequency-dependent field solutions require only the physical input data:

(a) microstrip geometry (w, h, and box dimensions, if applicable), and
(b) the permittivity of the substrate material.

Itoh and Mittra,[33, 34] in particular, have developed a 'spectral-domain' approach. This approach involves Fourier transforms to convert the field formulation from the actual space domain to a domain having new (transformed) variables.

These workers set up a superposition of TE and TM fields in terms of scalar potential functions:

$$E_{zi} = j\frac{k_i^2 - \beta^2}{\beta}\psi_i^{(e)}(x, y)\exp(-j\beta z) \qquad (4.23)$$

and

$$H_{zi} = j\frac{k_i^2 - \beta^2}{\beta}\psi_i^{(h)}(x, y)\exp(-j\beta z) \qquad (4.24)$$

where

$$k_1 = \omega\sqrt{(\epsilon_r\mu_r\epsilon_0\mu_0)} \qquad k_2 = \omega\sqrt{(\epsilon_0\mu_0)} = k_0 \qquad (4.25)$$

β is the unknown propagation constant to be determined. Superscript (e) means 'TE' field and superscript (h) means 'TM' field. The Fourier transform of the scalar potentials is defined by

$$\psi_i^{(p)}(n, y) = \int_{-a}^{a} \psi_i^{(p)}(x, y) \exp(j\hat{k}_n x)\, dx \qquad (4.26)$$

where

$$i = 1, 2$$

$$p = e \text{ or } h$$

$$\hat{k}_n = \left(\frac{\pi}{a}\right)\left(n - \frac{1}{2}\right) \text{ for } E_z \text{ even and } H_z \text{ odd modes}$$

or

$$\hat{k}_n = \frac{n\pi}{a} \text{ for } E_z \text{ odd and } H_z \text{ even modes}$$

This procedure leads to a more economical computer technique. The current distributions on the strip are the crucial features of this particular technique. The final solution is selected only when successively improved current distribution functions, inserted in the program, are observed to give converging sets of output results.

Jansen[35] provided an improved and faster solution of this general complete electromagnetic analytical approach.

4.7.2 Analysis of arbitrary planar configurations

Spectral domain approaches of similar forms to that of Itoh and Mittra have been used elsewhere to analyse the characteristics of MICs and MMICs on a three-dimensional basis.[36] In their work Ke Wu and Vahldieck develop a space-spectral domain method for analysing arbitrary planar configurations with or without open boundaries.

They express the transverse fields in each layer of a multi-layer structure in a form suitable for application of a Fourier transform and reduce the results of this transformation to a differential solution. Basis functions are introduced for rectangular resonator shapes and, following the application of Galerkin's technique, a characteristic matrix equation system is obtained which has to be solved for the zeros of a determinant. Some interesting microstrip structure shapes are considered and results compare well with

measurements. Calculations are reported for a range of rectangular strip-patch microstrip resonators (suspended, grounded and sandwiched), giving resonant frequency as a function of resonator length. Although it is not clear from the paper, it seems that these calculations refer to microstrip on a substrate having a permittivity of 3.82 (fused quartz) and resonant frequencies in the 12–37 GHz range are obtained.

Ke Wu and Vahldieck demonstrated that this space-spectral domain technique can be run on an IBM PC/AT desktop computer with an accelerator board.

4.7.3. Asymmetry effects

The effects of asymmetrically-disposed shielded microstrip is important in practice and Dumbell et al. have reported an analytical technique for this situation. They apply a variational method to analyse asymmetric shielded microstrip both in single-layer and overlaid dielectric forms. In fact, they show that the variation in propagation coefficient as a function of degree of asymmetry can be controlled by the introduction of a dielectric overlay to the extent that microstrip tracks may be located close to conducting box walls and such variations then practically neglected.

Some results obtained by Dumbell et al.[37] are shown in Figure 4.9 for the case of a substrate having a relative permittivity of 10.5. This family of curves refers to microstrip with an aspect ratio of 0.866 enclosed in a conducting box 11.43 mm high by 12.7 mm wide. 'Strip offset' is the amount by which the strip is offset from the centre line of the system.

Fig. 4.9. Effective permittivity for shielded microstrip (Reprinted, with permission, from Dumbell et al.[37])

A comparative family of curves, applicable to the same geometry but with a dielectric overlay having permittivity 7.0 and 1.27 mm thickness, is shown in Figure 4.10. Clearly, at the lower frequencies particularly, the variations with strip offset are substantially reduced. Families of curves for the characteristic impedances are also determined by Dumbell *et al.*, and these indicate that variations caused by offset are less effectively compensated by the introduction of an overlay.

4.7.4 Time-domain approaches

A length of dispersive transmission line can be used as a 'phase equalizer' to compensate for the phase distortion otherwise resulting in the transmission of asymmetric pulses. This equalisation results in the effective reshaping and compression of ultrashort pulses which originate, typically, from high-speed photoconductive switches.

Qian and Yamashita[38] have shown that an optimum length of microstrip or coplanar waveguide can be used to correct the phase distortion. They begin with a Fourier spectrum of a single-sided exponential pulse which suitably quantifies this distortion and they adapt a microstrip dispersion expression (see Section 4.4.2). The phase function associated with the single-sided exponential pulse is expanded as a polynomial in which the frequency component at which the amplitude of the spectrum is 10 per cent of its peak value is defined.

The microstrip dispersion expression is expanded using Taylor's series and used to formulate an expresssion for the phase delay. A comparison is

Fig. 4.10. Effective permittivity for shielded microstrip with dielectric overlay (Reprinted, with permission, from Dumbell *et al.*[37])

finally made between the polynomial derived from the exponential pulse and this dispersion expression series expansion. Qian and Yamashita show that, if the second and third order terms in each series are made equal, then the total phase of the signal after a specific distance has been transversed is close to a linear function of frequency. They report a straightforward formula for the length of line required.

As an example, Qian and Yamashita provide details of a simulation for a 35 ps single-sided exponential pulse propagating along a 50 Ω microstrip line on a GaAs substrate and the resulting waveforms are repeated in Figure 4.11 here.

A finite-difference time-domain analysis can be useful for the determination of the broadband frequency response of complex MIC or MMIC components, as reported by Rittweger and Wolff.[39] These workers show results for spiral inductors (beyond the scope of this text) and microstrip radial stubs obtained using this technique. This numerical technique is three-dimensional and essentially trades a relatively modest requirement for analytical operations against a large numerical demand.

The microstrip radial stub is important in some commercial filter and matching circuit designs and Rittweger and Wolff's results for a specific case are shown in Figure 4.12. It can be seen that, at frequencies below 19 GHz, the measured and calculated results compare favourably. In the 19 to 24 GHz range there is a greater discrepancy in the data although the pattern of variation still compares quite well.

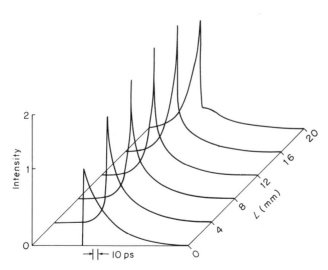

Fig. 4.11. Simulation results for the propagation of a 35 ps single-sided exponential pulse along 50 ohm microstrip lines on a GaAs substrate (© 1990 IEEE. Reprinted, with permission, from Qian and Yamashita.[38])

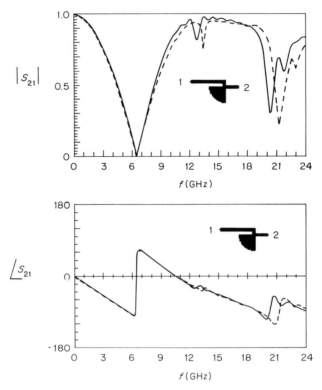

Fig. 4.12. Calculated (—) and measured (---) S-parameters of the radial stub. Al_2O_3, $\epsilon_r = 9.768$ (measured), height = 0.635 mm, width = 0.61 mm, gap = 0.61 mm, radius = 3.253 mm (© 1990 IEEE. Reprinted, with permission, from Rittweger and Wolff.[39])

4.8 FREQUENCY DEPENDENCE OF THE MICROSTRIP CHARACTERISTIC IMPEDANCE

There have been many different approaches to the dynamic problem of microstrip characteristic impedance and quite different functions of frequency have been predicted—extending even to opposing trends. With no definitive function available MIC designers have continued to use the static-TEM calculation for Z_0.

The problem of characteristic impedance as a function of frequency is not easy to satisfactorily solve, but some background theory and tentative recommendations are presented here in an effort to provide some insight and assistance for the design engineer.

4.8.1 Different definitions and different trends with increasing frequency

Fundamentally, characteristic impedance may be defined in the following three different ways:

$$Z_{0,a}(f) = \frac{V}{I} \tag{4.27}$$

$$Z_{0,b}(f) = \frac{P}{II^*} \tag{4.28}$$

$$Z_{0,c}(f) = \frac{VV^*}{P} \tag{4.29}$$

Different functions of frequency are obtained, depending on which of these definitions is selected. This is due mainly to different assumptions which have to be made regarding mode coupling as the frequency is increased. This mode coupling, it is recalled, is the fundamental dispersive mechanism for microstrip.

Bianco et al.[40] further define impedances in terms of mean voltage V and centre voltage V_c to yield a total of five definitions as follows:

$$Z_{0,1}(f) = \frac{V}{I} \tag{4.30}$$

$$Z_{0,2}(f) = \frac{V_c}{I} \tag{4.31}$$

$$Z_{0,3}(f) = \frac{P}{II^*} \tag{4.32}$$

$$Z_{0,4}(f) = \frac{V_c V_c^*}{P} \tag{4.33}$$

$$Z_{0,5}(f) = \frac{VV^*}{P} \tag{4.34}$$

The general theoretical conclusion is that both $Z_{0,2}$ and $Z_{0,4}$ always rise with increasing frequency whereas the remaining impedances all fall ($Z_{0,3}$ always exhibits the smallest variation).

In all calculations based upon eqns (4.27) to (4.34) the power P flowing in the microstrip is evaluated using Poynting's theorem.

Getsinger,[41] using the model previously described in this chapter, developed an expression which (like that of Denlinger[13]) has microstrip characteristic impedance varying inversely with the square root of $\epsilon_{\text{eff}}(f)$. The development is based upon Schelkunoff's definition of wave impedance

and Getsinger is very careful about the interpretation of the model for the analysis pursued. This derivation seems very plausible but it must be considered in the light of alternative definitions and trends which will now be examined.

4.8.2　Use of the planar waveguide model (Figure 4.13)

Owens[42, 43] developed a different approach with the effective microstrip width concept[44] as a principal feature, obtaining the following forms of expression:

$$Z_0(f) = \frac{h\eta}{w_{\text{eff}}(f)\sqrt{\epsilon_{\text{eff}}(f)}} \tag{4.35}$$

where, following the dispersion expressions given earlier (such as eqns 4.6 and 4.7), the effective microstrip width $w_{\text{eff}}(f)$ may be written as

$$w_{\text{eff}}(f) = w + \frac{w_{\text{eff}} - w}{1 + (f/f_p)^2} \tag{4.36}$$

and

$$f_p = \frac{c}{2w_{\text{eff}}\sqrt{\epsilon_{\text{eff}}}} \tag{4.37}$$

Also,

$$w_{\text{eff}} = \frac{h\eta}{Z_0\sqrt{\epsilon_{\text{eff}}}} \tag{4.38}$$

In all these expression η is the characteristic impedance of free space: $= \sqrt{\mu_0/\epsilon_0} = 376.7\ \Omega$. From eqn (4.36) it is observed that $w_{\text{eff}}(f)$ which appears in eqn (4.35) varies as a direct function of $\{1 + (f/f_p)^2\}^{-1}$, whereas $\sqrt{\epsilon_{\text{eff}}(f)}$ is also involved in eqn (4.35). Hence the frequency-dependent effect of $w_{\text{eff}}(f)$ dominates, and this quantity decreases as the frequency increases.

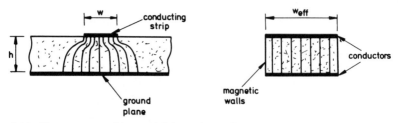

Fig. 4.13. Planar waveguide model for microstrip

Thus $Z_0(f)$ increases with increasing frequency. For a 50 Ω line the change is about 10 per cent over the 0–16 GHz range.

4.8.3 A further alternative expression

Another expression has been given by Bianco et al.,[45] who also utilize the form of the main Getsinger dispersion formula (eqn 4.6) to produce the following result:

$$Z_0(f) = Z_{0T} - \frac{Z_{0T} - Z_0}{1 + G(f/f_p)^2} \qquad (4.39)$$

In this formula Z_{0T} is *twice* the characteristic impedance calculated for a *triplate transmission line* having the same width w as the microstrip line but twice the substrate height (i.e. total triplate dielectric thickness is $2h$). Z_{0T} is readily determined from standard design curves[46] or from some good approximate expressions.[47]

Again we consider a 50 Ω line on an alumina substrate and find that this expression (eqn 4.39) yields an increase of only about 5 per cent over the 0–16 GHz frequency range. Now this order of magnitude (5 per cent) is in fair agreement with a more fundamental calculation by Bianco et al.[45] based upon eqn (4.20), the theoretical results due to Krage and Haddad[32] and Kowalski and Pregla,[31] and also the theoretical calculations produced by Knorr and Tufekcioglu.[48] It may be significant that the latter workers employed the spectral-domain techniques which have been demonstrably successful in the accurate prediction of $\epsilon_{eff}(f)$ (see Section 4.7). Mirshekar-Syahkal and Davies[49] also obtain similar results.

For these reasons the use of eqn (4.39) is, rather tentatively, recommended here. Further work is needed as follows:

(a) to test against the theoretical predictions, and
(b) to develop validating measurements.

Both (a) and (b) are needed to cover a wide range of 'characteristic impedance' (Z_0) and, preferably, for a range of useful substrates. This testing and validating should be performed in the light of recent theoretical work, referred to later (Section 4.12).

4.8.4 A design algorithm for microstrip width

It is very useful to have the facility for evaluating microstrip width with the aid of static-TEM expressions, such as those given in Chapter 3. The concept, and a routine for achieving this when the frequency-dependent characteristic impedance is known, was first suggested by Getsinger.[20] A modified form of this design algorithm is presented here. The design

specification includes ϵ_r, h, f, and $Z_0(f)$, and we require the microstrip width w.

Proceed as follows:

(a) As an initial approximation let $Z_0(f) \approx Z_0$.

(b) Find the approximate microstrip width w using this value of Z_0, with ϵ_r and h.

(c) Determine the frequency-dependent denominator $\{1 + G(f/f_p)^2\}$, as described in Sections 4.3.2 and 4.4, using this approximate value of Z_0. (It may eventually be found that the denominator function in eqn (4.11) gives improved results, but this requires further investigation.)

(d) Evaluate Z_{0T} for triplate having the same strip width w as the microstrip, but twice the dielectric thickness ($2h$). Design curves, or good approximations, may be used.

(e) Rearrange eqn (4.39) for Z_0, letting $A = \{1 + G(f/f_p)^2\}$. This is

$$Z_0 = A\{Z_0(f) - Z_{0T}\} + Z_{0T}$$

(f) With this approximate value of Z_0 calculate w/h using static-TEM expressions given in Chapter 3, and hence calculate w (knowing h).

(g) Repeat steps (b) to (f) until negligible difference is observed in the values of w obtained (e.g. it may be desired that the result should be accurate to within 1 per cent).

4.9 OPERATING FREQUENCY LIMITATIONS

As the signal frequency applied to an MIC is steadily increased some characteristic frequency may be reached at which undesirable effects occur. Two possible spurious effects restrict the desirable operating frequencies. These are

(a) the lowest-order TM mode
(b) the lowest-order transverse microstrip resonance.

In practice *one* of these modes will be experienced at some frequency lower than the other and will thus set the frequency limitation.

4.9.1 The TM mode limitation

Vendelin[50] has indicated that the most significant frequency forming modal limitations in microstrip is associated with strong coupling between the quasi-TEM mode and the lowest-order TM mode. In the cited paper Vendelin gives the main relationships quoted here (although the original analysis is due to Collin—ref. 6 of Vendelin's paper).

We start with the substrate viewed as a dielectric slab, having the coordinate notation shown in Figure 4.14. This slab is a fair approximation for the situation with narrow microstrip lines. Under these conditions the eigenvalue expressions are

$$0.5 \, ph = 0.5 \, qh \tan (0.5 \, qh) \tag{4.40}$$

and

$$(0.5 \, ph)^2 + (0.5 \, qh)^2 = (\epsilon_r - 1)(0.5 \, \beta_0 h)^2 \tag{4.41}$$

where β_0 is the free-space phase coefficient given in terms of the free-space wavelength λ_0 by

$$\beta_0 = \frac{2\pi}{\lambda_0} \tag{4.42}$$

p is the eigenvalue for the field in the air, and q is the eigenvalue for the field in the substrate (ϵ_r). It may be convenient to solve (4.40) and (4.41) graphically for p and q, as shown in Figure 4.15.

The following wave equation is then used to find the net phase coefficient for the TM wave

$$\beta^2 = \beta_0^2 + p^2 \tag{4.43}$$

A frequency for strong coupling between the quasi-TEM mode and the TM mode is identified when the associated two phase velocities are close. The result is given by

$$f_{\text{TEM 1}} = \frac{c \tan^{-1}(\epsilon_r)}{\sqrt{2}\pi h \sqrt{\epsilon_r - 1}} \tag{4.44}$$

This is an important relationship. Vendelin points out that the lowest-order TE wave has a cut-off frequency given by the better-known expression

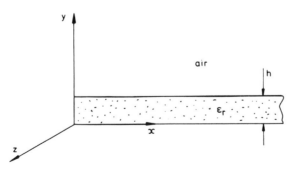

Fig. 4.14. An (isotropic) substrate viewed as a dielectric slab, showing the nomenclature

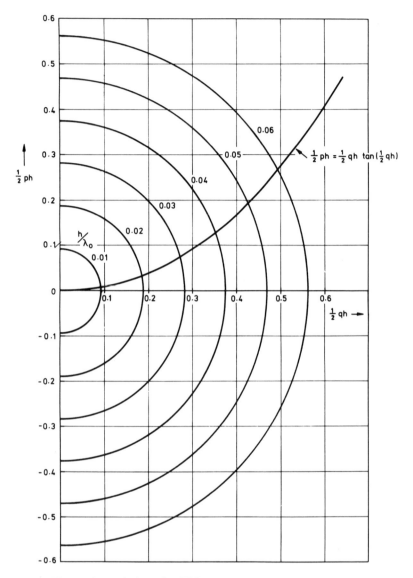

Fig. 4.15. Eigenvalue solutions for TM waves

$$f_c = \frac{c}{4h\sqrt{\epsilon_r - 1}}$$ (4.45)

This frequency merely happens to coincide with the $f_{\text{TEM 1}}$ result for wide microstrip lines, although the phenomena involved are different. In fact the

frequency for the lowest-order TM wave to couple strongly to wide microstrip line is given by

$$f_{TEM\ 1} = \frac{75}{h\sqrt{\epsilon_r - 1}} \tag{4.46}$$

where $f_{TEM\ 1}$ is in gigahertz and h is in millimetres.

Returning to the more important case of narrow microstrip lines, where the permittivity ϵ_r is high (say $\epsilon_r > 10$), eqn (4.44) tends towards

$$f_{TEM\ 1} = \frac{c}{2\sqrt{2}h\sqrt{\epsilon_r - 1}} \tag{4.47}$$

or

$$f_{TEM\ 1} = \frac{106}{h\sqrt{\epsilon_r - 1}} \tag{4.48}$$

where $f_{TEM\ 1}$ is in gigahertz and h is in millimetres.

The maximum restriction on usable substrate thicknesses is then easily obtained as

$$h = \frac{0.354\lambda_0}{\sqrt{\epsilon_r - 1}} \tag{4.49}$$

Care should be exercised in the use of these expressions, eqn (4.44) being the more general and applicable to lower-permittivity substrates. It is recommended that substrates should be as thin as possible (eqn 4.49) and that operating frequencies be kept below $f_{TEM\ 1}$ (eqn 4.44).

4.9.2 The lowest-order transverse microstrip resonance

For a sufficiently wide microstrip line a transverse-resonant mode can exist which can also couple strongly to the quasi-TEM microstrip mode. At the cut-off frequency for this transverse-resonant mode the equivalent circuit is a resonant transmission line of 'length' $(w + 2d)$, where d accounts for the microstrip side-fringing capacitance: $d = 0.2h$. The situation is illustrated in Figure 4.16. A half-wavelength must be supported by the 'length' $(w + 2d)$. Therefore the cut-off half-wavelength is

$$\frac{\lambda_{CT}}{2} = w + 2d$$

$$= w + 0.4h$$

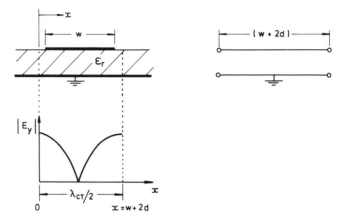

Fig. 4.16. Transverse resonance: standing wave ($|E_y|$) and equivalent transmission line of 'length' ($w + 2d$)

or

$$\frac{c}{2f_{CT}\sqrt{\epsilon_r}} = w + 0.4h$$

Hence

$$f_{CT} = \frac{c}{\sqrt{\epsilon_r}(2w + 0.8h)} \tag{4.50}$$

Vendelin[50] has indicated that slots, introduced into the metal strip, can suppress the transverse-resonant mode. However, this may not always be practicable (e.g. when short stubs are involved) and it should be checked, by calculating from eqn (4.50), whether the transverse resonance might be excited. Sometimes a change in the circuit configuration will enable the offending wide lines to be avoided altogether.

Under the assumption that transverse resonances are suppressed or avoided in some way, Figure 4.17 shows the cut-off frequency limitation imposed (eqn 4.44) as a function of substrate thickness h for a variety of substrate materials.

Microstrip, compared with media such as waveguide or coaxial lines, offers a rather low unloaded Q-factor. The details of power losses, leading to Q, are discussed shortly. Arbitrarily assuming that a Q-factor of at least a hundred is often desirable, the hatched region of Figure 4.18 then provides a 'useful operating region' knowing the substrate material and the intended operating frequencies.

As an example, consider a microstrip line on an alumina substrate of permittivity 9.6 and thickness 0.65 mm. A vertical ordinate drawn at the

Fig. 4.17. TM threshold excitation frequency $f_{\text{TEM 1}}$ as a function of substrate thickness h and relative permittivity ϵ_r, (Reproduced by permission of *Microwave Journal* from Vendelin.[50])

$h = 0.65$ mm position shows that we can safely operate at frequencies considerably above 10 GHz and with a 'good' Q-factor of several hundred.

Notice that thinner substrates, while allowing higher frequencies to be used before the onset of restrictions due to TM modes, cause serious degradation of the Q-factor, especially at the lower operating frequencies.

4.10 POWER LOSSES AND PARASITIC COUPLING

Four separate mechanisms can be identified for power losses and parasitic effects associated with microstrip lines:

(a) conductor losses
(b) dissipation in the dielectric of the substrate
(c) radiation losses
(d) surface-wave propagation.

The first two items are dissipative effects whereas radiation loss and surface-wave propagation are essentially parasitic phenomena. Thus, both conductor and dielectric losses can be lumped together for the purposes of calculation and embodied as the attenuation coefficient (α) for a microstrip line.

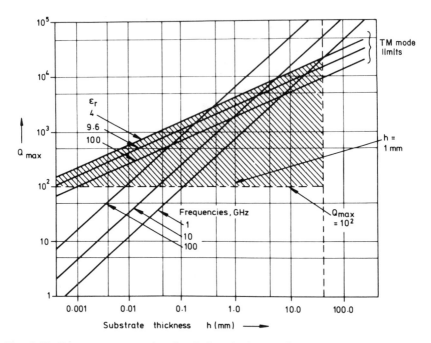

Fig. 4.18. Parameters governing the choice of substrate for any microstrip application: acceptable Q-factor, operating frequency, substrate thickness and relative permittivity ϵ_r, (Reproduced by permission of *Microwave Journal* from Vendelin.[50])

Considering a fictitious microstrip resonator, which does not radiate or propagate surface waves, the dissipative losses may also be interpreted in terms of a Q-factor as defined by the following expression:

$$Q = \frac{\omega_o U}{W} \tag{4.51}$$

where U is the stored energy and W is the average power lost per cycle.

We now provide the relationship between α and Q.

4.10.1 Q-factor and attenuation coefficient

With V_p and I_p, the peak voltage and current at some plane along the line, we may use eqn (4.51) to write, for unit length,

$$Q = \omega_o \frac{(LI_p^2/2 + CV_p^2/2)/2}{(RI_p^2 + GV_p^2)/2} \tag{4.52}$$

upon substituting average energy stored.

If $I_p = V_p/Z_0$ is now introduced, we get

$$Q = \omega_0 \frac{CV_p^2}{RV_p^2/Z_0^2 + GV_p^2}$$

But $Z_0 = \sqrt{L/C}$, yielding

$$Q = \frac{\omega_0\sqrt{LC}}{R/Z_0 + GZ_0}$$

Now $\beta = \omega\sqrt{LC}$ and $\alpha = R\sqrt{C/L} = R/Z_0 = GZ_0$ so that

$$Q = \frac{\beta}{2\alpha} \qquad (4.53)$$

The phase coefficient β is identical to $2\pi/\lambda_g$ and hence eqn (4.53) finally becomes

$$Q = \frac{\pi}{\alpha\lambda_g} \qquad (4.54)$$

In this important result it must be noted that α is in units of Nepers per metre (1 Np = 8.686 dB). Therefore, as α is usually given in decibels per metre, the expression must be divided by 8.686.

To clarify the meaning and use of this expression consider the following example. A Q-factor of 200 is obtained for a microstrip resonator which is $2\lambda_g$ in length and which involves negligible parasitic effects. We require the total loss in decibels for this resonator. From eqn (4.54):

$$\alpha = \frac{\pi}{Q\lambda_g} \qquad \text{N/m}$$

and hence, for length l,

$$\alpha(l) = \frac{8.686\pi l}{Q\lambda_g} \qquad \text{dB}$$

Substituting $Q = 200$ and $l = 2\lambda_g$, $\alpha(l) = 0.27$ dB for the $2\lambda_g$ length of line. If $\lambda_g = 10$ mm, the attenuation coefficient is then $\alpha = 0.27/20 \simeq 0.013$ dB/mm.

Measurements of this Q-factor are best made using ring resonators, where the parasitic effects can be kept very small. (See Chapter 7 for a discussion on measurements.)

4.10.2 Conductor losses

This topic has been treated in detail elsewhere,[51–56] perhaps the most useful expression has been given by Hammerstad and Bekkadal:[56]

$$\alpha_c = 0.072 \frac{\sqrt{f}}{w Z_0} \lambda_g \qquad \text{dB/microstrip wavelength} \qquad (4.55)$$

where the frequency f is in gigahertz and Z_0 is in ohms.

In practice eqn (4.55) yields somewhat low results and account must be taken of surface roughness. Hammerstad and Bekkadal[56] have given the following curve-fitted formula to accommodate this:

$$\alpha'_c = \alpha_c \left[1 + \frac{2}{\pi} \tan^{-1} \left\{ 1.4 \left(\frac{\Delta}{\delta_s} \right)^2 \right\} \right] \qquad (4.56)$$

where

Δ = r.m.s. surface roughness
$\delta_s = 1/(R_s \sigma)$ is the skin depth at the appropriate operating frequency
R_s = surface resistance
σ = metal film conductivity

In order to observe the typical magnitudes involved consider a copper microstrip, where the skin depth at a few gigahertz is 1 μm. Also assume the r.m.s. surface roughness to be of similar magnitude, i.e. 1 μm (typical of rutile, see Table 2.2. Equation (4.56) then shows that the attenuation coefficient becomes

$$\alpha'_c \simeq 1.6 \alpha_c$$

That is to say, the loss is approximately 60 per cent increased when surface roughness is taken into account. The increase will be still greater for alumina and other substrates where the roughness exceeds \sim 10 μm.

4.10.3 Dielectric loss

This is analysed by several workers, including Gupta et al.[57] and Hammerstad and Bekkadal.[56] The first group[57] derive the following expression for dielectric attenuation coefficient, α_d, per unit length:

$$\alpha_d = 27.3 \frac{\epsilon_r (\epsilon_{eff} - 1) \tan \delta}{\sqrt{\epsilon_{eff}} (\epsilon_r - 1) \lambda_0} \qquad \text{dB/unit length} \qquad (4.57)$$

where $\tan \delta$ is the loss tangent for the substrate material.

With the microstrip wavelength λ_g given by

$$\lambda_g = \frac{\lambda_0}{\sqrt{\epsilon_{eff}}}$$

eqn (4.57) can also be written

$$\alpha_d = 27.3 \frac{\epsilon_r(\epsilon_{eff} - 1)\tan \delta}{\epsilon_{eff}(\epsilon_r - 1)} \qquad \text{dB/microstrip wavelength} \qquad (4.58)$$

which is the expression given by Hammerstad and Bakkadal.[56]

For microstrip lines on alumina, $\epsilon_{eff} \to \epsilon_r$, at least very approximately, and the attenuation coefficient is, roughly, $\alpha_d \approx 27 \tan \delta$. In most cases $\tan \delta \approx 10^{-3}$ (or less) and therefore $\alpha_d \approx 0.027$ dB/microstrip wavelength. This is at least a factor of five smaller than the conductor loss which was calculated earlier.

Conductor losses greatly exceed dielectric losses for most microstrip lines on alumina or sapphire substrates. However, where plastic substrates are used this will by no means always be the case, and (especially) silicon or gallium arsenide substrates result in much larger dielectric losses (about 0.04 dB/mm, that is 0.4 dB/microstrip wavelength at 10 GHz for silicon).

4.10.4 Radiation

Microstrip is an asymmetric transmission line structure and it is often used in unshielded or poorly shielded circuits where any radiation is either free to propagate away or to induce currents in the shielding. Further power loss is the net result.

In particular, *discontinuities* such as abruptly open-circuit microstrip (i.e. 'open' ends), steps, and bends will all radiate to a certain extent. Such discontinuities form essential features of a microwave integrated circuit and radiation cannot therefore be avoided altogether. Efforts must be made to reduce such radiation and its undesirable effects. In circuits such as filters, amplifiers, etc., this radiation is an acknowledged nuisance.

Both radiation and surface-wave propagation may be represented as a shunt admittance at the end of an open-circuit microstrip stub or at the plane location associated with some other abrupt discontinuity. This equivalent admittance is

$$Y = G_r + G_s + jB \qquad (4.59)$$

James and Henderson[58] show that, at frequencies where the surface wave is highly trapped in the substrate and with h/λ_0, $w_{eff}/\lambda_0 \ll 1$, the radiation conductance G_r is approximated by

$$G_r Z_0 \approx \frac{4\pi h w_{eff}}{3\lambda_0^2 \sqrt{\epsilon_{eff}}} \qquad (4.60)$$

in which w_{eff} is the effective microstrip width defined earlier (eqn 4.35).

The remaining terms in eqn (4.59), G_s and B, represent conductance due to surface-wave propagation and susceptance arising from the many field influences around the discontinuity.

For a 50 Ω microstrip line on 0.65 mm thick alumina the equivalent radiation conductance due to an open circuit, at 10 GHz, is approximately 30 μS. Under similar conditions a 10 Ω line yields a conductance of about 120 μS.

4.10.5 Surface-wave propagation

Surface waves, trapped just beneath the surface of the substrate dielectric, will be propagated away from microstrip discontinuities in the form of a range of TE and radial TM modes. James and Henderson[58] have analysed the problem thoroughly and have given a complicated expression for G_s.

The important conclusion is that, for microstrips on alumina substrates up to frequencies around 10 GHz, $G_s \ll G_r$ and G_r dominates. However, at higher frequencies G_s also becomes a very significant component. The curves in James and Henderson's paper (Figures 4 and 6 particularly) are very illuminating in this respect and they should be studied for detailed further information.

When we are concerned with a rectangular microstrip 'patch' resonator, loosely coupled to some source of microwave excitation, there will be *two* equivalent end admittances, each given by eqn (4.59). In Appendix B it is shown that the 'external' Q-factor due to such end loads may be written as

$$Q_e = \frac{l\pi\rho}{1 - \rho^2} \tag{4.61}$$

where l is an integer and ρ is given by

$$\rho = \left| \frac{YZ_0 - 1}{YZ_0 + 1} \right| \tag{4.62}$$

With the aid of the expressions given in all the preceding sections James and Henderson calculate a *total* Q-factor of 300 falling to 180, and 60 falling to 20 and lower, for rectangular microstrip patch resonators on alumina and plastic (polyolefin) substrates respectively. The frequency ranges were 6 to 10 GHz for plastic. Agreement with Q-factor measurements was fairly good, although only 130 was measured for the resonator on alumina at 18 GHz. (Details of the resonators are given in the cited paper.)

Various techniques are possible to reduce radiation:

(a) Metallic shielding or 'screening'
(b) The introduction of a small specimen of lossy (i.e. absorbent) material near any radiative discontinuity[59]
(c) The utilization of compact, planar enclosed circuits (e.g. spur-line filters—see Chapter 8, hairpin resonators).
(d) Possibly shape the discontinuity in some way to reduce the radiative efficiency

Surface-wave propagation may be reduced by (b) above or by cutting slots into the substrate surface just in front of an open circuit.[60]

4.10.6 Parasitic coupling

It has been pointed out[58] that while screening the circuit removes the radiation it reinforces the surface-wave fields. Thus we are often left with the option (b in the foregoing discussions) of introducing a lossy element to absorb the waves. This may not always be practicable due to space limitations and in any case it further reduces the Q-factor.

In the absence of any such material or other technique the total power in the parasitic waves, $(P_{rad} + P_{sur})$, is related to the incident power in the microstrip P by the ratio

$$\xi = \frac{P}{P_{rad} + P_{sur}} = \frac{1}{1 - \rho^2} \tag{4.63}$$

For a high degree of isolation the circuit must be operated at a relatively low frequency so that $P_{sur} \ll P_{rad}$ and then, substituting eqns (4.60) and (4.62) into (4.63):

$$\xi = \frac{1}{4G_r Z_0} = \frac{Z_0 \epsilon_{eff}}{640} \left(\frac{\lambda_0}{\pi h}\right)^2 \tag{4.64}$$

Where possible, the following observations then apply for high isolation:

(a) Use relatively high permittivity substrates. Then ϵ_{eff} is relatively large (e.g. alumina is better than plastic).
(b) Use fairly thin substrates (λ_0/h as large as possible).
(c) Employ high impedance (Z_0) stubs, etc., wherever this is feasible.

4.10.7 Radiation and surface-wave losses from various discontinuities

Although microstrip discontinuities are the principal subject of Chapter 5, the losses originating from them should logically receive some consideration here. Horng et al.[61] have reported a generalized method for calculating and clearly distinguishing between radiation and surface-wave losses. They base their method on a rigorous Poynting vector analysis in which the current distribution around the entire discontinuity is a result of a complete electromagnetic solution by moments. These workers provide self-consistency checks of their results, based upon power conservation calculations.

The losses calculated by Horng et al., due to radiation and surface-wave propagation from the important open-end discontinuity, are shown as a function of frequency in Figure 4.19. This feature is taken up further in

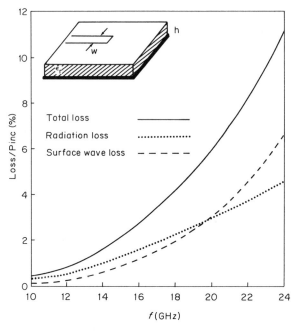

Fig. 4.19. Power losses versus frequency for open-end discontinuity (ϵ_r = 10.2, w = 24 mil, h = 25 mil) (© 1990 IEEE. Reprinted, with permission, from Horng et al.[61])

Chapter 10 where microstrip antennas are introduced, but it must be appreciated that such losses are also significant from the circuit design standpoint (usually they have to be minimized).

4.10.8 Losses in microstrip on semi-insulating GaAs

This specific and important subject has been addressed by several workers, most recently Goldfarb and Platzker[62] who compare data measured over the frequency range 0–40 GHz with predictions made using three proprietary CAE packages (see Appendix D). For the measurements microstrip resonators are formed on 0.1 mm thick GaAs substrates.

Goldfarb and Platzker's results for a 20-Ω line having such a structure are shown in Figure 4.20. Even with this relatively wide line operated at 27 GHz the radiation loss is negligible as a consequence of the high permittivity (12.9) and the small thickness of the substrate. So these results are dominated by conductor and dielectric losses. It is clear that all three CAE packages yield predictions of loss well within the range of the measurements, except for the losses as frequency approaches zero. At this limit Version 2.0 of SuperCompact™ does approach the correct value of

Fig. 4.20. Losses in microstrip on a GaAs substrate. (© 1990 IEEE. Reprinted, with permission, from Goldfarb and Platzker.[62])

this asymptotic loss whereas Version 1.95 did not. Likewise Hewlett-Packard's MDS approaches the correct value, but Eesof's Touchstone™ does not.

Where there is discrepancy between predicted and measured results Goldfarb and Platzker recommend a modification of the conductor resistivity model in the simulator, forcing it to follow the observed approximate square law relationship as a function of frequency.

4.11 SUPERCONDUCTING MICROSTRIPS

As pointed out in Chapter 2, a major disadvantage of most planar transmission lines is the dissipative loss (when compared with more traditional media such as all-metallic waveguide or coaxial cable). This forms an advantage for antenna applications but represents a distinct difficulty in the realisation of several types of circuits—notably filters.

Since the principal mechanism responsible for the power dissipation in microstrips is conductor loss it is clear that if this could be reduced then overall loss would also be significantly decreased. One direct method of

achieving this is to 'supercool' the structure, typically down to a specific temperature within the range 20–77 K.

By no means incidentally, the action of supercooling in this manner can also benefit the function of other circuit components such as switches by accelerating their performance and thereby rendering them capable of higher-frequency operation. By the end of the 1980s this area of microwave technology had expanded to the extent that commercial organizations were already in evidence, for example Superconducting Technologies Inc., USA.

Lee et al.[63] have reported conductor loss calculations for superconducting microstrips. They describe the complex conductivity of the superconductor approximately, using the 'two-fluid' model. The resulting modelled conductivity is used to determine the conductor loss by means of the phenomenological loss equivalent method (PEM). A problem with this type of analysis is that for complete penetration of the current into the conductor the current distribution is uniform inside the conductor. Contrastingly, where the current penetration is relatively moderate then the current distribution becomes distinctly non-uniform inside the conductor. In order to obtain an equivalent strip for calculations Lee et al. consider two different superconducting materials to which are separately ascribed the shallow and deep penetrations.

The curves of Figure 4.21 are repeated from the data presented by Lee et al. and they show results for both conventional (room temperature) microstrip having aluminium metallization and, alternatively, the superconducting line held at a temperature of 77 K (critical temperature 92.5 K).

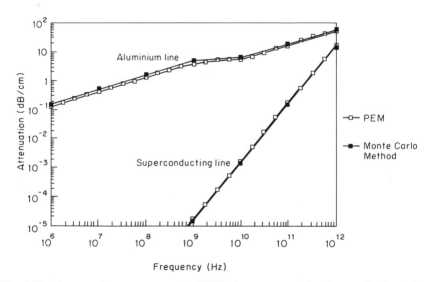

Fig. 4.21. Attenuation constants at 77 K for superconducting and aluminium microstrip lines. (Reprinted, with permission, from Lee, Kong and Itoh.[63])

The conventional microstrip line is fabricated on a quartz substrate with a permittivity of 3.9. In both cases results using the PEM are compared with results using an independent Monte Carlo method. Clearly, at all frequencies considered here the superconducting line offers a much lower loss regime than conventional (room temperature) metallic microstrip. Even at 100 GHz (10^{11} Hz) the superconducting line loses only 0.1 dB/cm in contrast to the conventional line for which a loss of 10 dB/cm is applicable at this frequency.

Phase velocity is also shown as a function of frequency in Figure 4.22. The behaviour associated with the aluminium line exhibits the usual variation of phase velocity, including the marked dispersion which, for this example, is pronounced about 1 GHz. In the case of the superconducting line, however, the phase velocity is clearly approximately constant with frequency over the entire band considered (d.c. to 1000 GHz). Any dispersive effects at the higher frequencies are masked by the scale of this graph.

Working in conjunction with another group (Bhasin et al.), Kong et al.[64] have also reported on the performance and modelling of superconducting ring resonators operating at 35 GHz (Chapter 7 provides information on ring resonators). They showed that, at temperatures below 60 K, the superconducting strip exhibited a performance superior to that of a conventional gold-metallized microstrip. In the superconducting case the unloaded-Q was higher by a factor of approximately 1.5. These workers also show results for resonant frequency as a function of temperature.

Clearly there is a substantial amount of work required to refine both the models and/or the quality of the metallizations used for the microstrips. The

Fig. 4.22. Phase velocities at 77 K for superconducting and aluminium microstrip lines. (Reprinted, with permission, from Lee, Kong and Itoh.[63])

difference between data obtained by theory and measurement approached a factor of two at the lower temperatures and remained about 25 per cent at the higher temperatures for the superconductive case.

Closely related work has also been performed by van Deventer et al.[65] This group used an integral equation approach to calculate the propagation characteristics of shielded superconducting microstrip lines. Phase and attenuation coefficients, as well as characteristic impedance, are presented as functions of frequency (d.c. to 10 GHz). It is shown that the attenuation coefficient remains below 0.5 dB/m in all cases, up to 10 GHz, except for the relatively extreme case where the substrate height is reduced to 25 microns, when the attenuation increases to 0.83 dB/m at 10 GHz. However, unlike all other situations reported, in this case the phase coefficient stays constant over the entire band i.e. at these lower microwave frequencies design is relatively straightforward and there is negligible dispersion. Characteristic impedance is also invariant with frequency over this range.

Hammond et al.[66] have reported the performance of superconducting microstrip resonators. Their measurements indicate that the loaded Q-factor of superconducting, open-ended S-shaped microstrip resonators is always much greater than that of conventional copper or even silver microstrip resonators. The Q-factors are, however, dependent upon the level of input power and significant reductions occur as this input power approaches 1 Watt. As the power level is increased from -65 to around 0 dBm, the Q-factor falls slowly from around 5500 at the lower power, to approximately 4000 at 0 dBm. Even so, across the entire range of power levels the Q-factors are about one order of magnitude higher than is achievable with conventional resonators held at room temperature.

In summary, the performance achievable with superconducting microstrip renders the approach attractive, at least in situations where the other circuit components also benefit.

4.12 MICROSTRIP CHARACTERISTIC IMPEDANCE—
COMMENTS ON RECENT THEORETICAL WORK

Brews[67] has provided a theoretical study of microstrip characteristic impedance. His paper is entirely theoretical but he points out that the 'very natural' requirement for complex power P is

$$P = \frac{1}{2} I^* V \qquad (4.65)$$

This leads to a preferred choice of impedance-defining expressions following eqns (4.32) to (4.34) given earlier, and it is based upon these definitions that calculations and measurements of $Z_0 (f)$ should be made and used.

4.13 DESIGN RECOMMENDATIONS

In the light of the techniques presented in Chapters 3 and 4 we can now summarize the main microstrip design recommendations.

Initially, system and network design procedures yield a specific value of characteristic impedance Z_0 and of electrical length θ for each microstrip line. We must determine the correct physical dimensions for the microstrip, that is w and l. It is recommended that the following sequence be used.

4.13.1

Select the *substrate* such that the TM mode effect is avoided. That is, the maximum intended operating frequency should be less than

$$f_{\text{TEM 1}} = c\tan^{-1}(\epsilon_r)/\sqrt{2\pi h}\sqrt{\epsilon_r - 1}$$

as given by eqn (4.53).

Also ensure that parasitic coupling is sufficiently low (eqn 4.53).

4.13.2

Carry out a 'first-cut' design on the static-TEM basis explained in Chapter 3. This yields

$$w \text{ and } \epsilon_{\text{eff}}$$

4.13.3

Check that the first-order transverse resonance cannot be excited at the highest frequencies to be used (eqn 4.39):

$$f_{\text{CT}} = c/\sqrt{\epsilon_r}(2w + 0.8h)$$

In the event that such a resonance might occur either redesign (using narrower microstrip lines) or, if possible, incorporate slots in the metal strips.

4.13.4

Calculate the total losses and the total Q-factor, and check that this meets the circuit requirements (Section 4.10 gives the methods for this). If the Q-factor is too low a major reappraisal of the design philosophy may be necessary.

However, as a general indication, many satisfactory microstrip circuits may be designed with Q-factors no greater than about 100.

4.13.5

Evaluate the frequency-dependent effective microstrip permittivity $\epsilon_{\text{eff}}(f)$, and hence the wavelength λ_{g} $(= c/f\sqrt{\epsilon_{\text{eff}}(f)}$.

For alumina-type substrates, where $10 \leq Z_0 \leq 100$ Ω and $2 \leq f \leq 18$ GHz the recommended expression is the relatively simple eqn (4.11):

$$\epsilon_{\text{eff}}(f) = \epsilon_{\text{r}} - \frac{\epsilon_{\text{r}} - \epsilon_{\text{eff}}}{1 + (h/Z_0)^{1.33}(0.43 f^2 - 0.009 f^3)}$$

where h is in millimetres and f is in gigahertz. The estimated accuracy is ± 0.8 per cent. (Note that this *includes* errors in the static-TEM expressions for ϵ_{eff} and Z_0, i.e. the ± 0.8 per cent is overall.)

Where higher frequencies, and/or substrates having permittivities differing substantially from 10, are involved the more elaborate set of expressions due to Kirschning and Jansen or Kobayashi are recommended. These were provided earlier in this chapter and will not be repeated here (see Section 4.4.2).*

Since the electrical length θ (degrees) is given in the circuit or antenna specification, the physical length l of the microstrip is

$$l = \frac{\theta\lambda_{\text{g}}}{360} \quad \text{(neglecting discontinuities)}$$

4.13.6 Characteristic impedance as a function of frequency

This problem was discussed in Section 4.8. The practical effects are fairly small, *mainly* amounting to a small degradation of VSWR in most cases. If it is considered that $Z_0(f)$ is significant in a particular design problem the algorithm given in Section 4.8.4 is recommended.

4.13.7 Computer-aided design

The techniques described in Chapters 3 and 4 are very appropriate for CAD implementation. Sometimes MIC designers prefer to set up their own CAD facility. This can incorporate the most accurate expressions available for static-TEM and dispersion computations. It can also embody formulae accounting for a number of microstrip discontinuity elements which are described in Chapter 5.

A selection of computer-aided design techniques are outlined in Appendix D at the back of this text.

*If sufficient time and computer power is available a technique such as that described in Section 4.7 might be used.

REFERENCES

1. Daly, P., 'Hybrid-mode analysis of microstrip by finite-element methods', *IEEE Trans.*, **MTT-19**, January 1971, pp. 19–25.
2. Gelder, D., 'Numerical determination of microstrip properties using the transverse field components', *Proc. IEE*, **117**, April 1970, 699–703.
3. Corr, D. G., and Davies, J. B., 'Computer analysis of the fundamental and higher order modes in single and coupled microstrip', *IEEE Trans.*, **MTT-20**, No. 10, October 1972, 669–677.
4. Corr, D G., 'Finite difference analysis of hybrid modes in microstrip structures', *PhD Dissertation*, Univ. of London, August 1970.
5. Grünberger, G. K., Keine, V., and Meinke, H. H., 'Longitudinal field components and frequency-dependent phase velocity in the microstrip transmission line', *Electron. Lett.*, **6**, No. 21, 15 Oct. 1970, 683–685.
6. Grünberger, G. K., and Meinke, H. H., 'A theory of the microstrip line including longitudinal components', *Proc. 1971 European Microwave Conference*, Stockholm, Sweden, August 1971, pp. C4/2:1 to C4/2:4.
7. Laloux, A., and Vander Vorst, A., 'The dispersive character of microstrip lines', *Proc. 1971 European Microwave Conference*, Stockholm, Sweden, August 1971, pp. C3/6:1 to C3/6:3.
8. Van de Capelle, A. R., and Luypaert, P. J., 'An investigation of the higher order modes in open microstrip lines', *Proc. of the Fifth Colloquium on Microwave Communication*, Akademiai, Kiado, Budapest, 24–30 June 1974, Vol. 3, pp. ET-32 to ET-31.
9. Schmitt, J. H., and Sarges, K. H., 'Wave propagation in microstrip', *Nachrichtentech NTZ*, **5**, May 1971, 260–264.
10. Hornsby, J. S., and Gopinath, A., 'Numerical solutions of inhomogeneously filled guides with symmetrical microstrip line', *Proc. 1969 European Microwave Conf.*, London, September 1969, IEE Conf. Publ. 58, pp. 114–117.
11. Hornsby, J. S., and Gopinath, A., 'Fourier analysis of a dielectric-loaded waveguide with a microstrip line', *Electron. Lett.*, **5**, 12 June 1969, 265–267.
12. Hornsby, J. S., and Gopinath, A., 'Numerical analysis of a dielectric-loaded waveguide with a microstrip line—Finite-difference methods', *IEEE Trans.*, **MTT-17**, September 1969, 684–690.
13. Denlinger, E. J., 'A frequency-dependent solution for microstrip transmission lines', *IEEE Trans.*, **MTT-19**, Jan. 1971, pp. 30–39.
14. Chudobiak, W. J., Jain, O. P., and Makios, V., 'Dispersion in microstrip', *IEEE Trans. (Correspondence)*, **MTT-19**, September 1971, 783–784.
15. Jain, O. P., Makios, V., and Chudobiak, W. J., 'The dispersive behaviour of microstrip transmission lines', *Proc. 1971 European Microwave Conf.*, Stockholm, Sweden, August 1971, pp. C3/5:1 to C3/5:4.
16. Jain, O. P., Makios, V., and Chudobiak, W. J., 'Coupled-mode model of dispersion in microstrip', *Electron. Lett.*, **7**, 15 July 1971, 405–407.
17. Schneider, M. B., 'Microstrip dispersion', *Proc. IEEE*, January **1972**, 144–146.
18. Carlin, H. L., 'A simplified circuit model for microstrip', *IEEE Trans.*, **MTT-21**, No. 9, September 1973, 589–591.
19. Getsinger, W. J., 'Microstrip dispersion model', *IEEE Trans.*, **MTT-21**, No. 1, January 1973, 34–39.
20. Getsinger, W. J., 'Microstrip dispersion model', *COMSAT Laboratories, Technical Memorandum* CL-9-72, 10 Mar. 1972.
21. Edwards, T. C., and Owens, R. P., '2–18 GHz dispersion measurements on 10–100 ohm microstrip lines on sapphire', *IEEE Trans.*, **MTT-24**, No. 8, August 1976, 505–513.

22. York, R. A., and Compton, R. C., 'Experimental evaluation of existing CAD models for microstrip dispersion' *IEEE Trans.*, **MTT-S**, No. 3, March 1990, 327–328.
23. Yamashita, E., Atsuki, K., and Ueda, T., 'An approximate dispersion formula of microstrip lines for computer-aided design of microwave integrated circuits' *IEEE Trans.* **MTT-27**, Dec. 1979, 1036–1038.
24. Kirschning, M., and Jansen, R. H., 'Accurate model for effective dielectric constant of microstrip with validity up to millimetre-wave frequencies'. *Electronics Letters*, 18 March 1982, 272–273.
25. Kobayashi, M., 'A dispersion formula satisfying recent requirements in microstrip CAD'. *IEEE Trans.*, **MTT-36**, August 1988, 1246–1250.
26. Atwater, H. A., 'Tests of microstrip dispersion formulas'. *IEEE Trans.*, **MTT-36**, March 1988, 619–621.
27. Kurushin, E. P., *et al.*, 'Dispersion in microstripline on a ferrite substrate', *Radio Physics and Quantum Electronics*, March 1982, 251–254.
28. Hsia, I. Y., Yang, H.-Y., and Alexopoulos, N. G., 'Basic Properties of Microstrip Circuit Elements on Nonreciprocal Substrate-Superstrate Structures, *IEEE MTT-S International Microwave Symposium* 1990, Digest of Papers, 665–668.
29. Kobayashi, M., 'Frequency dependent characteristics of microstrips on anisotropic substrates', *IEEE Trans.* **MTT-S**, No. 11, Nov. 1982, 2054–2059.
30. Mariki, G. E., and Yeh, C., Dynamic three-dimensional TLM analysis of microstriplines on anisotropic substrate', *IEEE Trans.*, **MTT-33**, No. 9, Sept. 1984, 789–798.
31. Kowalski, G., and Pregla, R., 'Dispersion characteristics of single and coupled microstrips', *Arch. Elek. Übertragung*, **26**, June 1972, 276–280.
32. Krage, M. K., and Haddad, G. I., 'Frequency-dependent characteristics of microstrip transmission lines', *IEEE Trans.*, **MTT-20**, No. 10, October 1972, 678–688.
33. Itoh, T., and Mittra, R., 'Spectral-domain approach for calculating the dispersion characteristics of microstrip lines', *IEEE Trans.*, **MTT-21**, No. 7, July 1973, 496–499.
34. Itoh, T., and Mittra, R., 'A technique for computing dispersion characteristics of shielded microstrip lines', *IEEE Trans.*, **MTT-22**, October 1974, 896–898.
35. Jansen, R. H., 'High-speed computation of single and coupled microstrip parameters including dispersion, high-order modes, loss and finite strip thickness', *IEEE Trans.*, **MTT-26**, No. 2, February 1978, 75–82.
36. Wu, Ke., and Vahldeick, R., 'A novel space-spectral domain approach and its application to three-dimensional MIC/MMIC circuits' *Proceedings of the 19th European Microwave Conference*, London, Sept. 1989, 751–756.
37. Dumbell, K. D., Pennock, S. R., and Rozzi, T. E., 'Effects of asymmetry in shielded microstrip transmission lines', *Proceedings of the 19th European Microwave Conference*, London, Sept. 1989, 761–766.
38. Qian, Y., and Yamashita, E., 'Compression and reshaping of picosecond electrical pulses using dispersive microwave transmission lines', *IEEE MTT-S International Microwave Symposium* 1990, Digest of Papers, 951–954.
39. Rittweger, M., and Wolff, I., 'Analysis of complex passive (M)MIC-components using the finite difference time-domain approach', *IEEE MTT-S International Microwave Symposium* 1990, Digest of Papers, 1147–1150.
40. Bianco, B., *et al.*, 'Some consideration about the frequency dependence of the characteristic impedance of uniform microstrips', *IEEE Trans.*, **MTT-26**, No. 3, March 1978, 182–185.

41. Getsinger, W. J., 'Microstrip characteristic impedance', *IEEE Trans.*, **MTT-27**, April 1979, 293.
42. Owens, R. P., 'Predicted frequency dependence of microstrip characteristic impedance using the planar waveguide model', *Electron. Lett.*, **12**, 1976, 269–270.
43. Owens, R. P., 'The electrical characteristics of transferred-electron device package and mounts, and their effects on oscillator performance', *PhD Thesis*, University of London, 1977.
44. Kompa, G., and Mehran, R., 'Planar waveguide model for calculating microstrip components', *Electron. Lett.*, **11**, 1975, 459–460.
45. Bianco, B. *et al.*, 'Frequency dependence of microstrip parameters', *Alta Frequenza*, **43**, 1974, 413–416.
46. Howe, Jr, H., *Stripline Circuit Design*, Artech House Inc., Dedham, Massachusetts, USA., 1974.
47. Bahl, I. J., and Garg, Ramesh, 'A designer's guide to stripline circuits', *MicroWaves*, January **1978**, 90–96.
48. Knorr, J. B., and Tufekcioglu, A., 'Spectral-domain calculation of microstrip characteristics impedance', *IEEE Trans.*, **MTT-23**, 1975, 725–728.
49. Mirshekar-Syahkal, D., and Davies, J. B., 'Accurate solution of microstrip and coplanar structures for dispersion and for dielectric and conductor losses', *IEEE Trans.*, **MTT-27**, No. 7, July 1979, 694–699.
50. Vendelin, G. D., 'Limitations on stripline *Q*', *Microwave Journal*, May **1970**, 63–69.
51. Pucel, R. A. Masse, D. J., and Hartwig, C. P., 'Losses in microstrip', *IEEE Trans.*, **MTT-16**, 1968, pp. 342–350. Also see correction in *IEEE Trans.*, **MTT-16**, 1968, 1064.
52. Schneider, M. V., 'Microstrip lines for microwave integrated circuits', *B.S.T.J.*, **48**, 1969, 1422–1444.
53. Wheeler, H. A., 'Formulas for the skin effect', *Proc. IRE*, **30**, 1942, 412–424.
54. Welch, J. D., and Pratt, H. J., 'Losses in microstrip transmission systems for integrated microwave circuits', *NEREM Rec*, **8**, 1966, 100–101.
55. Horton, R., Easter, B., and Gopinath, A., 'Variation of microstrip losses with thickness of strip', *Electron. Lett.*, **7**, 1971, 490.
56. Hammerstad, E. O., and Bekkadal, F., 'A microstrip handbook', *ELAB Report*, STF 44 A74169, University of Trondheim, Norway, 1975, pp. 98–110.
57. Gupta, K. C., Garg, Ramesh, and Bahl, I. J., *Microstrip Lines and Slotlines'*, Artech House, Inc., 1979, pp. 91–92.
58. James, J. R., and Henderson, A., 'High-frequency behaviour of microstrip open-circuit terminations', *IEE Journal on Microwaves, Optics and Acoustics*, **3**, No 5, September 1979, 205–218.
59. Armstrong, A. F., and Cooper, P. D., 'Techniques for investigating spurious propagation in enclosed microstrip', *Rad. & Electron Eng*, **48**, 1978, 64–72.
60. James, J. R., and Ladbrooke, P. H., 'Surface-wave phenomena associated with open-circuited stripline terminations', *Electron. Lett.*, **9**, No. 24, 29 Nov. 1973, 570–571.
61. Horng, T. S., Yang, H. Y., and Alexopoulos, N. G., 'A full-wave analysis of shielded microstrip line-to-line transitions' *IEEE MTT-S International Microwave Symposium* 1990, Digest of Papers, 251–254.
62. Goldfarb, M. E., and Aryeh Platzker, 'Losses in GaAs Microstrip', *IEEE MTT-S International Microwave Symposium* 1990, Digest of Papers, 563–565.
63. Lee, H.-Y., Kong, K.-S., and Itoh, T., 'Conductor loss calculation of superconducting microstrip line using a phenomenological loss equivalence method' *Proceedings of the 19th European Microwave Conference*, London,

Sept. 1989, 757–760.

64. Bhasin, K. B., *et al.*, Performance and modelling of superconducting ring resonators at millimeter-wave frequencies', *IEEE MTT-S International Microwave Symposium* 1990, Digest of Papers, 269–276.

65. van Deventer, T. E., *et al.*, 'High frequency characterization of high-temperature superconducting thin film lines', *IEEE MTT-S International Microwave Symposium* 1990, Digest of Papers, 285–288.

66. Hammond, R. B., *et al.*, 'Superconducting Tl-Ca-Ba-Cu-O thin film microstrip resonator and its power handling performance at 77K', *IEEE MTT-S International Microwave Symposium* 1990, Digest of Papers, 867–870.

67. Brews, 'Characteristic impedance of microstrip lines', *IEEE Trans.* **MTT-35**, January 1987, 30–34.

5 Discontinuities in Microstrip

5.1 THE MAIN DISCONTINUITIES

Virtually *all* practical distributed circuits, whether in waveguide, coaxial lines, or any other propagation structure, must inherently contain discontinuities. A straight, uninterrupted length of waveguide or microstrip line (strictly speaking, with a perfectly invariant cross-section) would be genuinely *continuous*, i.e. it would not contain any discontinuities. But in fact such straight lines, taken alone, are of little engineering use and in any case junctions are essential. In the region of bends, abruptly stopped open circuits, width changes, and transitions *discontinuities* occur in the transmission line.

Although such discontinuities give rise to only very small capacitances and inductances (often < 0.1 pF and < 0.1 nH) the reactances of these become particularly significant at the high microwave and into millimeter-wave frequencies. The performance of amplifiers, for example, has been shown to be considerably affected by the microstrip discontinuities.[1] As with dispersion, we can very often neglect these discontinuities when the frequencies involved do not exceed a few gigahertz. Above approximately 10 GHz they are definitely very significant.

Several forms of discontinuity emerge from circuit requirements:

(a) foreshortened open circuits

(b) series coupling gaps

(c) short circuits through to the ground plane

(d) right-angled corners or 'bends' (unmitred and mitred)

(e) step width changes

(f) transverse slit

(g) T-junction

(h) cross-junctions.

An example of a microwave transistor amplifier layout[2] is shown in Figure 5.1, and at least three of the discontinuities can be readily identified. Some

Fig. 5.1. Layout of a typical microwave amplifier using a GaAsFET, showing discontinuities in the microstrip lines (d.c. bias filters are *not* shown)

of them are numbered on the diagram, in accordance with the quoted list. Many other circuits, such as filters, mixers, and oscillators, involve several discontinuities.

It is beyond the scope of this treatment, being oriented towards circuit design, to consider in any detail the techniques which have been used to evaluate discontinuity capacitances and inductances.

In the first twelve sections of this chapter we consider discontinuity modelling based upon equivalent capacitances and inductances. Later, in Section 5.13, non-linear frequency-dependent effects are considered.

These techniques have received a good, comparative assessment in detail elsewhere (e.g. Gupta *et al.*)[3] and we shall only note the approaches briefly here, starting with capacitance evaluation techniques as follows:

(a) matrix inversion method
(b) variational method
(c) Galerkin's method in the spectral domain
(d) use of line sources with charge reversal.

Of all the techniques the last (d), adhered to by Silvester and Benedek,[4] appears to have been the most successful. This is probably because it avoids the subtraction of two large and similar numbers.

Discontinuity *inductance* evaluation has been performed by Thomson and Gopinath.[5] Their method starts with magnetic vector potential and uses Maxwell's equations and divergence. Inductance is finally obtained after integrating the current density distribution.

At frequencies much beyond 10 or 12 GHz the lumped C and L description of most discontinuities becomes less and less meaningful. Discontinuities at these higher frequencies are probably best described by means of scattering

matrices, although in many cases considerable success has been enjoyed by using the lumped *LC* approach at frequencies up to around 18 GHz.

Radiation from various discontinuities gives rise to equivalent resistive elements. These will not be discussed in any detail here since they enter mainly into the province of those interested in antennas rather than circuits, except where the reduction in *Q*-factor is significant, as discussed in Chapters 4, 7 and 10.

In some cases techniques have been developed which compensate, at least over a particular range of frequencies, the existing discontinuity effects. Where this is the case less detailed attention is given here to the properties of the uncompensated discontinuity.

Although 'quasi-static', the effectiveness of results obtained from these various aforementioned approaches has been adequately vindicated by incorporating the methods in several full circuit designs and checking with careful measurements. The *calculation* of the reactive effects due to most discontinuities will now be shown, as far as practicable.

5.2 THE FORESHORTENED OPEN CIRCUIT

There are essentially three phenomena associated with the 'open circuit':

(a) There will be fringing fields extending beyond the abrupt physical end of the metallic strip.
(b) Surface waves will be launched from the end of the strip.
(c) Energy will be radiated from the open end.

Item (a) may be accounted for by assuming some equivalent capacitance to be connected at the open end, and this feature nearly always dominates the 'end-effect'. Items (b) and (c) require equivalent shunt conductance at the open end of the line. In most cases in this work it will not be necessary to *quantify* these last two phenomena, but their minimization may be carried out on a practical basis in circuits involving microstrip. This has already been discussed in Chapter 4, and James and Ladbrooke[6] have indicated how substrate slots may be used to minimize surface-wave interactions.

The physical appearance of both a foreshortened open circuit and a series gap are shown in Figure 5.2, where equivalent lumped capacitor networks are also given.

Several quasi-static field calculations of the capacitance C_f have been documented. The most reliable, judged by comparison with many independent measurements, seems to be the work of Silvester and Benedek.[4] While being thorough, the resulting curves of results need further interpretation and a curve-fitting formula can be deduced for microwave circuit design. In their paper Silvester and Benedek give a formula which amounts to a series expansion fit to their theoretical data. The expression is

(a) Physical open circuits

(b) Equivalent networks

Fig. 5.2. Microstrip open circuits and their equivalent lumped capacitive networks

$$\frac{C_f}{w} = \exp\left\{2.2036 \sum_{i=1}^{5} K_\epsilon\left(\log\frac{w}{h}\right)^{i-1}\right\} \quad \text{pF/m} \tag{5.1}$$

where the coefficients K_ϵ are given in Table 5.1.

These results may readily be plotted out as a set of curves giving K_ϵ as a function of ϵ_r and with i as parameter. Thus coefficients K_ϵ for other materials may be found, within a fairly narrow range of permittivities. A set of first-order linear approximation equations can be simply obtained from these curves by curve fitting so that the formula (eqn 5.1) can be used in an iterative CAD routine.

Although this method works reasonably satisfactorily and accurately, especially on a minicomputer, it is not really suited to rapid pocket calculator or personal computer evaluation. An empirical formula which yields, quite simply, the *equivalent end-effect* length is given shortly, but first we must properly define this quantity.

Table 5.1. Coefficients for eqn (5.1) (©1972 IEEE. Reprinted, with permission, from Silvester and Benedek.[4])

ϵ_r	1.0	2.5	4.2	9.6	16.0	51.0
i						
1	1.110	1.295	1.443	1.738	1.938	2.403
2	−0.2892	−0.2817	−0.2535	−0.2538	−0.2233	−0.2220
3	0.1815	0.1367	0.1062	0.1308	0.1317	0.2170
4	−0.0033	−0.0133	−0.0260	−0.0087	−0.0267	−0.0240
5	−0.0540	−0.0267	−0.0073	−0.0113	−0.0147	−0.0840

5.2.1. Equivalent end-effect length

In many aspects of circuit design it is very useful to pretend that the microstrip line is rather longer than it actually is, to account for the end-effect. Then we can deal with fully distributed (microstrip) structures throughout and do not need to work separately in terms of lumped capacitance. The concept is illustrated in Figure 5.3.

We can now define the equivalent end-effect length, l_{eo}:

l_{eo} is the equivalent extra length of microstrip line, *having all the propagation parameters applicable to the main line*, continuing from the 'physical' end plane to the final *electrically open* end plane.

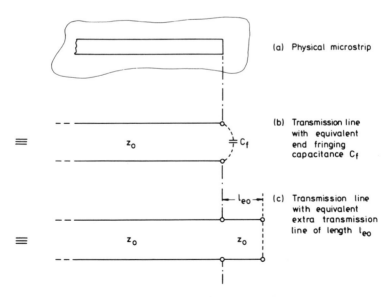

Fig. 5.3. Development of the equivalent end-effect length concept

Notice that the 'line' of length l_{eo} must be an exact continuation of the main transmission line—same Z_0 and hence same w/h and identical ϵ_{eff}.

It is a simple matter to obtain the end-effect length as a function of C_f, Z_0, and ϵ_{eff}. The input reactance to the extra length of line is given by a standard open-circuit terminated line result presented in Chapter 1:

$$X_{eo} = -jZ_0 \cot \beta l_{eo} \qquad (5.2)$$

Also, the capacitive reactance due to C_f is

$$X_f = \frac{1}{j\omega C_f} \qquad (5.3)$$

For equivalence we must equate (5.2) and (5.3), giving

$$\frac{1}{\omega C_f} = \frac{Z_0}{\tan \beta l_{eo}} \qquad (5.4)$$

Since $l_{eo} \ll \lambda_g$ we may apply the small-angle approximation: $(\tan \beta l_{eo}) \rightarrow \beta l_{eo}$, so that eqn (5.4) then becomes

$$\frac{1}{\omega C_f} \simeq \frac{Z_0}{\beta l_{eo}} \simeq \frac{Z_0 \lambda_g}{2\pi l_{eo}}$$

Writing c as the free-space velocity, this may be written as

$$\frac{1}{\omega C_f} = \frac{cZ_0}{\omega \sqrt{\epsilon_{eff}} \, l_{eo}}$$

Then, finally,

$$l_{eo} \simeq \frac{cZ_0 C_f}{\sqrt{\epsilon_{eff}}} \qquad (5.5)$$

Frequency-dependent effects are discussed in Section 5.13. For most microstrip lines, certainly those on alumina substrates less than 1 mm thick, the static-TEM expressions given here appear to work quite accurately over the approximate frequency range: $2 \leq f \leq 20$ GHz. An alternative empirical expression, yielding the length extension directly, has been given by Hammerstad and Bekkadal:[7]

$$l_{eo} = 0.412h \left(\frac{\epsilon_{eff} + 0.3}{\epsilon_{eff} - 0.258} \right) \left(\frac{w/h + 0.262}{w/h + 0.813} \right) \qquad (5.6)$$

It appears that, over a wide range of materials and microstrip width-to-height ratios (w/h), this expression can often give errors of 5 per cent or more. Where such errors are acceptable, which is frequently the case for this quantity, eqn (5.6) should be used because it involves less computing than the more accurate eqn (5.5).

5.2.2 Upper limit to end-effect length (quasi-static)

End-effect length is usually normalized to h, to obtain l_{eo}/h, which is the quantity usually employed as a design parameter. The graph of Figure 5.4. shows l_{eo}/h plotted as a function of w/h for the particular case of a substrate having $\epsilon_r = 11.0$. The maximum value, shown as a broken line, is determined on a static capacitance basis for a line embedded totally in a free-space environment.

Cohn[8] has given an exact expression for the fringing capacitance due to a foreshortened semi-infinite line situated at a distance $b/2$ above a ground plane, where the entire structure is completely filled with dielectric of relative permittivity ϵ_r. Cohn's expression may be reinterpreted for the microstrip case and calculated for l_{eo}/h. This yields the result

$$\left(\frac{l_{eo}}{h}\right)_{max} = \frac{2}{\pi}\ln(2) \qquad (5.7)$$

$$\simeq 0.441$$

Recent work has suggested that this statically derived 'limit' may not be absolute. It does, however, give an indication to the designer of the approximate maximum value likely for open-circuit equivalent length.

5.3 THE SERIES GAP

The physical appearance of the microstrip series gap was also given in Figure 5.2. This discontinuity is important in some forms of end-coupled filters, coupling to resonators, and other circuits. For convenience here the nomenclature is repeated in Figure 5.5.

For energy to be coupled across the gap the 'open-circuit' microstrip ends have to be equally and oppositely charged. Significant capacitance therefore

Fig. 5.4. Normalized end-effect length (l_{eo}/h) as a function of shape ratio w/h

(a) Physical arrangement

(b) Lumped capacitive equivalent circuit

Fig. 5.5. The series gap and its equivalent lumped circuit

exists across the gap and is denoted C_2. Grounded capacitors C_1 represent the fields fringing directly down to ground from each end of the line at a symmetrical gap.

An extensive analysis of the gap allowing for the necessary inversion to account for opposite potential signs on each side leads to comprehensive results which work for tightly coupled gaps. Since this is rarely encountered in MIC practice, only the result for loosely coupled gaps is given here.

By analogy with the development of eqn (5.5), for the foreshortened open circuit, the expression for the gap end-effect line extension may be written

$$l_{eg} \simeq \frac{cZ_0(C_1 + C_2)}{\sqrt{\epsilon_{eff}}} \qquad (5.8)$$

This now means that, where we *know* C_1 and C_2, we can determine the line extension l_{eg}. This is fairly satisfactory in the case of simple resonators and some other structures where the degree of coupling, leading to the gap separation g, has been rather arbitrarily set.

However, this situation is quite unsuitable for synthesis, e.g. of a filter. In such a case we require ultimately to determine g itself from a starting point of filter characteristics. (For present purposes suffice it to say that we *can* obtain the gap capacitance information from prototype filter characteristics.) Garg and Bahl[9] have given curve-fitted expressions for the gap capacitances as follows:

$$\frac{C_o}{w} = \left(\frac{g}{w}\right)^{m_o} \exp(k_o) \quad \text{pF/m} \tag{5.9}$$

$$\frac{C_e}{w} = \left(\frac{g}{w}\right)^{m_e} \exp(k_e) \quad \text{pF/m} \tag{5.10}$$

applicable for $\epsilon_r = 9.6$ and over the range

$$0.5 \leq \frac{w}{h} \leq 2$$

where

$$C_o = 2C_2 + C_1 \tag{5.11}$$

and

$$C_e = 2C_1 \tag{5.12}$$

The indices and arguments are defined by the following set of equations:

$$\left.\begin{aligned}
m_o &= \frac{w}{h}\left\{0.619 \log\left(\frac{w}{h}\right) - 0.3853\right\} \\
k_o &= 4.26 - 1.453 \log\left(\frac{w}{h}\right)
\end{aligned}\right\} 0.1 \leq \frac{g}{w} \leq 1.0 \tag{5.13a}$$

$$\left.\begin{aligned}
m_e &= 0.8675 \\
k_e &= 2.043 \left(\frac{w}{h}\right)^{0.12}
\end{aligned}\right\} 0.1 \leq \frac{g}{w} \leq 0.3 \tag{5.13b}$$

$$\left.\begin{aligned}
m_e &= \frac{1.565}{(w/h)^{0.16}} - 1 \\
k_e &= 1.97 - \frac{0.03}{w/h}
\end{aligned}\right\} 0.3 \leq \frac{g}{w} \leq 1.0 \tag{5.13c}$$

Garg and Bahl[9] also showed that values can be calculated for C_o and C_e for other values of substrate permittivity ϵ_r by using the following expressions:

$$C_o(\epsilon_r) = C_o(9.6)\left(\frac{\epsilon_r}{9.6}\right)^{0.8} \tag{5.14}$$

and

$$C_e(\epsilon_r) = C_e(9.6)\left(\frac{\epsilon_r}{9.6}\right)^{0.9} \tag{5.15}$$

These relations suit the range

$$2.5 \leq \epsilon_r \leq 15$$

and an accuracy of 7 per cent was quoted.

In order to obtain g we select eqn (5.9), because C_o involves both C_1 and C_2 gap capacitances, and rearrange as follows:

$$\ln\left(\frac{g}{w}\right) = \frac{1}{m_o}\left\{\ln\left(\frac{C_o}{w}\right) - k_o\right\} \tag{5.16}$$

Having obtained a first value for g, it is desirable that a second value be found using eqn (5.10). Should the two values differ significantly then the final value of g will be that which most closely satisfies the two relations.

5.3.1 Accuracy of gap capacitance calculations (C_1 and C_2)

Several approaches have been made to the problem of gap capacitance calculation and the results have differed significantly. In a critical paper by Costamagna[10] these various approaches were compared and recommendations were made. Costamagna argued that the results due to Benedek and Silvester[11] should be used—but should be reduced by a certain amount in the case of C_2 (Costamagna's C_b, shown in his Figure 2). In critical design situations great care should therefore be exercised over the determination of C_1 and C_2.

5.4 MICROSTRIP SHORT CIRCUITS

At lower microwave frequencies, up to 2 or 3 GHz, a short wire bonded to the microstrip and ground plane and passed through an unmetallized hole in the substrate is known to provide a fairly good short circuit. At these frequencies discontinuity parasitics (inductance and capacitance) present negligible reactive effects. As the frequency is increased beyond 2 or 3 GHz the equivalent reactance associated with such a wire becomes increasingly significant so that the structure departs noticeably from a short circuit.

This equivalent reactance is also frequency dependent and thus different effective lengths of the microstrip line are associated with the same structure at different frequencies. This means that the location, along the line, of the effective short-circuit plane also varies significantly with frequency.

Some interesting work has been performed by Owens,[12] primarily in an effort to characterize shunt holes in microstrip lines where active devices such as transferred-electron devices may be connected. One consequence of this work was the discovery that a hole can be produced which has a practically frequency-independent reactance, so that its effective short-circuit plane remains stationary over a wide frequency range.

Extensive microstrip resonator measurements were employed and it was established that the holes may either contain a brass rod or merely be metallized around their 'cylindrical' surfaces to obtain essentially identical results.

For fairly low impedance (18 Ω) microstrip lines on alumina substrates ($\epsilon_r = 10.1$, $h = 0.635$ mm), it was found that a 1.52-mm diameter hole down through these 3 mm wide lines gave

$$\text{Reactance} = X < \pm j0.5 \quad \Omega$$

over the frequency range:

$$4 \leq f \leq 18 \text{ GHz}$$

This represents a good *broad-band* short circuit. Holes yielding this performance had microstrip continuing on both sides of them.

Where such a hole appears at the otherwise foreshortened end of the strip Owens gives the following condition for a broad-band short-circuiting hole:

$$\ln\left(\frac{w_{\text{eff}}}{\pi d_e}\right) \simeq \left(\frac{\pi d_e}{w_{\text{eff}}}\right)^2$$

where d_e = 0.03 + 0.44d
 d = actual hole diameter
 w_{eff} = effective microstrip width (see Section 4.8.2)

In the case of the 3-mm wide line this leads to a hole diameter of 1.30 mm—substantially lower than the previous optimum value, when the hole was not at the end of a microstrip line. Figure 5.6 illustrates the appearance of the structure.

A 50 Ω line on this substrate material would have a width of approximately 0.6 mm, and many circuits, including filters, require such lines. Assuming, very roughly for present purposes, a pro rata reduction in the hole diameter

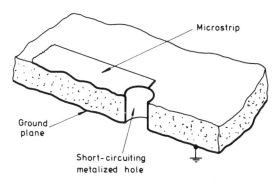

Fig. 5.6. A shunt metallized hole which may form a short circuit in a microstrip line

we find that 0.26 mm is required for a good broad-band short circuit. This would be extremely difficult, if not impossible, to drill (except, may be, using a laser).

It is quite difficult to accurately and repeatably locate these holes, or 'shunt parts', relative to the microstrip circuit. Microstrip-like structures can be precisely defined and located photolithographically, but this is considerably more difficult with shunt structures such as these short circuits.

However, with techniques such as computer-controlled laser drilling (or the use of vias in MMICs) these short circuits may become more important and some recent work is briefly described in Section 5.13.5.

5.5 FURTHER DISCONTINUITIES

The remaining sections deal with bends, steps, slits, T-junctions, and crossover junctions. In these cases fringing field effects, leading to equivalent capacitive elements, occur as we have already seen for other discontinuities. However, current flow disturbances are now produced and these give rise to the concept of equivalent circuit inductances.

For bends and T-junctions arrangements have been developed which, at least partially, compensate for the discontinuity effects. These are described here because they are significant to the MIC circuit designer, certainly up to around 18 GHz. Sometimes it may not be practical to apply the compensation and for this reason, for example, the uncompensated bend equivalent circuit parameters are studied.

5.6 THE RIGHT-ANGLED BEND OR 'CORNER'

In many cases where quite complex circuits are required on a single substrate it is necessary to feed, between circuits, using lengths of microstrip that include bends. These usually pass through an angle of 90°. Most often the line does not change width; the bend, together with its equivalent circuit, is shown in the diagrams of Figure 5.7.

The capacitance arises through additional charge accumulation at the corners—particularly around the outer point of the bend where electric fields concentrate. The inductances arise because of current flow interruption. This is considerable, especially bearing in mind that most of the current flows in the outer edges of microstrip.

An alternative equivalent circuit simply treats the inductances as two lengths of equivalent, inductive, microstrip lines. Capacitance data have been determined theoretically by Silvester and Benedek[13] and inductance data by Thomson and Gopinath.[5] Fairly extensive measurements have also been conducted.

(a) Structure and nomenclature (b) Equivalent circuit

Fig. 5.7. Right-angled microstrip bend. (© 1973 IEEE. Reprinted, with permission, from Silvester and Benedek.[13])

Gupta *et al.*[3] have given closed formulae for the evaluation of bend capacitance:
For $w/h < 1$:

$$\frac{C_{BEND}}{w} = \frac{(14\epsilon_r + 12.5)w/h - (1.83\epsilon_r - 2.25)}{\sqrt{w/h}} \quad \text{pF/m} \qquad (5.17a)$$

For $w/h > 1$:

$$\frac{C_{BEND}}{w} = (9.5\epsilon_r + 1.25)\frac{w}{h} + 5.2\epsilon_r + 7.0 \quad \text{pF/m} \qquad (5.17b)$$

and inductance:

$$\frac{L}{h} = 100\left\{4\sqrt{\left(\frac{w}{h}\right)} - 4.21\right\} \quad \text{nH/m} \qquad (5.18)$$

The accuracy of eqns (5.17) is quoted as within 5 per cent over the ranges: $2.5 \le \epsilon_r \le 15$ and $0.1 \le w/h \le 5$. The accuracy of eqn (5.18) is quoted as about 3 per cent for the range: $0.5 \le w/h \le 2.0$, compared with the results due to Thomson and Gopinath.

As an example, let us consider a microstrip line of 0.75 mm width and fabricated on a 0.5-mm thick substrate whose permittivity is 9.9. The bend equivalent circuit capacitance and inductance are 0.135 pF and 0.031 nH respectively, shown on the equivalent circuit of Figure 5.8. Although these components may have 'small' values, their reactances, even at 10 GHz, have a significant effect on the electrical properties of the complete network. For the inductance, at 10 GHz:

$$\omega L = 2\pi \times 10^{10} \times 0.031 \times 10^{-9}$$

$$= 2\,\Omega$$

and for the capacitance:

Fig. 5.8. Lumped equivalent circuit for the microstrip bend calculated in the text. (© 1973 IEEE. Reprinted, with permission, from Silvester and Benedek.[13])

$$\frac{1}{\omega C} = \frac{1}{2\pi \times 10^{10} \times 0.135 \times 10^{-12}}$$

$$= 120 \ \Omega$$

It should be observed that both values are significant compared with the microstrip line characteristic impedance—having a value of some tens of ohms.

5.7 'MATCHED' MICROSTRIP BENDS: COMPENSATION TECHNIQUES

Several techniques have been investigated for the compensation of microstrip bends, greatly reducing the effect of the capacitance and hence improving the VSWR. In particular, Anders and Arndt[14] have used a moment method to calculate the appropriate capacitances and inductances for both curved and mitred bends. Their results indicate that, at least up to a frequency of about 10 GHz, a mitred bend produces as good as, or better, performance than curved bends. This applies to a wide range of bend angles, from 30° up to 120°. At least 70 per cent mitring is recommended for an acute-angled bend of 120°, i.e. one which bends back upon itself. Guidance for the design of such mitred bends, but for an angle of 90°, is also available for the structure shown in Figure 5.9.

The equivalent circuit shown in Figure 5.9(b) is for the region between planes P and P′ and the curves of Figure 5.9(c) relate to *measured* results.[15] Although these curves apply to the specific instance defined, the considerable reduction of susceptance B (and therefore also capacitance C) would be expected in widely different substrates and structures. As can be seen, the equivalent line-length parameter l_c *increases* with enhanced chamfering. Because of this and extreme line narrowing in the centre of the bend, the degree of chamfer is generally restricted to around

$$1 - \frac{b}{\sqrt{2w}} \simeq 0.6$$

(b) Equivalent circuit

a) Structure and nomenclature

(c) Parameter trends

Fig. 5.9. Chamfered, right-angled bend together with its equivalent circuit and parameter variations (as a function of amount of chamfer). (Reproduced by permission of the Institution of Electronic and Radio Engineers from Easter *et al.*[15])

so that

$$b \simeq 0.57w$$

It is tentatively recommended that approximately this amount of chamfer is applied in the majority of practical cases where the substrate permittivity approximates that of alumina. This feature is re-addressed in Section 5.13.3.

5.8 STEP CHANGES IN WIDTH (IMPEDANCE STEPS)

It is necessary for microstrip lines to experience a width change in many circuits. This is clearly so even for the simple case of a quarter-wave transformer, and several such changes are demanded in transistor amplifier circuits such as that shown in Figure 5.1. Couplers and filters provide further examples.

These forms of discontinuities have been fairly widely investigated and characterized.[11] Like the microstrip bend, the shunt capacitance due to corners is the dominant equivalent circuit parameter and edge-current disturbances occur again which give rise to equivalent inductances.

5.8.1 The symmetrical microstrip step

This is illustrated in Figure 5.10(a), with an equivalent circuit in Figure 5.10(b) and parameter variations in Figure 5.11.

We previously interpreted a discontinuity capacitance in terms of an equivalent extra length of the microstrip (for the foreshortened open circuit see Section 5.2, Figure 5.3). This approach can be used again for the step,

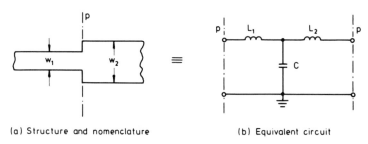

(a) Structure and nomenclature (b) Equivalent circuit

Fig. 5.10. Structure and equivalent circuit of the symmetrical microstrip step (change in width). (© 1972 IEEE. Reprinted, with permission, from Benedek and Silvester.[11])

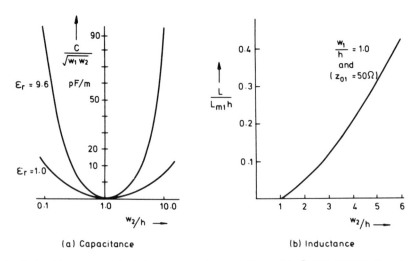

(a) Capacitance (b) Inductance

Fig. 5.11. The symmetrical step: parameter variations. (a) © 1972 IEEE. Reprinted, with permission, from Benedek and Silvester.[11] (b) Thomson and Gopinath.[5]

where the principal influence of the capacitance is to increase the effective length of the wider line (width w_2). Since the changes and values are all quite small, the fully foreshortened open circuit result can be used in a modified form as given by the following expression:

$$l_{es} \simeq l_{eo} \left(1 - \frac{w_1}{w_2}\right) \tag{5.19}$$

where l_{eo} is obtained from eqn (5.5) or eqn (5.6).

Gupta et al.[3] have shown how to account approximately for the inductances. Where L_{m1} and L_{m2} are the inductances per unit length of the appropriate microstrips, having widths w_1 and w_2, the discontinuity inductances are given respectively by the following relations:

$$L_1 = \frac{L_{m1}}{L_{m1} + L_{m2}} L \tag{5.20a}$$

and

$$L_2 = \frac{L_{m2}}{L_{m1} + L_{m2}} L \tag{5.20b}$$

The equivalent extra lengths of microstrip are then given by

$$\frac{l_{es1}}{h} = \frac{l_{es2}}{h} = \frac{L}{(L_{m1} + L_{m2})h} \tag{5.21}$$

Garg and Bahl[9] obtained design formulae for both C and L. The following expressions for C were derived by curve fitting earlier theoretical results:
For $\epsilon_r \leq 10$; $1.5 \leq w_2/w_1 \leq 3.5$:

$$\frac{C}{\sqrt{w_1 w_2}} = (10.1\log \epsilon_r + 2.33) \frac{w_2}{w_1} - 12.6\log \epsilon_r - 3.17 \quad \text{pF/m} \tag{5.22}$$

For $\epsilon_r = 9.6$; $3.5 \leq w_2/w_1 \leq 10$:

$$\frac{C}{\sqrt{w_1 w_2}} = 130\log(w_2/w_1) - 44 \quad \text{pF/m} \tag{5.23}$$

The error involved in applying eqn (5.22), which applies to relatively slight steps (w_2/w_1 maximum is 3.5), can be as high as 10 per cent. The error in eqn (5.23) is claimed to be less than 0.5 per cent.

Garg and Bahl's expression for the inductance is

$$\frac{L}{h} = 40.5\left(\frac{w_1}{w_2} - 1.0\right) - 75\frac{w_1}{w_2} + 0.2\left(\frac{w_1}{w_2} - 1.0\right)^2 \quad \text{nH/m} \tag{5.24}$$

This formula is quoted as having an accuracy of better than 5 per cent provided that $w_1/w_2 \leq 5.0$ and $w_2/h = 1.0$.

5.8.2 The asymmetrical step in microstrip

The geometry of this is shown in Figure 5.12, together with the appropriate equivalent circuit. As a first approximation the equivalent circuit parameter variations may be taken as the same as those for the symmetrical step. The actual parameter values will be about half of the values obtained for the symmetrical step.

5.9 NARROW TRANSVERSE SLIT

A narrow slit, cut transversely into the microstrip width, yields a predominantly series inductance effect. This structure has been investigated by Hoefer[16] and it may be used to compensate for excess capacitance at discontinuities or to fine-tune lengths of microstrip such as stubs. The appearance of the slit and its equivalent circuit are shown in Figure 5.13.

In Hoefer's analysis, dispersion and capacitive effects are neglected and a non-magnetic substrate is taken. A dipole field perturbation method, originally developed by Wheeler,[17] is used in the analysis. A parallel-plate waveguide model with magnetic side walls is also employed. The reader is recommended to the paper for further details since only the resulting expressions and their use will be described here. The final formula for the inductance is

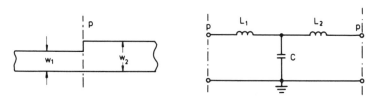

Fig. 5.12. Structure and equivalent circuit of the asymmetrical microstrip step. (© 1972 IEEE. Reprinted, with permission, from Benedek and Silvester.[11])

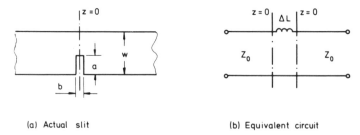

(a) Actual slit (b) Equivalent circuit

Fig. 5.13. The narrow transverse slit. (© 1977 IEEE. Reprinted, with permission, from Hoefer.[16])

$$\frac{\Delta L}{h} = \frac{\mu_0 \pi}{2} \left(\frac{a'}{A}\right)^2 \tag{5.25}$$

where

$$\frac{a'}{A} = 1 - \frac{Z_{0(air)}}{Z'_{0(air)}} \tag{5.26}$$

For design, the procedure is then as follows:

(a) Calculate the characteristic impedance of the air-filled microstrip line, width w ($Z_{0(air)}$).
(b) Calculate the characteristic impedance of an air-filled microstrip line of width $(w - a)$ ($Z'_{0(air)}$)
(c) Evaluate a'/A with these results, (a) and (b), using eqn (5.26) and hence calculate ΔL using eqn (5.25).

As an example a slit of relative penetration $(a/w) = 0.5$ can be calculated. Considering such a slit in a 50 Ω microstrip line on 0.635 mm thick alumina ($w/h = 1$ when $\epsilon_r = 9.6$), the inductive reactance of this slit is j3.7 Ω at 10 GHz.

There are two important limitations to this type of calculation. In the case of a relatively small slit width b the capacitance may be significant. Using the calculation techniques for *gaps* in microstrip, as described in Section 5.3, the capacitance of the slit may be approximately determined as $(a/w)C_2$. For the example calculated above the capacitive reactance is approximately $-$j320 Ω, which would represent a negligible shunt reactance effect in many applications.

Another limitation occurs as the slit is progressively widened. Eventually the structure looks like a short microstrip line of width $(w - a)$ asymmetrically stepped to the main microstrips of widths w. Hoefer suggests that a lumped inductance, independent of b, is approximately valid while $b < h$.

Experimental results for an RT-duroid 5880 substrate, with $w/h = 2.0$ and a range of a/w values, show good agreement with the theory.

5.10 THE MICROSTRIP T-JUNCTION

This junction necessarily occurs in a wide variety of microstrip circuits. Examples range through stub-matching elements, stub filters, branch-line couplers, and microstrip antenna element feeds, together with several other conceivable types of circuits. The appearance and a fairly elementary equivalent circuit are shown in Figure 5.14. The inductances and capacitance are included for identical reasons to those described previously in respect of other discontinuities. However, one new element has been introduced here: the transformer.

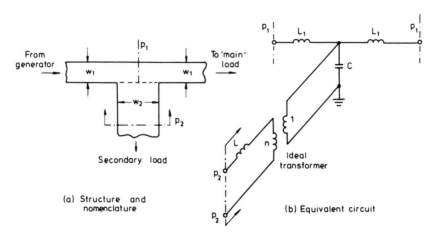

Fig. 5.14. The T-junction in microstrip. (© 1973 IEEE. Reprinted, with permission, from Silvester and Benedek.[13])

It is necessary to determine the effect, on the main circuit through from generator to main load, of a secondary load connected to the shunt branch arm (width w_2). At very low frequencies any such load will, of course, appear directly connected across the through circuit, i.e.

$$\text{when } f \to 0, n \to 1$$

At microwave frequencies a load connected at some point along the shunt branch will appear both transformed by the ratio n^2 and located at some specified electrical length from the physical junction. The inductances, the capacitance, and the transformation ratio have all been deduced experimentally and, in a number of cases, theoretically.[13,15] It is not intended that all the information be provided here and only the capacitance and transformation ratio results are given (Figure 5.15). Note that, for moderate-to-high impedance branch lines, the junction capacitance becomes reduced below that of either microstrip line capacitance per unit length. Thus, capacitance is 'taken out' of the junction and the equivalent circuit capacitance is negative.

The small difference of 0.1 in the permittivities should not matter much in practice if the two sets of results were both required when carrying out a design on an appropriate alumina substrate. It is observed that a substantial discrepancy exists between the measured and theoretical results for n^2, and it is not clear whether this is due mainly to experimental or to theoretical error. Judging by the given error bounds (only three points) it may be that Vogel's theory is not fully adequate here. The fact that the discrepancy grows with increasing frequency is also unfortunate, since design around 10 GHz and above is generally important.

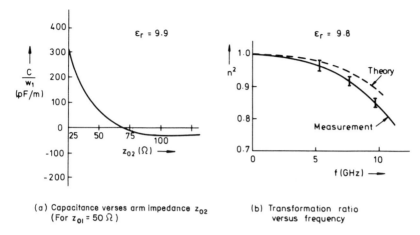

(a) Capacitance verses arm impedance z_{02}
(For $z_{01} = 50 \, \Omega$)

(b) Transformation ratio
versus frequency

Fig. 5.15. Parameter trends for the T-junction. (a) © 1973 IEEE. Reprinted, with permission, from Silvester and Benedek.[13] (b) Reproduced by permission of the Institution of Electronic and Radio Engineers from Easter *et al.*[15]

Formulae for the capacitance and for the inductances are given by Garg and Bahl,[9] but not for the transformation ratio. Semi-empirical expressions have also been given by Hammerstad and Bekkadal[7] for shifts in the locations of the electrical reference planes, as well as *n*, and it is these results that are more useful for circuit design than the equivalent lumped parameters.

The reference planes and hypothetical widths associated with Hammerstad and Bakkadal's method are shown in Figure 5.16, where the effective width (see eqn 4.38, Section 4.8.2) of either line is

$$w_{\mathrm{eff}1.2} = \frac{h\eta}{Z_{0(1.2)} \sqrt{\epsilon_{\mathrm{eff}}}} \tag{5.27}$$

The transformation ratio *n* is given by

Fig. 5.16. T-junction: reference planes and hypothetical widths. (Reproduced by permission of *MicroWaves* from Dydyk.[20])

$$n^2 = \left[\frac{\sin\{\pi(w_{\text{eff1}}/\lambda_{g1})\,(Z_{0(1)}/Z_{0(2)})\}}{\pi(w_{\text{eff1}}/\lambda_{g1})\,(Z_{0(1)}/Z_{0(2)})}\right]^2 \left[1 - \{\pi(w_{\text{eff1}}/\lambda_{g1})\,(d_2/w_{\text{eff1}})\}^2\right]$$

(5.28)

The displacement of the reference plane for the primary line (1) is

$$\frac{d_1}{w_{\text{eff2}}} = 0.05 \frac{Z_{0(1)}}{Z_{0(2)}} n^2$$

(5.29)

The displacement of the reference plane for the secondary arm (2) is

$$\frac{d_2}{w_{\text{eff1}}} = 0.5 - \left\{0.076 + 0.2\left(\frac{2w_{\text{eff1}}}{\lambda_{g1}}\right)^2 + 0.663\exp\left(-1.71\frac{Z_{0(1)}}{Z_{0(2)}}\right)\right.$$

$$\left. - 0.172\ln\left(\frac{Z_{0(1)}}{Z_{0(2)}}\right)\right\} \frac{Z_{0(1)}}{Z_{0(2)}}$$

(5.30)

Also, the shunt capacitance is determined by the following expressions:
For the condition $Z_{0(1)}/Z_{0(2)} \leq 0.5$:

$$\frac{\omega C\lambda_{g1}}{Y_{0(1)}w_{\text{eff1}}} = \left(\frac{2w_{\text{eff1}}}{\lambda_{g1}} - 1\right)\frac{Z_{0(1)}}{Z_{0(2)}}$$

(5.31a)

For the condition $Z_{0(1)}/Z_{0(2)} \geq 0.5$:

$$\frac{\omega C\lambda_{g1}}{Y_{0(1)}w_{\text{eff1}}} = \left(\frac{2w_{\text{eff1}}}{\lambda_{g1}} - 1\right)\left(2 - 3\frac{Z_{0(1)}}{Z_{0(2)}}\right)$$

(5.31b)

Actual accuracies for these expressions are not quoted, but their discrepancies rise when

$$\frac{2w_{\text{eff1}}}{\lambda_{g1}} > 0.3$$

i.e. when the wavelength becomes small enough (frequency large enough) for this condition to occur. Hall and James[18] have reported measurements at 17 GHz on T-junctions and they conclude that Hammerstad's numerical results, as he suggests, are too large at such high frequencies. In particular, a discrepancy of a factor of two or more was found for d_2, where the measurements always showed smaller results.

It would be useful to attempt to modify Hammerstad's expressions so that a better fit is obtained at higher frequencies. One might, for example, note that if w_{eff1} in eqn (5.30) is replaced by the frequency-dependent value given by eqn (4.36) then the magnitude of the term $(w_{\text{eff1}}/\lambda_{g1})$ rises more strongly with frequency, hence reducing d_2 as desired. Some further investigations are required in this area.

Another limitation to the accuracy of the expressions is imposed by the impedance ratio $Z_{0(1)}/Z_{0(2)}$. When this exceeds about 2.0 the calculated

values of d_2 become too large and an improved accuracy is obtained when the ratio $Z_{0(1)}/Z_{0(2)}$ is replaced by its inverse, that is $Z_{0(2)}/Z_{0(1)}$.

Some useful work has been carried out to compensate T-junctions and thereby improve the performance of circuits using them. This is discussed in the next section.

5.11 COMPENSATED T-JUNCTIONS

One simple way to at least compensate the capacitance of a T-junction is to introduce a slit across the width of the main-through microstrip line, opposite the branch arm. The characteristics of such a slit were discussed in Section 5.9. Kompa[19] has shown further that a comparatively wide 'slot' ($b \simeq w_2$) in this position, i.e. a rather wider slit, strongly affects the transmission of higher-order modes in the main line and leads to enhanced skirt sensitivity of filters composed of stubs with such modified T-junctions. To the best of the present writer's knowledge this kind of approach has not yet been developed further and more investigation could prove very fruitful.

Dydyk[20] has described a T-junction compensation technique which appears to work well in the case of a branch-line microstrip coupler circuit. Dydyk's aim was to modify the microstrip lines in the vicinity of the junction in order to compensate for reference plane shifts, at least over a specified range of frequencies. The treatment of the junction can exclude any discussion of radiation loss with little error in circuit performance results, at least up to a frequency of 17 GHz.

The compensated T-junction takes on the form shown in Figure 5.17. A microstrip branch-line coupler designed using these principles yielded the following performance over the frequency band 16–18 GHz:

Insertion loss	3.5–4.0 dB over entire band
Isolation	> 20 dB over entire band
Return loss	10 (17.5 GHz)–25 dB

The very 'flat', almost frequency-independent, insertion loss and isolation are attributed to the compensated junction design. Dydyk also describes an SPDT switch designed according to these principles.

5.12 CROSS-JUNCTIONS

There are occasions where microstrip lines are required to separate in the shape of a cross, forming a four-port circuit. Examples can be seen in the arrangement indicated in Figure 5.1. In general a cross-junction may be symmetrical or asymmetrical—where the lines forming the cross do not all have the same widths.

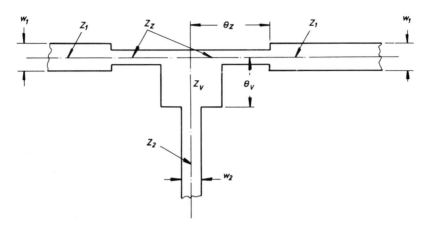

Fig. 5.17. A compensated form of T-junction. (Reproduced by permission of *MicroWaves* from Dydyk.[20])

In attempting to design a single stub to be fabricated in microstrip it can be found that the width of the stub line is wide enough to sustain transverse resonance modes at the higher frequencies of operation (see Section 4.9.2 and eqn 4.50). One way to overcome this problem is to use two stubs placed on each side of the microstrip instead of the original one. The impedance of each new stub will be twice that of the originally intended single stub, and thus the widths will be greatly reduced. A cross-junction is therefore formed.

Theoretical and experimental agreement is not good, especially for some inductance parameters associated with the equivalent circuits of cross-junctions. Because of this it does not appear sensible to give design formulae which merely fit theoretical curves. Instead, some useful engineering design information is given for the asymmetric cross.

The asymmetric cross is the most important form for practical microstrip circuits, and is shown in Figure 5.18 together with its equivalent circuit. Microstrip resonator measurement techniques, described in Chapter 7, have proved to be very powerful in the characterization of certain specific cross-junctions. Akello[21] has used such techniques to show that appropriate T-junction inductance parameters can be judiciously applied to the cross. The capacitance C_T presents prediction difficulties (theoretical treatment is available for symmetrical crosses *only*). Akello indicates that C_T can be approximated to within 8 per cent by using the simple relationship

$$C_T \simeq \tfrac{3}{4} C_m$$

where C_m is the capacitance per unit length of the (uniform) microstrip line on port 4 of the cross-junction. This appears to work quite well for the range:

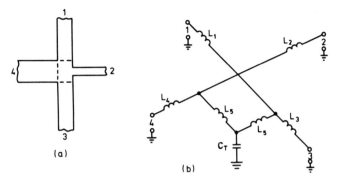

Fig. 5.18. The asymmetric crossover junction: (a) structure and (b) equivalent circuit. (Reproduced by permission of the Institution of Electrical Engineers from Akello *et al.*[21])

$$1.32 \leq \frac{w_4}{h} \leq 3.0$$

The capacitance C_m may be calculated from expressions given early in Chapter 3. Specifically,

$$Z_0 = \frac{1}{v_p C_m}$$

whence

$$C_m = \frac{c}{Z_0 \sqrt{\epsilon_{eff}}}$$

(Z_0 and ϵ_{eff} are well known at an early design stage).

Akello[21] has tabulated values of all the various equivalent circuit elements and has shown how to apply them in practical situations. In particular, computer simulation was used to indicate that discontinuity reactances attributable to steps and crossovers are significant and can be incorporated into response prediction programs. X-band (8–12 GHz), narrow-band, and wide-band single-stage and double-stage transistor amplifiers are considered. The details and results may be seen by consulting the cited reference.

It would be useful to see whether some compensation technique, perhaps of the type described in Section 5.11 (Dydyk), could be developed for cross-junctions. Further information on cross-junctions is given in Section 5.13.3.

5.13 FREQUENCY DEPENDENCE OF DISCONTINUITY EFFECTS

At frequencies much above X-band (10 or 12 GHz) it appears that the essentially quasi-static techniques described in this chapter begin to break

down significantly in a number of cases. Eventually, for millimetre and submillimetre design, it is unlikely that these techniques would provide *any* useful design data, not at least directly.

5.13.1 Open circuits and series gaps

We have already seen here that the quasi-static results provide quite accurate design information, particularly for the capacitive effects, at frequencies approaching 18 GHz. For example, the commonly encountered open circuit (see Section 5.2) has been fairly extensively investigated both theoretically and experimentally. More effort has been expended in this regard than for any other discontinuity—for the good reason that the open-circuit termination has many applications in both circuits and microstrip antennas. With an emphasis on the latter James and Henderson[22] have published a paper describing detailed research results. The values and trends shown in Figure 10 of their paper are of particular interest. There is a consistent trend for alumina substrates amounting to a peaked behaviour with frequency; two of the curves reported by James and Henderson are reproduced here as Figure 5.19, with the horizontal axis denormalized for the case of a 0.65 mm thick alumina substrate. Some experimental data are also shown which originate from unpublished results obtained by the present author.

The measured results were obtained by a resonator method which will be described in Chapter 7. They are very tedious to perform and require scrupulous attention to repeatability and high accuracy of manufacture. The 'scatter' in the experimental results is felt to be far from satisfactory, especially where the data are intended for use in critical circuits (e.g. filters). There is about 19 per cent scatter in the results obtained on sapphire substrates and about 15 per cent on alumina. Even so it is clear that the measured results differ quite markedly from the frequency-dependent theoretical predictions. Results from quasi-static calculations are shown as horizontal chain lines in Figure 5.19. These fit the measurements fairly well up to 18 GHz although the calculated results are slightly lower in both cases.

It would be interesting to investigate experimentally the behaviour in the 20–40 GHz range for microstrip open circuits on alumina. Resonator methods, to be described in Chapter 7, might be attempted, but these pose considerable difficulties at such high frequencies because of the presence of higher-order modes. It may even be possible for transverse-resonant modes to become excited. A better approach might be to model the structure for measurement at much lower frequencies.

Other workers, notably Itoh,[23] have theoretically deduced a non-linear l_{eo}/h variation with frequency. Itoh's results also show up-turns in l_{eo}/h at

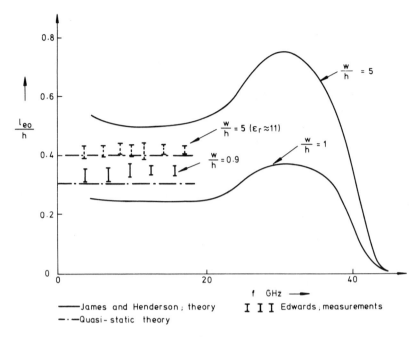

Fig. 5.19. Open-circuit normalized equivalent end-effect length (l_{eo}/h) as a function of frequency. The data refer to a 0.65 mm thick alumina substrate with a relative permittivity of 10, except for one case where the values were 0.5 mm and 11, as indicated. (Reproduced by permission of the Institution of Electrical Engineers from James and Henderson.[22])

both lower and higher frequencies, but the absolute values of his end-effect lengths were substantially lower than any other reported data. This is at least partially due to the shielded nature of the microstrip which was investigated, and the present writer has performed a number of computations using Itoh's technique with the top shielding cover steadily raised. The results show that the end-effect length then increases. However, a further source of possible error in this technique is probably the necessity for subtracting two large numbers (resonant frequencies) of similar magnitudes.

Katehi and Alexopoulos[24] have described the frequency-dependent characteristics of microstrip discontinuities on a theoretical basis—concentrating largely on open-circuited lines and also the series gap. These workers evaluate the current distributions in the vicinity of the discontinuities by means of Pocklington's integral equation, the dyadic Green's function and impedance matrices. They extend their calculations to 50 GHz and derive

results for radiation conductances, equivalent length extensions and gap capacitances.

Figures 5.20, 5.21, 5.22 and 5.23 show the results obtained by Katehi and Alexopoulos. The radiation conductances both rise with frequency, as should be anticipated from physical considerations. It is interesting that the equivalent length extension ('excess length') rises slightly, to peak at about 40 GHz, then falling with frequency as predicted elsewhere independently.

A frequency-dependent equivalent circuit for gap discontinuities has been reported by Ozmehmet.[25] This includes the radiation resistance and separates gap capacitances (series) into fringing field and 'plate' components. Therefore the series structure of the equivalent circuit shown as Figure 5.5(b) is enhanced by these additional elements. Ozmehmet shows that this slightly more complex circuit enables wideband MICs to be designed at frequencies up to at least 15 GHz.

A complete electromagnetic modelling of microstrip open end discontinuities has been reported by McLean et al.[26] They use a spectral domain method of similar general type to that described in Chapter 4 and they include space wave and surface wave radiation by implementing the exact spectral domain Green's function.

In this work longitudinal and transverse current components are included so that strips ranging from electrically narrow to electrically wide can be analysed. Results are shown for both magnitude and phase of the reflection

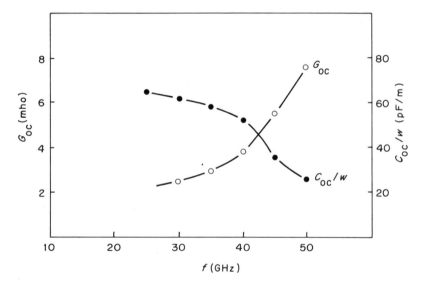

Fig. 5.20. Radiation conductance G_{oc} and normalized capacitance C_{oc}/w of an open-circuited microstrip line as functions of frequency ($\epsilon_r = 9.6$, $w/h = 1$, $h = 0.6$ mm). (© 1985 IEEE. Reprinted, with permission, from Alexopoulos.[24])

Fig. 5.21. Gap-discontinuity radiation conductances G_g, G_p, and normalized capacitances C_g/w, C_p/w as functions of frequency ($\epsilon_r = 9.6$, $w/h = 1$, $h = 0.6$ mm, $s/h = 0.3762$). (© 1985 IEEE. Reprinted, with permission, from Alexopoulos.[24])

Fig. 5.22. Open-circuit normalized excess length and normalized capacitances as functions of frequency ($\epsilon_r = 9.6$, $w/h = 1$, $h = 0.6$ mm). (© 1985 IEEE. Reprinted, with permission, from Alexopoulos.[24])

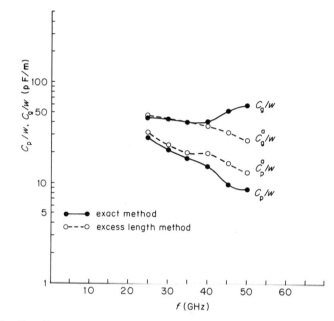

Fig. 5.23. Gap-discontinuity normalized capacitances C_p/w, C_g/w as functions of frequency ($\epsilon_r = 9.6$, $w/h = 1$, $h = 0.6$ mm, $s/h = 0.3762$). (© 1985 IEEE. Reprinted, with permission, from Alexopoulos.[24])

coefficient as functions of frequency. In the case of an electrically narrow microstrip line on an alumina substrate both the magnitude and phase of the reflection coefficient decrease with frequency—the magnitude slowly at first, then rapidly for frequencies approaching 30 GHz (about 0.75 at 31 GHz and falling). The phase is approximately 0.05 radians at 5 GHz, 0.25 radians at 25 GHz and then decreasing faster to about 0.5 radians at 31 GHz. These phase calculations are compared with results measured elsewhere and agreement is very good.

Results are also obtained, in the form of magnitude only as a function of frequency, for a range of line widths applicable to microstrip lines on a substrate having a permittivity of 12.8. Pronounced singularities are now evident for all lines at high frequencies. At first the reflection coefficient continues to decrease rapidly but this is followed by a further increase, a peak, and finally a resumption of the general fall in value. This behaviour occurs in the vicinity of cut-off for the first TE surface wave, due to the tendency of the transverse current to store as opposed to radiate energy and hence to increase the frequency sensitivity of the reflection coefficient. It is demonstrated that this effect is strengthened as the line width increases, which is consistent with the above argument.

With their significance in, for example, filter realisation some work has been reported by Achkar *et al.* on open end discontinuities of suspended striplines.[27] The analysis uses eigenmodes to determine the dispersion characteristics of suspended substrate striplines and a modal analysis to find the open end discontinuity characteristics. For this type of structure, fabricated on a substrate of permittivity 2.2, it is shown that the amplitude of the reflection coefficient is essentially zero for all frequencies over the 26–40 GHz range. Its phase, however, increases from about −0.028 to −0.015 radians over this frequency range. Hence also the equivalent end-effect length increases slowly with frequency—although it is still only around one micron at 40 GHz for this structure. The behaviour of the transmission coefficient (S_{12}) was also quantified, but this parameter remains less than 130 dB down on the incident signal power over the entire band.

5.13.2 Other discontinuities

Easter[15] has performed measurements which indicate that the values of the inductive series equivalent elements for discontinuities can be expected to increase considerably with frequency. Over the frequency range 5–11 GHz Easter's measurements reveal a 40 per cent increase in the inductance values for lines having $w/h > 1.0$. For lines with smaller width-to-height ratios the increase is greatly reduced.

Alternative approaches to the problem of frequency dependence of microstrip discontinuities have been presented by Menzel and Wolff,[28] Kompa,[29] and Mehran.[30] The complete validity of their magnetic-wall method of modelling microstrip has, however, been questioned by Easter. These workers present their data in scattering (S) parameter form, which is often difficult to reinterpret as an equivalent circuit.

The work reported by Kompa[29] is, however, particularly interesting, even though most of his computations relate to plastic substrates ($\epsilon_r = 2.32$). He reports S parameters for steps in microstrip, using the planar waveguide model to analyse the structure. It is clear that the transmission parameter $|S_{12}|$ is substantially constant up to at least 12 GHz for steps produced on an alumina substrate. A dramatically different behaviour is noted for the plastic substrate, where $|S_{12}|$ decreases very rapidly for moderate-to-high width ratios at high frequencies. For example, the transmission drops to zero at just over 9 GHz for a width ratio of approximately 5.0, whereas when the ratio is only 3.0 there is little change in transmission up to 12 GHz. Kompa describes measurements, a one-section quarter-wavelength transformer (including a field-matching solution to this), and the analysis of a straight-tapered, six-stepped microstrip line.

We now consider analysis and modelling results, in terms of scattering parameters, for a selection of discontinuities.

The use of microstrip circuits fabricated on multilayer substrates is important in many MMIC designs and antenna array feed networks. To meet the circuit design needs discontinuities have been characterized for these types of structures.

In particular, Harokopus and Katehi[31] report results for unchamfered 90° bends. They employ a space domain method of moments solution of Pocklington's integral equation which is suitable for application to most discontinuities. These workers present results for various combinations of substrates, generally incorporating high-alumina/low-permittivity Duroid (i.e. permittivities of 10.2 and 2.2).

Over the frequency range 10–20 GHz, for a 90° bend fabricated on a single layer substrate, it is shown that the magnitude of the reflection coefficient rises slowly from approximately 0.08 to 0.15 (roughly following the prediction of Eesof's Touchstone™ software). In contrast, the magnitude of the transfer parameter (S_{12}) falls slowly with frequency—from about 0.98 at 10 GHz to approximately 0.88 at 20 GHz. This differs greatly from the predictions of Eesof's Touchstone™ which indicates a value very close to 1.0 over the entire frequency range. For multilayer bends the general trends of the parameters are the same but the actual values differ considerably. For example, in the case of a low-permittivity Duroid substrate topped by an alumina substrate (equal thicknesses) the reflection coefficient increases from about 0.2 to 0.4 whilst the transfer parameter decreases from 0.97 to 0.74. This is due to the enhancement of radiation fields where the microstrip itself is held above a ground plane separated by a relatively low-permittivity substrate (Duroid).

The important case of 90° bends, direct, chamfered and double-chamfered (see Figure 5.24), has also been investigated by Zheng and Chang.[32] These workers employed an integral equation approach (with the Galerkin method) to evaluate the performance of these bends.

They found similar trends and values for the coefficients to those reported by Harokopus and Katehi[31] for the unchamfered bend. In common with the well known results they also noted that both S_{11} and S_{12} were considerably lower for the optimally chamfered bend than for the unchamfered one, over the frequency range 8–40 GHz at least (see also Section 5.7). The double-chamfered case was, however, particularly interesting since it provided a substantially lower reflection coefficient over the frequency range 22–38 GHz and a lower value of S_{21} over the entire frequency range (much lower above about 18 GHz). This response is due to interference effects operating between the chamfer ends—i.e. a distinctly frequency-dependent phenomenon.

5.13.3 Cross and tee junctions

The frequency-dependent behaviour of cross and tee junctions has been investigated by Wu *et al.*[33] They use an electromagnetic spectral-domain

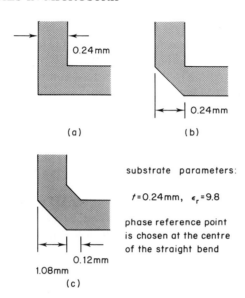

substrate parameters:

t = 0.24 mm, ϵ_r = 9.8

phase reference point
is chosen at the centre
of the straight bend

Fig. 5.24. Right-angled single bend: unchamfered and chamfered. (© 1990 IEEE. Reprinted, with permission, from Zheng and Chang.[32])

analysis to develop the scattering parameters of these types of junctions. An N-port junction approach is used together with a concept whereby a 'module' around the junction region contains that section where higher order modes are generated as a consequence of the discontinuity. The current distributions inside the module are expanded and a 'nearly Galerkin' method is applied to transform the integral equation into a matrix equation. The scattering parameters are generated from submatrices within the major current, voltage and reflection coefficient interrelating matrices.

Wu *et al.* compare their results, to which experimental data generally fit very well, with quasi-static predictions generated using Eesof's Touchstone™ software over the frequency range 4–24 GHz. For a basic T-junction the magnitude of the transfer parameters S_{21} and S_{31} indicates a greater inequality in the division of power between arms than the Touchstone™ software predicts. In general the disparities amount to some 10 to 15 per cent—being most marked as frequency increases. Phase performance is more sensitive to the accuracy of the prediction solution and, again for a basic T-junction, the phase of S_{33} is found to differ from the Touchstone™ result by up to 45° (at 24 GHz). Wu's method, supported by measurements, shows the phase of S_{33} varying only between 170° and 175°.

In the case of a right-angled cross junction the use of this approach is essential since, for example, Touchstone™ predicts simple equal power splitting over a wide frequency range. Wu's approach shows that the power

separation is distinctly unequal and indeed varies markedly with frequency. Curves applicable to a right-angled cross junction in microstrip on a high-alumina substrate are shown in Figure 5.25.

This method of analysis is also applied to a compensated T-junction—as originally suggested by Dydyk (see Section 5.11). In the example analysed by Wu the top slot is omitted and the intention is to extend the frequency range over which power is equally split between the coupled arms. For such a compensated microstrip T-junction, fabricated on a similar substrate to that described above, it is found that the power is split equally when the structure is driven at any frequency from d.c. to 16 GHz (compared with d.c. to 6 GHz for a basic T-junction).

Chang et al.[34] have reported an analytical CAD program applicable to a wide variety of discontinuities. Their method is based on the effective width principle (see Section 4.8.2) following which an impedance is derived based on the boundary integral method. The scattering matrix is obtained directly from the impedance matrix.

Chang et al. show results for right-angled bends, T-junctions and Y-junctions. In the case of the right-angled bends a family of curves is computed showing S_{11} as a function of frequency (see Figure 5.26) and these clearly indicate an optimum chamfer or 'cut' ratio of approximately 0.8 for a 0.65 mm thick plastic substrate of permittivity 2.53. It is also clear that this optimum continues for frequencies up to at least 29 GHz. Although the definition contrasts somewhat the absolute magnitude of this chamfer

Fig. 5.25. Magnitude of S-parameters of a right-angle crossing junction ($\epsilon_r = 10.2$, $h = 25$ mil, $w = 24$ mil) Note that in Touchstone™ $|S_{11}| = |S_{21}| = |S_{31}|$. (© 1990 IEEE. Reprinted, with permission, from Wu, Yang and Alexopoulos.[33])

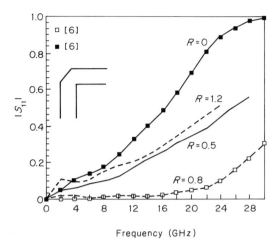

Fig. 5.26. Variations in $|S_{11}|$ of a right-angle bend for cut ratio $R = 0, 0.5, 0.8, 1.2$. $W1_{eff} = W2_{eff} = 0.1882$ cm, $d = 0.065$ cm, $\epsilon_{re} = 2.53$. (Reprinted, by permission, from Chang, Huang and Itoh.[34])

fraction differs considerably from that given in Section 5.7 here—perhaps because of the much lower permittivity in this case.

Reflection coefficient data are also shown for a T-junction with a vee-cut opposite the leg of the T. With a cut ratio definition similar to that used for the bend, it is clear that when this ratio is 0.4 an approximate optimum exists in that S_{22} is minimized over the frequency range 0 to 16 GHz. This applies to a substrate of the type applicable to the case of the bend above.

An entirely different approach from those described above for the modelling of microstrip cross junctions has been developed by Giannini et al.[35] These workers showed how a lumped model can be used to represent the characteristics of cross junctions over a broad band (2 to 12 GHz on alumina substrates). In contrast to most other approaches Giannini et al. derive sophisticated Z-parameter expressions for a set of equivalent elements. The overall configuration is shown in Figure 5.27, where the impedance elements (actually totally reactive) are related to the physical microstrip parameters, permittivity and of course frequency—as well as complicated forms of infinite series. The term $b/2$ is the half-width of the entire cross-junction. The full details will not be set down here and the interested reader is recommended to follow the original paper.

Like the approach due to Wu and co-workers, Giannini et al. also find that when compared with Eesof's Touchstone™ data obtained using their expressions provide a much better fit to experimental results. They justify this with calculations using a version of Touchstone™ modified ('enhanced') with their families of formulae.

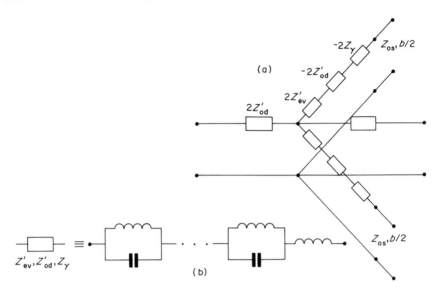

Fig. 5.27. Lumped equivalent model for the cross-junction discontinuity. (Reprinted, by permission, from Giannini *et al.*[35])

5.13.4 Radial bends

Most work concerning discontinuities has centered on abrupt changes in the width or direction of the transmission line and relatively little research appears to have been conducted on more continuously changing structures such as radial bends. Reported research by three groups in different locations (the first being headed by Weisshaar at Oregon State University) has, however, redressed the balance somewhat.[36] The nomenclature, equivalent circuit and waveguide model are indicated in Figure 5.28.

In the waveguide model the effective mean radius of curvature R_c is related to the effective microstrip width w_c as shown in Figure 5.28(c). R_c is chosen such that there is always a positive effective inner radius regardless of actual inner and outer radii.

The frequency-dependent behaviour of the radial bend is determined using a second-order perturbation analysis of the equivalent waveguide model together with a mode-matching technique which includes the higher-order modes. Scattering parameters and radiation losses are calculated for alumina (MIC) and GaAs (MMIC) versions of the structure and these are shown, for 50 Ω lines, on Figures 5.29 and 5.30.

Weisshaar *et al.* also report reflection coefficient results for radial bends in 30 Ω lines, and the scattering transfer parameter S_{21} for a T-junction with an open stub leg having curved bends.

Some workers have suggested the use of rounding and tapering techniques, as shown in Figure 5.31 to reduce the effects of discontinuities and these

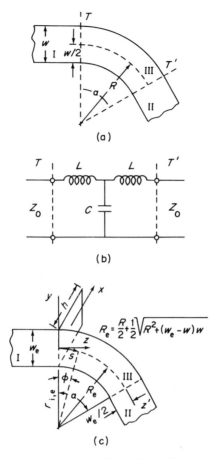

Fig. 5.28. The radial microstrip bend: (a) physical, (b) equivalent lumped model, (c) waveguide model. (© IEEE 1990. Reprinted, with permission, from Weisshaar et al.[36])

may well be effective and worth considering at least at the lower microwave frequencies. At higher frequencies, however, it would be difficult to accurately quantify the excess conductances and reactances associated with these complex shapes and it is recommended that the better understood characterization of the conventional discontinuities described in this chapter continue to be used.

5.13.5 Frequency dependence of shunt post parameters

Finch and Alexopoulos[37] have reported a theoretical study of the frequency dependence of parameters relating to shunt posts in microstrip. They use the planar waveguide model and a multiple expansion method within

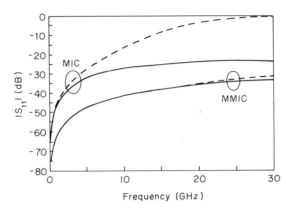

Fig. 5.29. S_{11} for $50\Omega, 90°$ bends with $R = 2w$ on MIC \rightarrow 25 mil Alumina and MMIC \rightarrow 4 mil GaAs substrate (—— waveguide model; ------ quasi-static model). (© IEEE 1990. Reprinted, with permission, from Weisshaar *et al.*[36])

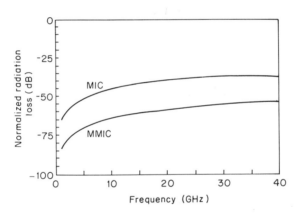

Fig. 5.30. Normalized radiation loss for 50Ω bends with $R = 2w$ on MIC \rightarrow 25 mil Alumina and MMIC \rightarrow 4 mil GaAs substrate. (© IEEE 1990. Reprinted, with permission, from Weisshaar *et al.*[36])

algorithms for the computation of Schloemilch-type series to analyse the behaviour of these structures. Their approach applies to perfectly conducting, imperfectly conducting (i.e. 'lossy') and composite shunt posts.

These workers determine scattering parameters (S_{11} and S_{21}) for this type of structure. In the case of a 1.22 mm diameter conducting post in microstrip, having a width of 3 mm, using an alumina substrate of thickness 0.635 mm and permittivity 10, they show that $|S_{11}|$ decreases with frequency. Starting with perfect short circuit behaviour at d.c., the value of $|S_{11}|$ falls to 0.92 at 12 GHz. Doubtless the method used by Finch and Alexopoulos could be extended to almost any of these types of structures.

Fig. 5.31. Rounding and tapering techniques that may be applied to reduce the discontinuity effects associated with bends, T-junctions and steps in width

5.14 RECOMMENDATIONS FOR THE CALCULATION OF DISCONTINUITIES

The following recommendations generally apply with reasonable accuracy for frequencies up to about 10–12 GHz, where the microstrip is on alumina-type substrates. As indicated, some of the formulae also apply at higher frequencies and on quite different, e.g. plastic, substrates.

5.14.1 Foreshortened open-circuits

Fundamental considerations were presented in Section 5.2.1, and eqn (5.5) gives the open-end length extension as a function of the lumped fringing capacitance C_f and the microstrip line parameters:

$$l_{eo} \simeq \frac{c Z_0 C_f}{\sqrt{\epsilon_{eff}}}$$

It is difficult to specify the accuracy of this expression, but empirical relationships can be set up for C_f, and both Z_0 and ϵ_{eff} are readily known parameters.

The alternative empirical formula is

$$l_{eo} = 0.412h \left(\frac{\epsilon_{eff} + 0.3}{\epsilon_{eff} - 0.258} \right) \left(\frac{w/h + 0.262}{w/h + 0.813} \right)$$

This expression is useful over a wide range of materials and microstrip lines, but errors can exceed 5 per cent and this can be significant—especially in filter design.

Independent theoretical calculations and measurements verify that either expression can be used satisfactorily at frequencies up to around 18 GHz for alumina-type substrates of less than 1 mm thickness.

See also Section 5.13.1 for frequency-dependent effects.

5.14.2 Series gaps

These are usually only of significance in resonator design for measurement or filter realisation purposes, where the value of g required for a given order-of-magnitude of coupling (i.e. 'loose', 'tight',. . .) can often be determined experimentally. In case such a discontinuity is required in other circuits some design equations are given in Section 5.3, but they will not be repeated here.

5.14.3 Short circuits

With the present state of technology short circuits are unusual and undesirable in microstrip for a variety of reasons, discussed in Section 5.4. However, this situation may change, and therefore the diameter of the 'optimum metallized hole' (Owens[12]) is obtainable using

$$\ln\left(\frac{w_{\mathrm{eff}}}{\pi d_{\mathrm{e}}}\right) \simeq \left(\frac{\pi d_{\mathrm{e}}}{w_{\mathrm{eff}}}\right)^2$$

where $\quad d_{\mathrm{e}} \quad = 0.03 + 0.44d$
$\qquad d \quad = $ actual hole diameter
$\qquad w_{\mathrm{eff}} \quad = $ effective microstrip width

This expression yields the diameter of such a hole situated at the end of foreshortened microstrip which provides a broad-band short circuit up to about 18 GHz.

It may be noted that the foregoing expression in w_{eff} and d_{e} has the form

$$\ln(x) \simeq x^{-2}$$

and this can be solved iteratively. In fact a total of around six or seven iterations reveal that $x = 1.532$ satisfies the equation quite accurately. The values of d_{e} and hence d are then easily found (w_{eff} can be found from eqn 4.38, Section 4.8.2).

5.14.4 Right-angled bends: mitring

Expressions were given in Section 5.6 for the equivalent circuit capacitance and inductance associated with a right-angled bend in microstrip.

The compensating techniques referred to in Section 5.7 are, however, probably of more practical significance to the circuit designer—especially the simple device known as the *mitred* or chamfered bend. It was indicated

that the approximately optimum degree, or extent, of chamfer (Figure 5.9a) should be

$$b \simeq 0.57w \qquad \text{for alumina-like substrates}$$

(but see also Section 5.13.2 and, particularly, 5.13.3).

5.14.5 Steps in width

Rather approximately, the equivalent length extension of the *wider line* (width w_2) can be determined as a function of the fully open-circuit value l_{eo}:

$$l_{es} \simeq l_{eo}\left(1 - \frac{w_1}{w_2}\right)$$

More accurate expressions can be found by incorporating the inductance and capacitance for the equivalent circuit of the bend itself; these are taken directly from the work of Garg and Bahl:[9]

$$\frac{L}{h} = 40.5\left(\frac{w_1}{w_2} - 1.0\right) - 75\frac{w_1}{w_2} + 0.2\left(\frac{w_1}{w_2} - 1.0\right)^2 \qquad \text{nH/m}$$

This formula (eqn 5.24) is quoted as having an accuracy of better than 5 per cent provided that $w_1/w_2 \le 5.0$ and $w_2/h = 1.0$.

The following expressions for C were derived by curve fitting earlier theoretical results:

For $\epsilon_r \le 10$; $1.5 \le w_2/w_1 \le 3.5$:

$$\frac{C}{\sqrt{w_1 w_2}} = (10.1 \log \epsilon_r + 2.33)\frac{w_2}{w_1} - 12.6 \log \epsilon_r - 3.17 \qquad \text{pF/m}$$

For $\epsilon_r = 9.6$; $3.5 \le w_2/w_1 \le 10$:

$$\frac{C}{\sqrt{w_1 w_2}} = 130 \log\left(\frac{w_2}{w_1}\right) - 44 \qquad \text{pF/m}$$

The error involved in applying the first equation, which applied to relatively slight steps (w_2/w_1 maximum is 3.5), can be as high as 10 per cent. The error in applying the second equation is claimed to be less than 0.5 per cent.

Appropriate extra equivalent lengths of the microstrip line can then be determined in terms of L and C.

5.14.6 Transverse slit

As outlined in Section 5.9, this yields the effect of a lumped inductance, ΔL, in the plane of the slit, given by

$$\frac{\Delta L}{h} = \frac{\mu_0 \pi}{2} \left(\frac{a'}{A}\right)^2$$

where

$$\frac{a'}{A} = 1 - \frac{Z_{0(air)}}{Z'_{0(air)}}$$

For design, the procedure is then as follows:

(a) Calculate the characteristic impedance of the air-filled microstrip line, width w ($Z_{0(air)}$).
(b) Calculate the characteristic impedance of an air-filled microstrip line of width $(w-a)$ ($Z'_{0(air)}$).
(c) Evaluate a'/A with these results, (a) and (b), using the second expression, and hence calculate ΔL using the first expression.

The accuracy of these equations is very difficult to specify, but the main application appears to be in trimming microstrip stubs, etc., for correct electrical length. That is, the slit could be deliberately introduced at a late manufacturing stage.

The expressions tend to break down as the 'slit' becomes too narrow (capacitive) or too wide. In the latter case the 'slit' tends towards a length of (distributed) microstrip line of reduced width.

5.14.7 The T-junction

This was fully described in Section 5.10 and further in Section 5.13.3.

Although it is clear that substantial further work is required, especially for frequencies much in excess of 10 GHz, the following expressions hold approximately, provided $2w_{eff1}/\lambda_{g1} > 0.3$, with

$$w_{eff1.2} = \frac{h\eta}{Z_{0(1.2)}\sqrt{\epsilon_{eff}}}$$

The transformation ratio n is given by

$$n^2 = \left[\frac{\sin\{\pi(w_{eff1}/\lambda_{g1})\,(Z_{0(1)}/Z_{0(2)})\}}{\pi(w_{eff1}/\lambda_{g1})\,(Z_{0(1)}/Z_{0(2)})}\right]^2 \left[1 - \{\pi(w_{eff1}/\lambda_{g1})\,(d_2/w_{eff1})\}^2\right]$$

The displacement of the reference plane for the primary line (1) is

$$\frac{d_1}{w_{eff2}} = 0.05 \frac{Z_{0(1)}}{Z_{0(2)}} n^2$$

The displacement of the reference plane for the secondary arm (2) is

$$\frac{d_2}{w_{eff1}} = 0.5 - \left\{ 0.076 + 0.2 \left(\frac{2w_{eff1}}{\lambda_{g1}} \right)^2 + 0.663 \exp \left(-1.71 \frac{Z_{0(1)}}{Z_{0(2)}} \right) \right.$$

$$\left. -0.172 \ln \left(\frac{Z_{0(1)}}{Z_{0(2)}} \right) \right\} \frac{Z_{0(1)}}{Z_{0(2)}}$$

(Replace $Z_{0(1)}/Z_{0(2)}$ by $Z_{0(2)}/Z_{0(1)}$ when $Z_{0(1)}/Z_{0(2)} \gtrsim 2.0$.) Also, the shunt capacitance is determined by the following expressions:
For the condition $Z_{0(1)}/Z_{0(2)} \leq 0.5$:

$$\frac{\omega C \lambda_{g1}}{Y_{0(1)} w_{eff1}} = \left(\frac{2w_{eff1}}{\lambda_{g1}} - 1 \right) \frac{Z_{0(1)}}{Z_{0(2)}}$$

For the condition $Z_{0(1)}/Z_{0(2)} \geq 0.5$:

$$\frac{\omega C \lambda_{\hat{g}1}}{Y_{0(1)} w_{eff1}} = \left(\frac{2w_{eff1}}{\lambda_{g1}} - 1 \right) \left(2 - 3 \frac{Z_{0(1)}}{Z_{0(2)}} \right)$$

Also refer to Section 5.13.3 for frequency-dependent effects.

5.14.8 The asymmetric cross-junction

This structure was considered in Section 5.12, and a fairly comprehensive equivalent circuit was shown which is again applicable for frequencies up to about 12 GHz. The capacitance is given simply by

$$C_m = \frac{c}{Z_0 \sqrt{\epsilon_{eff}}}$$

Alternatively, and probably more powerfully at somewhat higher frequencies, the cross-junction may be considered as *two* T-junctions operating from the same plane in the main microstrip line. Some elaboration of the equivalent circuits may be necessary if interactions between the branch arms are to be taken into account.

For virtually all the discontinuities it is clear that substantial further work is necessary before techniques are available for circuit designers working at frequencies beyond about 10–12 GHz. Up to these frequencies, and occasionally beyond, the expressions given may be used in design.

Further design information, including dispersive effects, is provided in Section 5.13.3. This should prove used in support of design at frequencies up to 25 GHz.

REFERENCES

1. Akello, R. J., Easter, B., and Stephenson, I. M., 'Effects of microstrip discontinuities on GaAs MESFET amplifier gain performance', *Electron. Lett.*, **13**, No. 6, 17 March 1977, 160–162.

2. Slaymaker, N. A., Soares, R. A., and Turner, J. A., 'GaAs MESFET small-signal X-band amplifiers', *IEEE Trans.*, **MTT-24**, No. 6, June 1976, 329–337.
3. Gupta, K. C., Garg, Ramesh, and Bahl, I. J., *Microstrip Lines and Slotlines*, Artech House, Inc., Dedham, Massachusetts, 1979.
4. Silvester, P., and Benedek, P., 'Equivalent capacitances of microstrip open circuits', *IEEE Trans.*, **MTT-20**, No. 8, August 1972, 511–576.
5. Thomson, A. F., and Gopinath, A., 'Calculation of microstrip discontinuity inductances', *IEEE Trans.*, **MTT-23**, No. 8, August 1975, 648–655.
6. James, J. R., and Ladbrooke, P. H., 'Surface-wave phenomena associated with open-circuited stripline terminations', *Electron. Lett.*, **9**, No. 24, 29 Nov. 1973, 570–571.
7. Hammerstad, E. O., and Bekkadal, F., 'A microstrip handbook', *ELAB Report*, STF 44 A74169, N7034, University of Trondheim–NTH, Norway, 1975.
8. Cohn, S. B., 'Problems in strip transmission lines', *IRE Trans.*, **PG MTT-3**, No. 2, March 1955, 119–126.
9. Garg, Ramesh, and Bahl, I. J., 'Microstrip discontinuities', *Int. J. Electronics*, **45**, No. 1, 1978, 81–87.
10. Costamagna, E., 'Some notes on a method of calculating gap capacitances in microstrip structures', *Alta Frequenza*, **XLIII**, No. 6, June 1974, 362–364.
11. Benedek, P., and Silvester, P., 'Equivalent capacitances for microstrip gaps and steps', *IEEE Trans.*, **MTT-20**, No. 11, November 1972, 729–733.
12. Owens, R. P., 'The electrical characteristics of transferred-electron device packages and mounts, and their effects on oscillator performance', *PhD Thesis*, University of London, 1977 (work carried out at RMCS).
13. Silvester, P., and Benedek, P., 'Microstrip discontinuity capacitances for right-angle bends, T-junction and crossings', *IEEE Trans.*, **MTT-21**, No. 5, May 1973, 341–346.
14. Anders, P., and Arndt, F., 'Moment method of designing matched microstrip bends', *1979 European Microwave Conference, Brighton, England, Digest of Papers*, 430–434.
15. Easter, B., *et al.*, 'Theoretical and experimental methods for evaluating discontinuities in microstrip', *Radio and Electronic Engineer*, **48**, No. 1/2, January/February 1978, 73–84.
16. Hoefer, W. J. R., 'Equivalent series inductivity of a narrow transverse slit in microstrip', *IEEE Trans.*, **MTT-25**, No. 10, October 1977, 822–824.
17. Wheeler, H. A., 'Coupling holes between resonant cavities or waveguides evaluated in terms of volume ratios', *IEEE Trans.*, **MTT-12**. March 1964, 231–244.
18. Hall, P. S., and James, J. R., Private communication, December 1978.
19. Kompa, G., 'Reduced coupling aperture of microstrip stubs provides new aspects in stub filter design', *European Microwave Conference, Digest of Papers*, Rome, 1976, 39–43.
20. Dydyk, M., 'Master the T-junction and sharpen your MIC designs', *MicroWaves*, May 1977, 184–186.
21. Akello, R. J., Easter, B., Stephenson, I. M., 'Equivalent circuit of the asymmetric crossover junction', *Electron. Lett.*, **13**, No. 4, 17 Feb. 1977, 117–118.
22. James, J. R., and Henderson, A., 'High-frequency behaviour of microstrip open-circuit terminations', *IEEE Journal on Microwaves, Optics and Acoustics*, **3**, No. 4, September 1979, 205–218.
23. Itoh, T., 'Analysis of microstrip resonators', *IEEE Trans.*, **MTT-22**, November 1974, 946–952.
24. Katehi, P. B., and Alexopoulos, N. G., 'Frequency-dependent characteristics

of microstrip discontinuities in millimeter-wave integrated circuits', *IEEE Trans.*, **MTT-33**, October 1985, 1029–1035.

25. Ozmehmet, K., 'New frequency dependent equivalent circuit for gap discontinuities of microstriplines', *IEEE Proceedings No. 3*, June 1987, 333–335.

26. McLean, J., Ling, H., and Itoh, T., 'Full wave modeling of electrically wide microstrip open end discontinuities via a deterministic spectral domain method', *IEEE MTT-S International Microwave Symposium* 1990, Digest of Papers, 1155–1158.

27. Achkar, J. *et al.*, 'Frequency-dependent characteristics of an asymmetric suspended stripline and its open end discontinuity for millimeter wave applications', *Proceedings of the 19th European Microwave Conference*, London, September 1989, 767–772.

28. Menzel, W., and Wolff, I., 'A method for calculating the frequency-dependent properties of microstrip discontinuities', *IEEE Trans.*, **MTT-25**, February 1977, 107–113.

29. Kompa, G., 'Design of Stepped microstrip components', *Radio and Electronic Engineer*, **48**, January/February 1978, 53–63.

30. Mehran, Reza, 'Computer-aided design of microstrip filters considering dispersion, loss and discontinuity effects', *IEEE Trans.*, **MTT-27**, No. 3, March 1979, 239–245.

31. Harokopus, W. P. Jr., and Katehi, P. B., 'Analysis of multilayer irregular microstrip discontinuities', *Proceedings of the 19th European Microwave Conference*, London, September 1989, 745–750.

32. Zheng, J. X., and Chang, D. C., 'Numerical modelling of chamfered bends and other microstrip junctions of general shape in MMICs', *IEEE MTT-S International Microwave Symposium*, 1990, Digest of Papers, 709–712.

33. Wu, S.-C., Yang. H.-Y., and Alexopoulos, N. G., 'A rigorous dispersive characterization of microstrip cross and tee junctions', *IEEE MTT-S International Microwave Symposium*, 1990, Digest of Papers, 1151–1154.

34. Chang, S. S., Huang, W. X., and Itoh, T., 'Computer aided design for planar microstrip circuit analysis by boundary-integral method', *Proceedings of the 19th European Microwave Conference*, London, September 1989, 975–978.

35. Giannini, F., Bartolucci, G., and Ruggieri, M., 'An improved equivalent model for microstrip cross-junction', *Proceedings of the 19th European Microwave Conference*, London, September 1989, 1226–1231.

36. Weisshaar, A., *et al.*, 'Modelling of radial microstrip bends', *IEEE MTT-S International Microwave Symposium*, 1990, Digest of Papers, 1051–1054.

37. Finch, K. L. and Alexopoulos, N. G., 'Shunt posts in microstrip transmission lines', *IEEE* **MTT-38**, November 1990, 1585–1594.

6 Parallel-coupled Microstrip Lines and Directional Couplers

6.1 STRUCTURE AND APPLICATIONS

The arrangement shown in Figure 6.1 illustrates the basic structure under consideration in this chapter.

It will be assumed that both microstrip lines have the same widths, which is nearly always the case in practical applications. Some work has also been carried out on multiple arrays of such parallel, edge-coupled lines but we can usefully restrict the considerations to just two lines mainly here.

There are two general areas of application for these structures:

(a) directional couplers—for use in a variety of circuits including balanced mixers, balanced amplifiers, phase shifters, attenuators, modulators, discriminators, and measurement bridges

(b) filters, delay lines, and matching networks—often using arrays of parallel-coupled microstrips as resonant elements.

In the first instance, (a), a prescribed amount of the incident power is required to be coupled out of the system. Thus, for example, a '3 dB coupler' is one in which half of the power input is coupled from one microstrip line into another and then on to separate circuitry. We shall see

Fig. 6.1. A pair of parallel, edge-coupled microstrip lines

later in this chapter that good coupler performance is achievable if special modifications are made to the basically simple structure consisting of two parallel microstrips.

Microstrip filters (b) employing the parallel-coupled structure are usually of the band-pass or band-stop types. An example of a design approach employing resonators coupled in this manner is given in Chapter 8.

Apart from direct applications, the calculations described here are also useful for estimating the degree of unwanted coupling between runs of microstrip lines on the same substrate. These calculations may also be judiciously applied to the problem of estimating the degree of crosstalk between such microstrip lines when they are used in high-speed pulse and digital systems.

In this chapter we principally consider the following features:

(a) directional coupler specification
(b) calculation of cross-sectional dimensions (for a desired amount of coupling)
(c) calculation of appropriate coupled-region physical length
(d) frequency response characteristics
(e) coupler directivity
(f) coupler compensation techniques
(g) 'special' designs (including Lange)
(h) thickness effects, losses, and fabrication tolerances
(i) crosstalk between microstrips used in high-speed digital systems.

Features (b), (c), (d), and (h) are not only applicable to directional couplers, they are also important in some filter design techniques and in crosstalk calculations (i).

6.2 PARAMETERS AND INITIAL SPECIFICATION

As described in Section 1.10, the even- and odd-mode characteristic impedances (Z_{0e} and Z_{0o}) are major design parameters for any parallel-coupled transmission line configuration—whatever its application. These impedances are obtained at an early design stage and they are functions of the degree of coupling (C) and the single-line terminating characteristic impedance (Z_0). The relationships between Z_{0e} and Z_{0o} and the physical dimensions of the coupled structure (including the substrate permittivity) are therefore of prime significance to the designer. As with single microstrip lines so also in this coupled situation we can determine Z_{0e} and Z_{0o} from known physical dimensions, which amounts to analysis. Alternatively, with greater difficulty, we may synthesize the physical structure from starting values of the impedances. Both of these procedures are useful in practice.

For coupler design the important starting parameters were defined in section 1.10; they are coupling factor C, transmission factor T, directivity D, and isolation I. Microstrip couplers, unless compensated by one of the methods to be described, exhibit a performance which can be far removed from the ideal in some cases. This is due to the different phase velocities associated with the even and odd modes, the main effect of which is to degrade the directivity.

The initial specification for such a coupler will include:

(a) coupling factor C at the centre frequency (usually in decibels)
(b) permittivity and thickness of the substrate
(c) terminating characteristic impedance Z_0 (usually 50 Ω)
(d) bandwidth and centre frequency
(e) coupling factor tolerance over the band (sometimes)
(f) lowest acceptable directivity D in decibels.

From this information the designer must ultimately determine the widths of the microstrip lines, the separation between them, and the length of the coupled region.

In the design of filters or matching networks using parallel-coupled lines we usually begin with either insertion loss as a function of frequency or VSWR requirements over some band. In either event we still arrive at desired values of Z_{0e} and Z_{0o} and the final synthesis problem is then similar to that of a coupler.

6.3 COUPLED MICROSTRIP LINES

In Chapter 1 we considered the basic properties of TEM lines in a parallel-coupled configuration. Just as single microstrip lines are quasi-TEM so coupled microstrip lines also exhibit a field pattern which is strictly quasi-TEM. We may, however, examine even-mode and odd-mode excitation (Figure 6.2) for the coupled microstrips.

(a) Even mode (b) Odd mode

Fig. 6.2. Field distributions resulting from (a) even-mode and (b) odd-mode excitation of parallel-coupled microstrip lines

Notice that the relative proportions of fields shared between the dielectric substrate and the air are different. Thus the phase velocities are also different, a fact which represents an extra design complication.

It is important to note that all parallel-line couplers, whether true TEM or not, have the odd- and even-mode property which always results in Z_{0o} and Z_{0e}. However, true TEM couplers yield equal phase velocities for each mode whereas microstrip (quasi-TEM) and certain other structures yield odd-mode and even-mode phase velocities, v_{po} and v_{pe}.

Referring to Appendix A this means that, strictly speaking, $\theta_e \neq \theta_o$ and eqn (A.16) cannot be evoked to exactly satisfy the equality relationship given (because only a single θ was present there). There is no reason why we should not continue to define the coupling factor as eqn (A.33) but the remaining relationships, for V_2, V_3, and V_4 are all more complicated. We shall return to this problem from a design viewpoint shortly.

The cross-sectional dimensions which we require from a design procedure are w and s, as indicated in Figure 6.3.

Both the even and the odd modes exist under the normal circuit excitation conditions associated with a microstrip coupler. However, coupler calculation and design proceeds from consideration of the effects due to each mode taken separately.

Analysis yielding Z_{0o} and Z_{0e} as functions of the shape ratio of w/h, s/h, and ϵ_r has been carried out by several groups—notably Bryant and Weiss.[1] Although this computer-orientated work is subject to some debate regarding accuracy (especially for tight coupling) it is a fundamental approach and graphs of results are very readily available in microwave handbooks.

The method of analysis due to Bryant and Weiss[1] amounts to a rigorous solution of the electrostatic problem for coupled microstrip lines. (Incidentally, single strips are also solved.) They determine capacitances on a static TEM basis. Longitudinal components of the fields are neglected and the thickness of the microstrips is also neglected. A formulation based essentially on Laplace's equation is employed with the proper boundary conditions inserted. Iterative numerical methods are used in the computer solution and the impedances are finally obtained as follows (either mode).

We start with the effective permittivity (see eqns 3.8 and 3.9):

$$\epsilon_{\text{eff}} = \frac{C}{C_1} \tag{6.1}$$

Fig. 6.3. Cross-sectional dimensions: nomenclature

where C is the microstrip capacitance per strip with the substrate present and C_1 is the microstrip capacitance per strip with the substrate absent (i.e. air spaced). The velocity of propagation v_p is then (from eqn 3.8)

$$v_p = \frac{c}{\sqrt{\epsilon_{eff}}} \tag{6.2}$$

and the characteristic impedance is (eqn 3.2b)

$$Z_0 = \frac{1}{v_p C} = \frac{1}{cC_1 \sqrt{\epsilon_{eff}}} \tag{6.3}$$

For odd and even modes we therefore have, respectively,

$$Z_{0o} = \frac{1}{cC_{1o} \sqrt{\epsilon_{effo}}} \tag{6.4}$$

and

$$Z_{0e} = \frac{1}{cC_{1e} \sqrt{\epsilon_{effe}}} \tag{6.5}$$

The calculations required, leading to $C_{1o,e}$ and $\epsilon_{effo,e}$, are carried out in a manner identical to that described in Section 3.6. Some numerical data obtained using this analysis technique are presented graphically here in Figure 6.4.

In design, Z_{0e} and Z_{0o} are marked on the vertical axis of a graph for the appropriate substrate material (ϵ_r). It is then necessary to read these values across horizontally until two points are found lying vertically above one another, and at identical values of s/h. Dropping a vertical ordinate through these points then also yields the required value of w/h. Interpolation and other difficulties with the graphs makes this an approximate and rather clumsy design technique. However, it may be useful in certain instances and a specific microstrip coupler design, using these curves, will be outlined shortly.

6.4 CHARACTERISTIC IMPEDANCES IN TERMS OF THE COUPLING FACTOR (C')

First, we repeat eqns (A.42), (A.16), (A.43a) and (A.43b) observing that several of these expressions are approximate in the microstrip case due to the unequal phase velocities. We have:

Coupling factor:
$$C' = 20\log \left| \frac{Z_{0e} - Z_{0o}}{Z_{0e} + Z_{0o}} \right| \quad \text{dB} \tag{6.6}$$

Fig. 6.4. Numerical results for the even- and odd-mode characteristic impedances of parallel-coupled microstrip lines: (a) air spaced and (b) with a substrate having relative permittivity ϵ_r. (© 1968 IEEE. Reprinted, with permission, from Bryant and Weiss.[1])

Impedance relationship: $Z_0^2 \simeq Z_{0e}Z_{0o}$ \qquad (6.7)

and, from these expressions, the impedances required are

$$Z_{0e} \simeq Z_0 \sqrt{\frac{1 + 10^{C'/20}}{1 - 10^{C'/20}}} \qquad (6.8)$$

and

$$Z_{0o} \simeq Z_0 \sqrt{\frac{1 - 10^{C'/20}}{1 + 10^{C'/20}}} \qquad (6.9)$$

The last two equations are the basic initial design expressions. The designer will know the characteristic impedance of the external lines connecting to the coupler (Z_0), often 50 Ω, and will also know the value of the mid-band coupling factor C' in decibels—or some quantities inherently involving C'.

As the coupling increases the approximation involved in eqn (6.7) worsens steadily. An accurate relationship can be obtained by setting θ_e and θ_o appropriately into eqns (A.2) and (A.3) and adding the matrices to obtain

$$\begin{bmatrix} V_1 \\ I_1 \end{bmatrix} = \begin{bmatrix} \cos\theta_e + \cos\theta_o & j(Z_{0e}\sin\theta_e + Z_{0o}\sin\theta_o) \\ j(Y_{0e}\sin\theta_e + Y_{0o}\sin\theta_o) & \cos\theta_e + \cos\theta_o \end{bmatrix} \begin{bmatrix} V_2 \\ I_2 \end{bmatrix} \quad (6.10)$$

The condition for a perfect input match and perfect isolation can be written down directly from this because diagonal terms involving impedances must be made equal; thus

$$\frac{Z_{0e}}{Z_0}\sin\theta_e + \frac{Z_{0o}}{Z_0}\sin\theta_o = \frac{Z_0}{Z_{0e}}\sin\theta_e + \frac{Z_0}{Z_{0o}}\sin\theta_o \qquad (6.11)$$

or

$$Z_0^2 = Z_{0e}Z_{0o}\frac{Z_{0e}\sin\theta_e + Z_{0o}\sin\theta_o}{Z_{0e}\sin\theta_o + Z_{0o}\sin\theta_e} \qquad (6.12)$$

This is a frequency-dependent result because θ_e and θ_o are both functions of frequency.

Having made a 'first-cut' design θ_e and θ_o may be evaluated and the value of Z_0 can then be obtained from eqn (6.12). If Z_0 is found to differ too greatly from the specified value then the geometry (w and s) might be varied slightly to correct the result.

In many cases, certainly for 10 dB couplings and looser, the approximation offered by eqn (6.7) is sufficient.

6.5 SEMI-EMPIRICAL ANALYSIS FORMULAE AS A DESIGN AID

Analysis formulae are available for the determination of characteristic impedances to an accuracy of within 3 per cent.[2,3] These could be useful to the designer who might start with an approximate synthesis and employ the analysis formulae to provide an improvement in accuracy using an iterative technique. Air-line and substrate-line capacitances form the basis of the expressions, and the capacitances are broken down as shown in Figure 6.5. The accuracy of better than 3 per cent holds over the ranges: $0.2 \leq w/h \leq 2$; $0.05 \leq s/h \leq 2$, and $\epsilon_r \geq 1$.

The total capacitances for each mode can then be written

$$C_e = C_p + C_f + C_f' \qquad (6.13)$$

and

$$C_o = C_p + C_f + C_{ga} + C_{gd} \qquad (6.14)$$

The capacitance C_p simply relates to the parallel-plate line value given by

$$C_p = \epsilon_o \epsilon_r \frac{w}{h} \qquad (6.15)$$

(a) Even-mode capacitances

(b) Odd-mode capacitances

Fig. 6.5. Separation of capacitances for use in the analysis. (Reproduced by permission of Artech House Books, Inc., Dedham, Mass., from Gupta *et al.,*[20] Figure 8.15.)

Also, C_f is simply the fringing capacitance due to each microstrip taken alone, as if for a single strip. This is given by

$$2C_f = \frac{\sqrt{\epsilon_{eff}}}{cZ_0} - C_p \tag{6.16}$$

where c is the free-space velocity and ϵ_{eff} and Z_0 are obtained by single-strip static-TEM methods.

An empirical expression for C_f' is given as follows:

$$C_f' = \frac{C_f}{1 + (A\,h/s)\tanh(8s/h)}\sqrt{\frac{\epsilon_r}{\epsilon_{eff}}} \tag{6.17}$$

where

$$A = \exp\{-0.1\exp(2.33 - 2.53\,w/h)\} \tag{6.18}$$

As can be seen from Figure 6.5, C_{ga} and C_{gd} represent, respectively, odd-mode fringing field capacitances for the air and dielectric regions across the coupling gap. C_{ga} was obtained by using an equivalent coplanar strip geometry calculation, yielding

$$C_{ga} = \epsilon_0 \frac{K(k')}{K(k)} \tag{6.19a}$$

where

$$k = \frac{s/h}{s/h + 2w/h}$$

$$k' = \sqrt{1 - k^2}$$

and the ratio of the elliptic functions is
For $0 \le k^2 \le 0.5$:

$$\frac{K(k')}{K(k)} = \frac{1}{\pi}\ln\left(2\frac{1 + \sqrt{k'}}{1 - \sqrt{k'}}\right) \tag{6.19b}$$

For $0.5 \le k^2 \le 1$:

$$\frac{K(k')}{K(k)} = \frac{\pi}{\ln\{2(1 + \sqrt{k})/(1 - \sqrt{k})\}} \tag{6.19c}$$

C_{gd} was determined differently. The capacitance for coupled striplines (triplate) was modified to give

$$C_{gd} = \frac{\epsilon_0\epsilon_r}{\pi}\ln\left\{\coth\left(\frac{\pi}{4}\frac{s}{h}\right)\right\} + 0.65C_f\left(\frac{0.02}{s/h}\sqrt{\epsilon_r} + 1 - \epsilon_r^{-2}\right) \tag{6.20}$$

The even-mode and odd-mode characteristic impedances are then

$$Z_{0e} = (c\sqrt{C_e C_{e1}})^{-1} \qquad (6.21)$$

and

$$Z_{0o} = (c\sqrt{C_o C_{o1}})^{-1} \qquad (6.22)$$

where c is the velocity in free space and the second subscript 1 refers to a free-space ('air') line.

The effective microstrip permittivities are

$$\epsilon_{effe} = \frac{C_e}{C_{e1}} \qquad (6.23)$$

and

$$\epsilon_{effo} = \frac{C_o}{C_{o1}} \qquad (6.24)$$

This concludes the presentation of the semi-empirical analysis. We next consider an approximate *synthesis* technique.

6.6 AN APPROXIMATE SYNTHESIS TECHNIQUE

In this potentially powerful procedure the shape ratios w/h and s/h are determined from Z_{0o} and Z_{0e} information either using just two universal graphs or with the aid of a relatively simple computer routine.

In both cases there are two distinct stages. One of these comprises a determination of *equivalent single* microstrip shape ratios $(w/h)_s$. The second stage relates the required w/h and s/h for the coupled structure to the equivalent single microstrip shape ratios. The final results are made independent of the substrate permittivity for a considerable range and are given by Akhtarzad *et al.*,[4] not only in graphical form but also as closed formulae. A summary of the design stages is thus:

(a) Determine shape ratios for equivalent *single* microstrip lines.
(b) Obtain the shape ratio w/h and the spacing ratio s/h for the desired coupled microstrip structure using the single-line shape ratios found at stage (a).

For design stage (a) we use the following relationships:

$$Z_{0se} = \frac{Z_{0e}}{2} \quad \text{(for single microstrip shape ratio } (w/h)_{se}) \qquad (6.25)$$

and

$$Z_{0so} = \frac{Z_{0o}}{2} \quad \text{(for single microstrip shape ratio } (w/h)_{so}) \qquad (6.26)$$

The appropriate shape ratios are then determined directly from a graph or from closed-form expressions which are readily available. The graphical technique described in Chapter 3 is recommended or the approximate static-TEM formulae given there may be used for more comprehensive CAD.

Akhtarzad et al.[4] give a graph for determining w/h and s/h from the single-line values now known. This graph is suitable for many approximate designs and therefore it is repeated here (Figure 6.6) with some additional information to guide the user.

To perform design using this graph, we start with stage (a) and evaluate $(w/h)_{se}$ and $(w/h)_{so}$ using *single-line* static-TEM methods. We then select the *two* curves of Figure 6.6 which have the appropriate w/h parameter for each mode as just evaluated (solid curves for odd mode, broken lines for even mode). The point of intersection of these two selected lines is the 'design point'. Reading across horizontally yields the required w/h for each of the coupled lines, and reading down vertically gives their normalized spacing s/h. A numerical example of design using this graph will be given shortly.

Difficulties may be encountered in crowded regions of the graph, such as low w/h combined with low s/h, and if the single-line-derived w/h values do not happen to fit the values 2, 3, 4, 6, 8, or 10, then some fairly awkward interpolation is necessary. To overcome these problems Akhtarzad et al.[4]

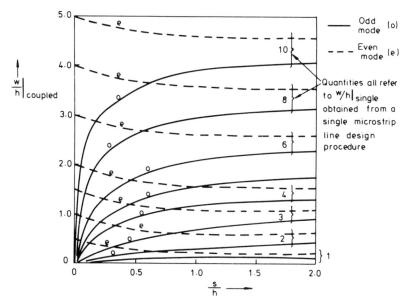

Fig. 6.6. 'General' curves for use in the synthesis technique. (© 1975 IEEE. Reprinted, with permission, from Akhtarzad et al.[4])

have given a family of useful expressions, the most complete of which are quoted here first:

$$\left(\frac{w}{h}\right)_{se} = \frac{2}{\pi} \cosh^{-1}\left(\frac{2d - g + 1}{g + 1}\right) \tag{6.27}$$

$$\left.\begin{array}{l} \left(\dfrac{w}{h}\right)_{so} = \dfrac{2}{\pi} \cosh^{-1}\left(\dfrac{2d - g - 1}{g - 1}\right) \\[4mm] \qquad + \dfrac{4}{\pi(1 + \epsilon_r/2)} \cosh^{-1}\left(1 + 2\dfrac{w/h}{s/h}\right) \end{array}\right\} \epsilon_r \le 6 \qquad (6.28a)$$

or

$$\left(\frac{w}{h}\right)_{so} = \frac{2}{\pi} \cosh^{-1}\left(\frac{2d - g - 1}{g - 1}\right) + \frac{1}{\pi} \cosh^{-1}\left(1 + 2\frac{w/h}{s/h}\right) \quad \epsilon_r \ge 6 \quad (6.28b)$$

where

$$g = \cosh\left(\frac{\pi s}{2h}\right)$$

and

$$d = \cosh\left(\pi\frac{w}{h} + \frac{\pi s}{2h}\right)$$

Equation (6.27) has to be solved simultaneously with either (6.28a) or (6.28b) as appropriate for the particular permittivity range. The curves in Figure 6.6 were originally obtained by Akhtarzad for the median value ($\epsilon_r = 6$) of the permittivity. As an approximation the second terms may be neglected in both eqns (6.28a) and (6.28b). Then an explicit formula for s/h is obtained:

$$\frac{s}{h} = \frac{2}{\pi} \cosh^{-1}\left[\frac{\cosh\{(\pi/2)\,(w/h)_{se}\} + \cosh\{(\pi/2)(w/h)_{so}\} - 2}{\cosh\{(\pi/2)(w/h)_{so}\} - \cosh\{(\pi/2)(w/h)_{se}\}}\right] \tag{6.29}$$

In any case the functional calculations are quite complicated whether one uses the more accurate expressions or starts with this formula, (6.29).

Also, when compared with the results of Bryant and Weiss,[1] an error of the order of 10 per cent has been quoted for this synthesis technique.[2] If this magnitude of error is actually found in practice then the synthesis technique may still be used to obtain an initial design, after which the more accurate analysis of Garg and Bahl (Section 6.5) could be employed to correct the design dimensions. The sequence would be:

(a) With the required Z_{0e} and Z_{0o} determine $(w/h)_1$ and $(s/h)_1$ by approximate synthesis.

(b) Using these values, $(w/h)_1$ and $(s/h)_1$, recalculate Z_{0e} and Z_{0o} by means of the semi-empirical analysis formulae given in Section 6.5.

(c) Compare the new values thus calculated for Z_{0e} and Z_{0o} with those originally required. A suitable programmable calculator or computer routine will then enable iterations to be made, incrementing w/h and s/h until the impedances agree within a specified tolerance.

Although this procedure can yield coupled-line design to approximately 3 per cent accuracy there is still clearly scope for significant improvement to aid the circuit designer.

In particular, either the synthesis technique due to Akhtarzad might be modified to obtain greater accuracy or a new technique might be developed.

6.7 A SPECIFIC EXAMPLE: DESIGN OF A −10 dB MICROSTRIP COUPLER

We will approach this by *two* methods:

(a) using Bryant and Weiss curves
(b) using Akhtarzad synthesis curves.

For the moment, in this section, we shall only determine the cross-sectional dimensions. Evaluation of the length of a coupler is considered shortly.

The specification is:

coupling coefficient $C' = -10$ dB
single microstrip feedline characteristic impedances $Z_0 = 50\ \Omega$
substrate permittivity $\epsilon_r = 9.0$; thickness $= 1$ mm
system centre frequency (mid-band for the coupler) $= 5$ GHz.

This is low enough for dispersion effects (dealt with later) to be neglected to a first approximation.

Design calculations:
Z_{0e} and Z_{0o} are obtained using eqns (6.8) and (6.9). This yields

$$Z_{0e} = 69.5\ \Omega \quad \text{and} \quad Z_{0o} = 36\ \Omega$$

Next, always check that the relation $Z_0 \simeq \sqrt{Z_{0e}Z_{0o}}$ holds, although the approximation is not good for tight (around −3 dB or more) coupling. Hence

$$\sqrt{69.5 \times 36} = 50.02\ \Omega$$

6.7.1 Bryant and Weiss' curves

With Bryant and Weiss' method (Section 6.3) choose the $\epsilon_r = 9$ family of curves. Select two points, one of the 'even-mode' set of curves and the other on the 'odd-mode' set of curves. These points are controlled by the necessity for them to lie on the same w/h ordinate value, and on curves with an identical s/h parameter, i.e. they must both have the same w/h and s/h. Then w/h and s/h are read off as 0.87 and 0.25. The procedure is shown in the sketched graph of Figure 6.7.

From the values obtained we can determine w and s easily, since the substrate thickness h was given as 1 mm:

$$w = 0.87 \times 1 = 0.87\,\text{mm}$$

and
$$s = 0.25 \times 1 = 0.25\,\text{mm}$$

6.7.2 Synthesis (Akhtarzad)

First we must calculate 'single' microstrip characteristic impedances, using eqns (6.25) and (6.26):

$$Z_{0se} = \frac{Z_{0e}}{2} \simeq 35\,\Omega$$

and

$$Z_{0so} = \frac{Z_{0o}}{2} \simeq 18\,\Omega$$

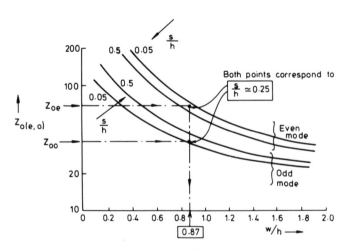

Fig. 6.7. Graphical design procedure yielding the 'shape ratios' for a $-10\,\text{dB}$ directional coupler

Details of the actual calculations will not be given here because they follow the same routine that was given in Chapter 3. Approximate values are

$$\left(\frac{w}{h}\right)_{se} \simeq 2.0$$

and

$$\left(\frac{w}{h}\right)_{so} \simeq 5.0$$

Now, using Figure 6.6, we first select the broken (even-mode) curve which has parameter $(w/h)_{se} = 2$. Next, select the solid (odd-mode) curve which has parameter $(w/h)_{so} = 5$. The point where these two curves intersect is the 'design point' and its coordinates give the results

$$\frac{w}{h} \simeq 0.85$$

and

$$\frac{s}{h} \simeq 0.25$$

Hence

$$w \simeq 0.85 \text{ mm and } s \simeq 0.25 \text{ mm.}$$

6.7.3 Comparison of methods

Although only graphical techniques have been used it can be seen that the values obtained compared well in this case. Notice that, for the synthesis method, no special set of curves was required to be selected—the curves used (Figure 6.6) are 'universal' in that they apply over a wide range of substrate permittivities.

6.8 COUPLED-REGION LENGTH

The requirement for coupled-region length is determined and given in Appendix A following eqn (A.37). It is seen that the maximum degree of coupling occurs when the length of the coupled region is made $\lambda_{gm}/4$, where λ_{gm} is the mid-band wavelength. At other wavelengths the coupling decreases.

As a rough first approximation, only really applicable to loose amounts of coupling (perhaps less than -10 dB), it is possible to consider a *single*, i.e. uncoupled, microstrip of width w and to evaluate the mid-band

wavelength as described in Chapter 3. However, the error involved in this can rise to 10 per cent or more and the mean of the even-mode and odd-mode wavelengths should be taken for greater accuracy.

It is best, for design purposes, to work in terms of characteristic impedances. We first recollect the three general forms in which characteristic impedance may be written (eqns 1.10, 1.11a and 1.11b):

$$Z_0 = \sqrt{\frac{L}{C}} = \frac{1}{v_p C} = v_p L \qquad (6.30)$$

The last form of this expression is the most convenient one to use here, since we may assume that the inductance L remains constant independent of whether a substrate is present or absent. (This is really another way of stating that only the electric flux is affected by changing the dielectric material on the substrate, *for a given mode*.)

Consider the odd mode on microstrip. For the actual situation with a substrate present (Figure 6.8a), we have

$$Z_{0o} = v_{po} L_o \qquad (6.31)$$

When the substrate is removed, as in Figure 6.8(b), the velocity is simply the free-space limit value c. For that situation the characteristic impedance is given by

$$Z_{01o} = c L_o \qquad (6.32)$$

Equations (6.31) and (6.32) can be simply solved for v_{po} giving

$$v_{po} = c \frac{Z_{0o}}{Z_{01o}} \qquad (6.33)$$

By a similar consideration for the even mode:

$$v_{pe} = c \frac{Z_{0e}}{Z_{01e}} \qquad (6.34)$$

The odd- and even-mode wavelengths are then easily found at any frequency f by using

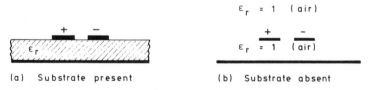

(a) Substrate present (b) Substrate absent

Fig. 6.8. Parallel-coupled microstrips with the substrate (a) present and (b) absent (only the odd mode is indicated)

$$\lambda_{go} = \frac{v_{po}}{f} \quad \text{and} \quad \lambda_{ge} = \frac{v_{pe}}{f} \tag{6.35}$$

The characteristic impedances Z_{0e} *and* Z_{0o} are known early on in the design calculations, as explained in Section 6.4. Then the cross-sectional dimensions w/h and s/h are obtained. Once these are known the air-spaced characteristic impedance Z_{01e} and Z_{01o} can either be obtained from Bryant and Weiss' curves applicable to $\epsilon_r = 1$, or by calculations using the analysis given in Section 6.5.

We can finally combine eqns (6.32), (6.34), and (6.35), and set frequency F (in gigahertz) to give

$$\lambda_{ge} \simeq \frac{300}{F} \frac{Z_{0e}}{Z_{01e}} \quad \text{mm} \tag{6.36}$$

and

$$\lambda_{go} \simeq \frac{300}{F} \frac{Z_{0o}}{Z_{01o}} \quad \text{mm} \tag{6.37}$$

Returning to the numerical design example of Section 6.7, we now apply the values of w/h and s/h to the Bryant and Weiss curves ($\epsilon_r = 1$) and obtain

$$Z_{01e} = 185\ \Omega \quad \text{and} \quad Z_{01o} = 75\ \Omega$$

With the given frequency of 5 GHz, eqns (6.36) and (6.37) then yield

$$\lambda_{go} = 28.8\ \text{mm} \quad \text{and} \quad \lambda_{ge} = 22.54\ \text{mm}$$

The mean of these two wavelengths (λ_{gm}) is 25.67 mm and thus the length of the coupled region ($\lambda_{gm}/4$) is 6.42 mm. With the approximations involved, and also considering that we have neglected dispersion, the value 6.4 mm is sensible. The final complete microstrip coupler design is illustrated, in plan view, in Figure 6.9.

Fig. 6.9. Final numerical dimensional values for the -10 dB coupler design shown in the text

In some cases, especially at higher frequencies, the length may become very small and difficult to define in manufacture. Then the length can be made a convenient odd number of quarter-wavelengths:

$$l = (2n - 1)\lambda_{gm}/4 \qquad (6.38)$$

It is, however, important to appreciate that the coupler would then exhibit n responses with frequency.

6.9 FREQUENCY RESPONSE

6.9.1 Overall effects and Getsinger's model

So far we have concentrated on 'mid-band' design. The full frequency-dependent (i.e. angle-dependent) coupling coefficient developed in Appendix A as eqn (A.36) is reproduced here, with phase velocity differences neglected:

$$C(\theta) = C(j\omega) = \frac{jC \sin \theta}{\sqrt{1 - C^2} \cos \theta + j \sin \theta} \qquad (6.39)$$

where C is the mid-band coupling factor given by eqn (A.33) and $\theta = \beta l = 2\pi l/\lambda_g$.

Accepting the validity of the result for the loosely coupled frequency response of the *ideal matched coupler*, the actual shape would be similar to that given in Figure 6.10.

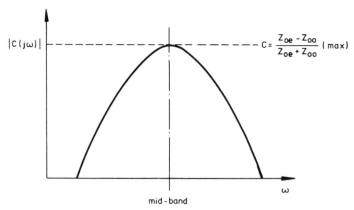

Fig. 6.10. The characteristic shape of the frequency response for a loosely coupled, ideal matched coupler

This general form of response is typical for couplers of this type, since they are difficult to construct for even moderately tight couplings: C' is generally smaller than -3 dB in designs with this geometry.

In several applications such a poor frequency response is intolerable, and a number of techniques have been developed to 'flatten' the response. (The phase responses are also significant, and there are also techniques to improve the directivity by near-equalization of the even- and odd-mode phase velocities.) Some of these techniques will be dealt with shortly. As with single microstrip lines, *dispersion* is also present in the behaviour of these coupled microstrips; this problem is dealt with first.

For calculations on parallel-coupled microstrips, especially at frequencies exceeding a few gigahertz, dispersion is significant and it is essential to have quantitative information on this dispersion. Only then can wavelengths be at all accurately calculated. The problem of dispersion in *single* microstrip lines was discussed in Chapter 4. Coupled microstrips suffer dispersion for precisely the same physical reasons as single microstrip, but the amount of dispersion depends upon the mode under consideration. The effective microstrip permittivities are now both somewhat different functions of frequency which involve the parameters indicated as follows:

$$\epsilon_{\text{effe}}(f, h, \epsilon_r, \epsilon_{\text{effe}}, Z_{0e})$$

and

$$\epsilon_{\text{effo}}(f, h, \epsilon_r, \epsilon_{\text{effo}}, Z_{0o})$$

Getsinger[5] has shown how his expressions, which were originally derived for single microstrip, can be adapted to suit the coupled microstrip problem (Getsinger's single-line expressions were given in Chapter 4). For use with coupled lines the following approximate substitutions are made where Getsinger's method is adopted:

(a) In the even mode the two strips are at the same potential and the total current is twice that on a single strip. Thus the total mode impedance is half that of a single strip and the dispersion for even-mode propagation is calculated by substituting

$$Z_0 \simeq \frac{1}{2} Z_{0e}$$

(b) In the odd mode the two strips are at opposite potentials and the voltage between strips is twice that of a single strip to ground. Thus the total mode impedance is twice that of a single strip and the dispersion for odd-mode propagation is calculated by substituting

$$Z_0 \simeq 2 Z_{0o}$$

in Getsinger's single-line equations.

So, for example, the even-mode dispersion expression runs as follows:

$$\epsilon_{\text{effe}}(f) = \epsilon_r - \frac{\epsilon_r - \epsilon_{\text{effe}}}{1 + G_e(f/f_{pe})^2} \qquad (6.40)$$

with

$$f_{pe} = \frac{Z_{0e}}{4\mu_0 h}$$

and

$$G_e = 0.6 + 0.0045\,Z_{0e} \qquad (6.41)$$

Some caution is necessary regarding the accuracy of using the substitution (b) since Easter and Gupta[6] have indicated that an overestimation of odd-mode dispersion is obtained. This assessment is based on comparisons with the theoretical treatment due to Krage and Haddad,[7] and with the measurements due to Richings and Easter[8] (measurements are dealt with in Chapter 7).

It is suggested that empirical modifications might be made to Getsinger's odd-mode substitution to provide more accurate design formulae but an approach yielding greater accuracy is available and is described in the next section.

6.9.2 More accurate design expressions, including dispersion

Kirschning and Jansen[9] have given design expressions for the modal dispersion in parallel coupled microstrip lines. These workers develop a coupled model for the structure, in which they use normalized parameters as follows:

$$u = w/h, \qquad g = s/h \qquad \text{and} \qquad f_n = f/h \qquad \text{(in GHz/mm units)}$$

They quote the ranges of validity for which the design expressions remain accurate:

$$0.1 \leqslant u \leqslant 10, \qquad 0.1 \leqslant g \leqslant 10 \qquad \text{and} \qquad 1 \leqslant \epsilon_r \leqslant 18$$

The static-TEM formula for the even mode effective microstrip permittivity is provided by Hammerstad and Jansen as follows:

$$\epsilon_{\text{effe}}(0) = 0.5\,(\epsilon_r + 1) + 0.5\,(\epsilon_r - 1)\,(1 + 10/v)^{-\{a_e(v)\,b_e(\epsilon_r)\}} \qquad (6.42)$$

where

$$v = u\,(20 + g^2)/(u + g^2) + g\exp(-g)$$

$$a_e(v) = 1 + \ln\{(v^4 + (v/52)^2)/(v^4 + 0.432)\}/49 + \ln\{1 + (v/18.1)^3\}/18.7$$

$$b_e(\epsilon_r) = 0.564\,\{(\epsilon_r - 0.9)/(\epsilon_r + 3)\}^{0.053} \qquad (6.43)$$

In contrast the static-TEM formula applicable to the odd mode effective microstrip permittivity was remodelled by Kirschning and Jansen so that an improved accuracy of 0.5 per cent applies over the whole range of parameters. This permittivity is given by

$$\epsilon_{\text{effo}}(0) = \{0.5(\epsilon_r + 1) + a_o(u, \epsilon_r) + \epsilon_{\text{eff}}(0)\} \exp(-c_o g^{d_o}) + \epsilon_{\text{eff}}(0) \qquad (6.44)$$

where

$$a_o(u, \epsilon_r) = 0.7287 \{\epsilon_{\text{eff}}(0) - 0.5(\epsilon_r + 1)\} \{1 - \exp(-0.179u)\}$$

$$b_o(\epsilon_r) = 0.747 \epsilon_r/(0.15 + \epsilon_r)$$

$$c_o = b_o(\epsilon_r) - \{b_o(\epsilon_r) - 0.207\} \exp(-0.414u)$$

$$d_o = 0.593 + 0.694 \exp(-0.562u) \qquad (6.45)$$

(where, as elsewhere in this text, $\epsilon_{\text{eff}}(0)$ refers to zero-thickness single microstrip line).

The functional form in which modal dispersion is introduced for both modes is similar to that shown for single microstrip, described in Chapter 4. Based upon this the general expression then takes the form

$$\epsilon_{\text{effe,o}}(f_n) = \epsilon_r - \{\epsilon_r - \epsilon_{\text{effe,o}}(0)\}/\{1 + F_{e,o}(f_n)\} \qquad (6.46)$$

In this expression it is of course F that determines the detailed frequency dependence.

The even-mode result for F, given by Kirschning and Jansen, is

$$F_e(f_n) = P_1 P_2 \{(P_3 P_4 + 0.1844 P_7) f_n\}^{1.5763} \qquad (6.47)$$

where

$$P_1 = 0.27488 + \{0.6315 + 0.525/(1 + 0.0157 f_n)^{20}\}u$$
$$\qquad - 0.065683 \exp(-8.7513u)$$

$$P_2 = 0.33622 \{1 - \exp(-0.03442 \epsilon_r)\}$$

$$P_3 = 0.0363 \exp(-4.6u) [1 - \exp\{-(f_n/38.7)^{4.97}\}]$$

$$P_4 = 1 + 2.751 [1 - \exp\{-(\epsilon_r/15.916)^8\}]$$

$$P_5 = 0.334 \exp\{-3.3(\epsilon_r/15)^3\} + 0.746$$

$$P_6 = P_5 \exp\{-(f_n/18)^{0.368}\}$$

and

$$P_7 = 1 + 4.069 P_6 g^{0.479} \exp(-1.347 g^{0.595} - 0.17g^{2.5}) \qquad (6.48)$$

Also, the result for F applicable to the odd mode is

$$F_0(f_n) = P_1 P_2 \{(P_3 P_4 + 0.1844) f_n P_{15}\}^{1.5763} \qquad (6.49)$$

with

$$P_8 = 0.7168 \left[1 + 1.076/\{1 + 0.0576\,(\epsilon_r - 1)\}\right]$$

$$P_9 = P_8 - 0.7913 \left[1 - \exp\{-(f_n/20)^{1.424}\}\right] \tan^{-1}\{2.481\,(\epsilon_r/8)^{0.946}\}$$

$$P_{10} = 0.242\,(\epsilon_r - 1)^{0.55}$$

$$P_{11} = 0.6366\,\{\exp(-0.3401 f_n) - 1\}\tan^{-1}\{1.263\,(u/3)^{1.629}\}$$

$$P_{12} = P_9 + (1 - P_9)/(1 + 1.183\,u^{1.376})$$

$$P_{13} = 1.695\,P_{10}/(0.414 + 1.605\,P_{10})$$

$$P_{14} = 0.8928 + 0.1072 \left[1 - \exp\{-0.42\,(f_n/20)^{3.215}\}\right]$$

$$P_{15} = \mathrm{abs}\,\{1 - 0.8928\,(1 + P_{11})\,P_{12}\exp(-P_{13}\,g^{1.092})/P_{14}\} \qquad (6.50)$$

The upper frequency limitation for these expressions depends upon the substrate parameters. The limit for retention of accuracy approaches 40 GHz for 0.635 mm thick substrates, which is fairly typical for alumina. The accuracy is also preserved for thin substrates with a permittivity of 12.9 (GaAs) and a normalized frequency of 30 GHz/mm, which corresponds to high millimeter-wave frequencies for gallium arsenide substrates some tens of microns thick. In all cases the maximum error bound is 1.4 per cent. It is important to appreciate that these design expressions hold accuracy only where the microstrip circuit is relatively undisturbed, i.e. effectively unshielded and with no affecting adjacent conductors or dielectrics. In situations where such practical disturbances exist various design corrections can be used (e.g. similar to those described in Chapter 3) but in practice it is generally found that, provided the lines are reasonably tightly coupled, the unshielded results provide a good design basis.

Kirschning and Jansen have also given design expressions for the frequency-dependent characteristic impedances. They showed that earlier formulae given by Hammerstad and Jansen for the static-TEM impedances yielded small errors that would accumulate for these purposes and they reported improved design expressions. For the even mode the static-TEM impedance is given by

$$Z_{0e}(0) = Z_0(0)\,\{\epsilon_{\mathrm{eff}}(0)/\epsilon_{\mathrm{effe}}(0)\}^{0.5}/[1 - \{Z_0(0)/377\}\,\{\epsilon_{\mathrm{eff}}(0)\}^{0.5}\,Q_4] \qquad (6.51)$$

where

$$Q_1 = 0.8695\,u^{0.194}$$

$$Q_2 = 1 + 0.7519\,g + 0.189\,g^{2.31}$$

$$Q_3 = 0.1975 + \{16.6 + (8.4/g)^6\}^{-0.387} + \ln[g^{10}/\{1 + (g/3.4)^{10}\}]/241$$

$$Q_4 = (2Q_1/Q_2)\,[\exp(-g)\,u^{Q_3} + \{2 - \exp(-g)\}\,u^{-Q_3}]^{-1} \qquad (6.52)$$

and the static-TEM odd mode impedance is given by

$$Z_{0o}(0) = Z_0(0) \{\epsilon_{eff}(0)/\epsilon_{effo}(0)\}^{0.5}/[1 - \{Z_0(0)/377\} \{\epsilon_{eff}(0)\}^{0.5} Q_{10}] \tag{6.53}$$

where

$$Q_5 = 1.794 + 1.14 \ln[1 + 0.638/\{g + 0.517 g^{2.43}\}]$$

$$Q_6 = 0.2305 + \ln[g^{10}/\{1 + (g/5.8)^{10}\}]/281.3 + \ln(1 + 0.598 g^{1.154})/5.1$$

$$Q_7 = (10 + 190 g^2)/(1 + 82.3 g^3)$$

$$Q_8 = \exp\{-6.5 - 0.95 \ln(g) - (g/0.15)^5\}$$

$$Q_9 = \ln(Q_7)(Q_8 + 1/16.5)$$

$$Q_{10} = Q_2^{-1}[Q_2 Q_4 - Q_5 \exp\{\ln(u) Q_6 u^{-Q_9}\}] \tag{6.54}$$

These design formulae are accurate to within 0.6 per cent over the full range of validity given here.

The frequency-dependent expressions then follow:

$$Z_{0e}(f_n) = Z_{0e}(0)[0.9408 \{\epsilon_{eff}(f_n)\}^{C_e} - 0.9603]^{Q_0}$$
$$/[(0.9408 - d_e)\{\epsilon_{eff}(f_n)\}^{C_e} - 0.9603]^{Q_0} \tag{6.55}$$

where

$$C_e = 1 + 1.275[1 - \exp\{-0.004625 p_e \epsilon_r^{1.674} (f_n/18.365)^{2.745}\}]$$
$$- Q_{12} + Q_{16} - Q_{17} + Q_{18} + Q_{20}$$

$$d_e = 5.086 q_e \{r_e/(0.3838 + 0.386 q_e)\}$$
$$\times \{\exp(-22.2 u^{1.92})/(1 + 1.2992 r_e)\}[(\epsilon_r - 1)^6/\{1 + 10(\epsilon_r - 1)^6\}]$$

$$p_e = 4.766 \exp(-3.228 u^{0.641})$$

$$q_e = 0.016 + (0.0514 \epsilon_r Q_{21})^{4.524}$$

$$r_e = (f_n/28.843)^{12} \tag{6.56}$$

and

$$Q_{11} = 0.893[1 - 0.3/\{1 + 0.7(\epsilon_r - 1)\}]$$

$$Q_{12} = 2.121[(f_n/20)^{4.91}/\{1 + Q_{11}(f_n/20)^{4.91}\}] \exp(-2.87 g) g^{0.902}$$

$$Q_{13} = 1 + 0.038(\epsilon_r/8)^{5.1}$$

$$Q_{14} = 1 + 1.203(\epsilon_r/15)^4/\{1 + (\epsilon_r/15)^4\}$$

$$Q_{15} = 1.887 \exp(-1.5 g^{0.84}) g^{Q_{14}} \{1 + 0.41(f_n/15)^3 u^{2/Q_{13}}/(0.125 + u^{1.626/Q_{13}})\}^{-1}$$

$$Q_{16} = [1 + 9/\{1 + 0.403\,(\epsilon_r - 1)^2\}]\,Q_{15}$$

$$Q_{17} = 0.394\,[1 - \exp\{-1.47\,(u/7)^{0.672}\}]\,[1 - \exp\{-4.25\,(f_n/20)^{1.87}\}]$$

$$Q_{18} = 0.61\,[1 - \exp\{-2.13\,(u/8)^{1.593}\}]/(1 + 6.544\,g^{4.17})$$

$$Q_{19} = 0.21\,g^4\,[(1 + 0.18\,g^{4.9})\,(1 + 0.1u^2)\,\{1 + (f_n/24)^3\}]^{-1}$$

$$Q_{20} = \mathrm{abs}\,\{1 - 42.54\,g^{0.133}\exp(-0.812g)\,u^{2.5}/(1 + 0.033\,u^{2.5})\} \qquad (6.57)$$

In these expressions $\epsilon_{\mathrm{eff}}\,(f_n)$ denotes the effective microstrip permittivity as a function of frequency, given in Chapter 4 here. Similarly, the appropriate design formulae applicable to the frequency-dependent odd mode impedance are

$$Z_{0o}\,(f_n) = Z_0\,(f_n) + [Z_{0o}\,(0)\,\{\epsilon_{\mathrm{effo}}\,(f_n)/\epsilon_{\mathrm{effo}}\,(0)\}^{Q_{22}}$$

$$- Z_0\,(f_n)\,Q_{23}]\,\{1 + Q_{24} + (0.46\,g)^{2.2}\,Q_{25}\}^{-1} \qquad (6.58)$$

where

$$Q_{22} = 0.925\,(f_n/Q_{26})^{1.536}/\{1 + 0.3\,(f_n/30)^{1.536}\}$$

$$Q_{23} = 1 + 0.005\,f_n\,Q_{27}\,[\{1 + 0.812\,(f_n/15)^{1.9}\}\,(1 + 0.025\,u^2)]^{-1}$$

$$Q_{24} = 2.506\,Q_{28}\,u^{0.894}\,\{(1 + 1.3u)\,f_n/99.25\}^{4.29}\,(3.575 + u^{0.894})^{-1}$$

$$Q_{25} = \{0.3\,f_n^2/(10 + f_n^2)\}\,[1 + 2.333\,(\epsilon_r - 1)^2/\{5 + (\epsilon_r - 1)^2\}]$$

$$Q_{26} = 30 - 22.2\,[[\{(\epsilon_r - 1)/13\}^{12}/1 + 3\,\{(\epsilon_r - 1)/13\}^{12}]] - Q_{29}$$

$$Q_{27} = 0.4\,g^{0.84}\,[1 + 2.5\,(\epsilon_r - 1)^{1.5}/\{5 + (\epsilon_r - 1)^{1.5}\}]$$

$$Q_{28} = 0.149\,(\epsilon_r - 1)^3/\{94.5 + 0.038\,(\epsilon_r - 1)^3\}$$

$$Q_{29} = 15.16/\{1 + 0.196\,(\epsilon_r - 1)^2\} \qquad (6.59)$$

Here again the range of accurate applicability of these expressions is the same as that defined earlier here. The maximum error is 2.5 per cent up to $f_n = 30$ GHz/mm but, if the specified maximum substrate permittivity is reduced from 18 to 12.9 (GaAs), then these expressions can be used up to $f_n = 25$ GHz/mm with the accuracy remaining guaranteed. It is noted again that this corresponds to an actual operating frequency of approximately 40 GHz in the case of a 0.635 mm thick substrate.

In design the principal sets of expressions are those for the effective microstrip permittivities since these ultimately form the major quantity in the determination of physical lengths. The frequency-dependent characteristic impedances should be used within dynamically interactive CAD routines to set the requirements for matching the termination line impedance (usually 50 Ω).

Since the determination of physical lengths is of paramount importance we also need to know semi-open end equivalent lengths for the even and

odd modes. Kirschning and Jansen give the following expressions for these equivalent lengths, beginning with the length for the even mode

$$\Delta l_e = \{\Delta l\,(2u,\,\epsilon_r) - \Delta l\,(u,\,\epsilon_r) + 0.0198\,h\,g^{R_1}\}\,\exp\,(-0.328\,g^{2.244}) + \Delta l\,(u,\,\epsilon_r)$$
(6.60)

where

$$R_1 = 1.187\,\{1 - \exp\,(-0.069\,u^{2.1})\}$$
(6.61)

and

$$\Delta l_o = \{\Delta l\,(u,\,\epsilon_r) - h\,R_3\}\,\{1 - \exp(-R_4)\} + h\,R_3$$
(6.62)

where, in this expression

$$R_2 = 0.343\,u^{0.6187} + \{0.45\,\epsilon_r/(1 + \epsilon_r)\}$$

$$\times\,u^{\{1.357\,+\,1.65/(1\,+\,0.7\,\epsilon_r)\}}$$

$$R_3 = 0.2974\,\{1 - exp\,(-R_2)\}$$

$$R_4 = (0.271 + 0.0281\,\epsilon_r)\,g^{\{1.167\,\epsilon_r/(0.66\,+\,\epsilon_r)\}}$$

$$+\,\{1.025\,\epsilon_r/(0.687 + \epsilon_r)\}\,g^{\{0.958\,\epsilon_r/(0.706\,+\,\epsilon_r)\}}$$
(6.63)

Although the accuracy of these expressions is approximately 5 per cent over the range defined above, with a slight increase also with frequency, the equivalent end-effect lengths usually only amount to a second-order contribution in determining the final physical lengths of the structure. For this reason it is expected that these expressions will be sufficiently accurate for most requirements of effective parallel-coupled microstrip designs.

6.9.3 Complete coupling section response

Once $\epsilon_{effe}(f)$ and $\epsilon_{effo}(f)$ have been evaluated then the appropriate wavelengths can be calculated in a manner similar to that employed with single microstrips. To solve for the *full* frequency response of the coupler, in microstrip form, it is necessary to return to the full voltage expressions developed in Appendix A. It is also necessary to incorporate the foregoing dispersion calculations so that the non-linearly varying wavelengths are taken into account. This is a fairly onerous task that is best set up on a computer. To get some idea of the magnitude of the calculations consider the coupling factor, as determined from eqns (A.17) and (A.21), but with θ appropriately replaced by θ_e and θ_o:

$$C = V_{1e} - V_{1o} = \frac{Z_{0e}Z_0\cos\theta_e + jZ_{0e}^2\sin\theta_e}{2Z_{0e}Z_0\cos\theta_e + j(Z_{0e}^2 + Z_{0e}^2)\sin\theta_e}$$

$$-\,\frac{Z_{0o}Z_0\cos\theta_o + jZ_{0o}^2\sin\theta_o}{2Z_{0o}Z_0\cos\theta_o + j(Z_{0o}^2 + Z_0^2)\sin\theta_o}$$
(6.63)'

6.10 COUPLER DIRECTIVITY

This quantity has already been defined as the undesirable coupling which can occur to port 4 (see Figure 1.15). The directivity D was also quoted as

$$D = \left| \frac{V_4}{V_3} \right|$$

Now, for a pure TEM-mode (ideal) coupler, eqn (A.40) shows that $V_4 = 0$ and the directivity would also be zero because V_3 is finite. With microstrip, the differing field patterns associated with the odd and even modes give rise to different phase velocities. We have already seen that this feature greatly complicates the coupling factor expression and a similar complication arises for the voltage appearing at port 4 (see Figure 1.15). In fact, whereas V_4 cancels to zero ideally, this is now far from true.

Expressions for V_3 and V_4 are obtained by using eqns (A.4) and (A.8), and also eqns (A.6) and (A.10), as follows:

$$V_3 = V_{1e} - V_{1o} = \frac{Z_{0e}Z_0 \cos \theta_e + j Z_{0e}^2 \sin \theta_e}{2 Z_{0e} Z_0 \cos \theta_e + j(Z_{0e}^2 + Z_0^2) \sin \theta_e}$$

$$- \frac{Z_{0o}Z_0 \cos \theta_o + j Z_{0o}^2 \sin \theta_o}{2 Z_{0o} Z_0 \cos \theta_o + j(Z_{0o}^2 + Z_0^2) \sin \theta_o} \tag{6.64a}$$

$$V_4 = V_{2e} - V_{2o} = \frac{Z_0 Z_{0e}}{2 Z_{0e} Z_0 \cos \theta_e + j(Z_{0e}^2 + Z_0^2) \sin \theta_e}$$

$$- \frac{Z_0 Z_{0o}}{2 Z_{0o} Z_0 \cos \theta_o + j(Z_{0o}^2 + Z_0^2) \sin \theta_o} \tag{6.64b}$$

Dividing eqn (6.64b) by eqn (6.64a) and taking the modulus yields

$$D = \left| \frac{\chi Z_0 Z_{0e} - \psi Z_0 Z_{0o}}{\chi(Z_{0e} Z_0 \cos \theta_e + j Z_{0e}^2 \sin \theta_e) - \psi(Z_{0o} Z_0 \cos \theta_o + j Z_{0o}^2 \sin \theta_o)} \right| \tag{6.65}$$

where

$$\chi = 2 Z_{0o} Z_0 \cos \theta_o + j(Z_{0o}^2 + Z_0^2) \sin \theta_o$$

and

$$\psi = 2 Z_{0e} Z_0 \cos \theta_e + j(Z_{0e}^2 + Z_0^2) \sin \theta_e$$

Calculation using these expressions is lengthy and a simpler alternative expression has been given for the directivity[10]

$$D = \left\{ \frac{4|\zeta|}{\pi \Delta (1 - |\zeta|^2)} \right\}^{-2} \tag{6.66}$$

in which

$$\zeta = \left(\frac{\rho_e}{1 + \rho_e^2}\right) - \left(\frac{\rho_o}{1 + \rho_o^2}\right)$$

$$\rho_e = \frac{Z_{0e} - Z_0}{Z_{0e} + Z_0}$$

$$\rho_o = \frac{Z_{0o} - Z_0}{Z_{0o} + Z_0}$$

and

$$\Delta = \frac{\lambda_{go}}{\lambda_{ge}} - 1 \qquad\qquad (6.67)$$

Equation (6.66) clearly demonstrates that, when $\lambda_{go} = \lambda_{ge}$, the directivity goes to infinity.

Values much better than 12 or 14 dB are difficult to achieve for directivity with this basic form of microstrip coupler. (The basic structure is still very useful in filter design and as a starting point for couplers of improved performance.) We shall now examine some techniques aiming at:

(a) equalizing, or approximately equalizing, the phase velocities so that the *directivity* (at least) is improved;
(b) modifying the overall coupler structure so that the frequency response is made much flatter, i.e. the *bandwidth* is greatly increased; and, finally,
(c) completely different structures where, for example, coupling factors much greater than -3 dB can be achieved (-3 dB is approaching the limit with parallel-coupled microstrips).

6.11 SPECIAL COUPLER DESIGNS WITH IMPROVED PERFORMANCE

6.11.1 The 'Lange' coupler

This structure owes its name to its inventor Julius Lange,[11] who first reported the strictly empirical concepts and development in 1969. Lange's aim was to produce a nearly octave-bandwidth coupler having a coupling factor of around -3 dB. In his design, true quadrature coupling over an octave is realized as a consequence of the interdigital coupling section *which compensates for even- and odd-mode phase velocity dispersion over the wide frequency range*. It is necessary to bond directly, via very short wires, between transversely interposed 'fingers' of the coupler. In this way power is coupled optimally to the desired port, as well as the remaining power

(after losses) still being fed to a direct port. The arrangement is illustrated in Figure 6.11.

The bonding wires should look, electrically, as close as possible to short circuits—or at least very small lumped inductances. This means that their lengths l_s must be kept as short as possible: $l_s \ll \lambda_{gm}/4$, where λ_{gm} is the mid-band wavelength. Notice that the input-to-direct output link meanders through the structure.

The physically short finger elements are approximately $\lambda_g/4$ at the highest frequency in the band desired, whereas the entire length within the central region of the coupler is approximately $\lambda_g/4$ at the lowest frequency of the band desired. These features, together with properly phased travelling-wave reinforcement due to the bonding wires, give the excellent performance. One really significant (and often important) characteristic of this device is the very good relative phase shift over the band. This is generally better than $\pm 2°$ on the nominal $90°$ phase shift. Also, the VSWR is generally less than 1.15 over an octave bandwidth.

In practice, especially at higher frequencies, several bonding wires are required in parallel connection in place of each single one shown in Figure 6.11 in order to reduce inductance and therefore enhance performance. This results in extra manufacturing expense, since the wires must be judiciously bonded in *by hand*. The spacings are about 75 μm, compared with only 10 μm for the simple, parallel, microstrip -3 dB coupler (on alumina).

An early design procedure was reported by Presser.[12] A coupling factor design repeatability of ± 0.05 dB was generally obtained and a gap-variation sensitivity of approximately 0.03 dB/μm was observed. This means that a ± 2.5 μm tolerance on the coupling gap would yield approximately ± 0.1 dB

Fig. 6.11. Basic form of the 'Lange' coupler. (© 1969 IEEE. Reprinted, with permission, from J. Lange, Interdigitated strip-line quadrature hybrid, *IEEE Trans.*, **MTT-17**, December 1969, 1150–1151.)

coupling factor variation. (Parallel-coupled microstrip tolerances are dis-
cussed in more detail in Section 6.12.3.) The couplers designed and measured
by Presser showed a minimum directivity of approximately -24 dB,
corresponding to an isolation of about -30 dB.

A design procedure, together with recommendations, is now given for
Lange couplers. This follows the work of Ou,[13] and also Osmani,[14] whose
approach has previously been consolidated and described by Fusco.[15]

Ou developed the following elegant result:

$$Z_0^2 = \frac{Z_{0e} Z_{0o} (Z_{0e} + Z_{0o})^2}{\{Z_{0e} + (k - 1) Z_{0o}\} \{Z_{0o} + (k - 1) Z_{0e}\}} \tag{6.68}$$

where k, the number of lines within the coupler, has to be an even integer.

A further important result derived by Ou is that the voltage coupling
ratio may be expressed as:

$$C = 10^{-\{(\text{voltage coupling ratio in dB}/20\}} = \frac{(k - 1) Z_{0e}^2 - (k - 1) Z_{0o}^2}{(k - 1) (Z_{0e}^2 + Z_{0o}^2) + 2 Z_{0e} Z_{0o}} \tag{6.69}$$

As described previously here it is essential to know the even- and odd-mode
characteristic impedances in order to synthesize the coupled microstrip cross-
section. Osmani showed that, by combining the simple expression in terms
of C: $(1 + C)/(1 - C)$, with the above expression for C and multiplying the
result by the expression for Z_0^2, Z_{0e} and Z_{0o} can be determined as follows:

$$Z_{0e} = Z_{0o} \frac{C + q}{(k - 1) (1 - C)} \tag{6.70}$$

and

$$Z_{0o} = Z_0 \left(\frac{1 - C}{1 + C}\right)^{\frac{1}{2}} \frac{(k - 1) (1 + q)}{(C + q) + (k - 1) (1 - C)} \tag{6.71}$$

where

$$q = \{C^2 + (1 - C^2) (k - 1)^2\} \tag{6.72}$$

In a Lange coupler the even-mode and odd-mode phase velocities are ideally
equal and this is approximately the case in practice. Therefore the overall
length, which must be close to a quarter of a wavelength, can be based
upon either phase velocity. The inherently broadband performance of this
type of coupler means that this overall length must actually be a quarter of
a wavelength calculated at the lowest frequency in the band under
consideration (e.g. 10 GHz for a 10–40 GHz coupler).

By contrast the finger length must, for the same reason of maintaining
broadband operation, be set to one quarter wavelength calculated at the
highest frequency in the band under consideration (e.g. 40 GHz for a
10–40 GHz coupler).

However, largely capacitive end-effect discontinuities tend to slightly reduce the actual physical lengths in each case and the expressions given in Section 6.9.2 should be used to estimate the corrections required as a result. The bond wires or other direct connection links also have an effect on the design values but, as explained earlier, by paralleling several bonds the overall inductance is greatly reduced.

Two examples will now be given showing the results of using these expressions as first steps in the design of Lange couplers.

Example 1

Assume that we are required to design a four-finger Lange coupler to provide 10 dB of coupling in a 50 Ω system operating at 10 GHz. In this case a soft plastic substrate is to be used with a thickness of 0.234 mm and a permittivity of 2.23.

We first note that, since $C = 10^{-(10/20)}$, $C = 0.3162$. This, together with the specification of 4 fingers, leads to q:

$$q^2 = [0.3162^2 + (1 - 0.3162^2)(4 - 1)^2]$$

which yields $q = 2.864$

Then, calculating Z_{0o} next:

$$Z_{0o} = 50 \left(\frac{1 - 0.3162}{1 + 0.3162}\right)^{\frac{1}{4}} \left\{ \frac{3(3.864)}{(0.3162 + 2.864) + 3(1 - 0.3162)} \right\}$$

yielding $Z_{0o} = 80 \ \Omega$.

We now have sufficient quantities to determine Z_{0e}:

$$Z_{0e} = 80 \left\{ \frac{0.3162 + 2.864}{3(1 - 0.3162)} \right\}$$

yielding $Z_{0e} = 124 \ \Omega$

Applying the coupled line synthesis approach as described in Section 6.6. here we find that these impedances are satisfied by a structure whose cross-sectional parameters are:

$$w/h = 1.103 \qquad \text{and} \qquad s/h = 0.39$$

Since the substrate thickness is 0.254 mm, we require coupled microstrips having widths of 0.28 mm and separations (gaps) of approximately 0.1 mm.

We now consider briefly the contrasting situation if we had, instead of choosing a Lange coupler, attempted to design this coupler as a single, simple parallel-coupled element. In this case, using the design procedure

described earlier in this chapter, we would require the following cross-section parameters:

$$w/h = 2.335 \qquad \text{and} \qquad s/h = 0.095$$

These parameters yield actual width and gap values as follows:

$$w = 0.593 \text{ mm} \qquad \text{and} \qquad s = 0.024 \text{ mm}$$

The width is just over twice and the gap is approximately one quarter the value found for the Lange coupler. The difference in width requirement is not generally serious although it does have implications for 'real estate' and infers ultimate operating frequency limitations. However, the much smaller gap results in very significant manufacturing and operational problems. In particular, the small gap (only 24 μm) leads to difficulties for manufacturing tolerances, repeatability and yield. Also, even if these difficulties are largely overcome, there remains the limitation of voltage handling capability.

The final design requirements, centring on the calculation of the coupled microstrip lengths, must follow the use of expressions such as those given earlier here (Kirschning and Jansen[9]). In this way the physical lengths are determined but this is left as an exercise for the reader.

Example 2

For this example only the main specification and the resulting design values are given here.

In Example 1 we considered a 10 dB coupler, but 3 dB couplers are commonly required for use in balanced amplifiers, attenuators and phase shifters. Balanced amplifiers are considered in more detail in Chapter 9. Let us assume we require an octave-bandwidth Lange coupler to be built on an alumina substrate 0.5 mm thick. Experience in using the above design expressions shows that a good first guess for the design iteration is to set the width of the coupled microstrips 38 μm and the separating gap 25 μm.

Many practical amplifier applications demand a bandwidth much greater than one octave—often nearer two octaves or more. A significant increase in bandwidth can be obtained by slightly overcoupling the coupler. In this example we can transform the design into one covering two octaves by reducing the width and separation by about 30 per cent and 40 per cent respectively (again as first approximations at the start of a design and simulation routine). This of course results in $w = 25$ μm and $s = 15$ μm.

Waugh[16] has reported a sensitivity analysis of the Lange coupler. He points out that, conventionally, designers impose ± 0.25 μm tolerances on their designs to ensure that specifications are met. This, however, leads to a corresponding deterioration in yield of the MICs. Waugh shows that for a 2.7 dB octave-bandwidth coupler designed on 0.635 mm alumina

dimensional errors of *6 μm* in both gap and line widths *can* still enable the coupler to function according to specification.

Most of the major microwave CAEE packages, including Eesof's Touchstone and the SuperCompact software, embody Lange coupler models.

Designs using multiconductor configurations yielding coupling factors as tight as 1.5 dB have also been described.[17] Quadrature-coupling arrangements using these principles are particularly significant in some microwave transistor amplifier circuits and 2–18 GHz bandwidths are obtainable.[18]

6.11.2 The 'unfolded Lange' coupler

This is a development of the original Lange coupler just described and was reported by Waugh and LaCombe in 1972.[19] In Figure 6.12, which shows this arrangement, it can be seen that some line elements and two straps are saved. The two straps are saved essentially by avoiding the strapped feed across to the direct port and, instead, retaining a 'straight-through' direct port connection. Slightly less coupling is obtained, but the most serious trade-off appears to be the phase-shift variation over the band. This now has a worst value of −8° departure away from the nominal 90°. The fact of this phase departure and the lack of a d.c. path between input and opposite port have probably represented the main reasons why this version of the Lange coupler seems to have been largely avoided.

Both forms of 'Lange' coupler require the connection of short wire straps, and this can pose repeatability problems in production (although there *may*, of course, be other short wires associated with active devices). For this reason, among others, alternative techniques have been developed.

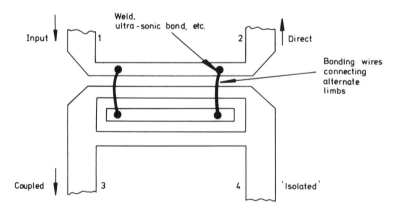

Fig. 6.12. The 'unfolded' Lange coupler. (© 1972 IEEE. Reprinted, with permission, from Waugh and LaCombe.[19])

6.11.3 Shielded parallel-coupled microstrips

This type of structure also tends to nearly equalize even- and odd-mode phase velocities. The geometry is shown in the drawing of Figure 6.13.

In this arrangement there is a considerable proportion of the total field in the air region above the coupler, whatever the mode. Thus the *proportional* redistribution of electric field into air space for the odd mode can be made much less than that for the unshielded structure. As a consequence the phase velocities can easily be within approximately 1 per cent of one another, as shown by Weiss.[20]

In order to obtain these advantages it is necessary for the shield-spacing d to be around or less than the substrate height h:

$$d \leq h$$

Although closely shielded circuits impose some manufacturing and other difficulties, they do yield certain specific design advantages because the structures are electrically well defined.

Several workers have reported studies on parallel-coupled microstrips including shielded structures and multiconductor geometries. For example, Kouki *et al.*[21] describe the finite element modelling of such structures analysed on a two-dimensional basis using an asymptotic boundary condition. Their method can accept conductors with arbitrary cross-sections in addition to arbitrarily inhomogeneous dielectrics (e.g. multilayer substrates). They demonstrate capacitance matrix results for shielded coupled pairs as well as six-conductor arrays of microstrips.

6.11.4 The use of a dielectric overlay

In this approach a layer of substrate-type dielectric material is fixed in the local region above the coupled (conventional) microstrip lines. The fixing is usually carried out with stycast epoxy, and the result is shown in Figure 6.14. Some very effective results can be achieved when the thickness of the overlay is only about 50 μm on an alumina substrate of 0.5–0.65 mm thickness.

The main principle is that electric field in the region immediately above the coupling gap experiences a permittivity similar to that of the substrate

Fig. 6.13. Parallel-coupled microstrips with a grounded shielding top plate

Fig. 6.14. Parallel-coupled microstrips with overlaid dielectric compensation. (Reproduced by permission of the Institution of Electrical Engineers from Shelad and Spielman.[22])

(instead of air when no overlay is present). This is most effective for the odd mode, when the field in the gap region is greatest. Thus the odd-mode phase velocity is reduced to approach, within a few per cent, the value of the even-mode phase velocity. If the overlay thickness is made approximately the same as that of the substrate and it overlaps the outer edges of the microstrips by at least half a strip width, then the phase velocity disparity is less than 1 per cent.

It is likely that production repeatability problems would be experienced with this technique, and epoxy bonding of the overlay is difficult to keep free of air bubbles. However, some excellent results have been reported for a -10 dB coupler.

6.11.5 The incorporation of lumped capacitors

It has been shown that the connection of lumped capacitors across a pair of coupled microstrip lines, as indicated schematically in Figure 6.15, will equalize the phase velocities over quite a broad band.[23]

An accurate method for the determination of these compensating capacitances, applicable over a wide range of coupling, has been derived by Dydyk.[24] In his approach Dydyk uses the standard *ABCD* matrix to represent both the even and odd modes for this structure, noting that the capacitance is only effective for the odd mode. The matrices therefore take the forms

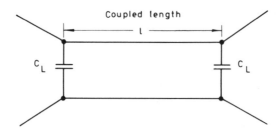

Fig. 6.15. Schematic illustration of parallel-coupled lines with lumped capacitors bridging the ends to provide compensation

$$\begin{bmatrix} A & B \\ C & D \end{bmatrix}_e = \begin{bmatrix} \cos \theta_e & j Z_{0e} \sin \theta_e \\ j Y_{0e} \sin \theta_e & \cos \theta_e \end{bmatrix} \qquad (6.73)$$

and

$$\begin{bmatrix} A & B \\ C & D \end{bmatrix}_o = \begin{bmatrix} (\cos \theta_o - Z_{0o} 2 \omega C \sin \theta_o) \\ j \{4 \omega C \cos \theta_o + (Y_{0o} - Z_{0o} (2 \omega C)^2 \sin \theta_o\} \end{bmatrix}$$

$$\begin{matrix} j Z_{0o} \sin \theta_o \\ (\cos \theta_o - 2\omega C Z_{0o} \sin \theta_o) \end{matrix} \Bigg] \qquad (6.74)$$

Note is then taken of the fact that, in the ideal case of equal even and odd mode electrical lengths, the $ABCD$ matrix for this case becomes

$$\begin{bmatrix} A & B \\ C & D \end{bmatrix}_{oi} = \begin{bmatrix} \cos \theta_e & j Z_{0oi} \sin \theta_e \\ j Y_{0oi} \sin \theta_e & \cos \theta_e \end{bmatrix} \qquad (6.75)$$

where Z_{0oi} and Y_{0oi} are respectively the odd-mode characteristic impedance and admittance in this ideal case.

This leads to the required capacitance by first equating the A and B element entries in the odd-mode matrices, giving

$$\cos \theta_e = \cos \theta_o - 2 \omega C_L Z_{0o} \sin \theta_o \qquad (6.76)$$

$$Z_{0o} \sin \theta_o = Z_{0oi} \sin \theta_e \qquad (6.77)$$

Combining these expressions and recognizing that the even mode electrical length is 90° at the centre frequency leads simply to the desired formula

$$C_L = \cos \theta_o / 2 \omega Z_{0oi} \qquad (6.78)$$

The electrical length can also be written in terms of the effective microstrip permittivities

$$\theta_o = \frac{2 \pi l}{\lambda g_o} = \frac{\pi}{2} \sqrt{\frac{\epsilon_{effo}}{\epsilon_{effei}}} \qquad (6.79)$$

so that the compensating capacitance, as a function of frequency, is given by

$$C_L = \frac{\cos \left(\dfrac{\pi}{2} \sqrt{\dfrac{\epsilon_{effo}}{\epsilon_{effei}}} \right)}{2 Z_{0oi} \omega} \qquad (6.80)$$

From the foregoing considerations it also follows that the actual odd mode characteristic impedance may be expressed as

$$Z_{0oa} = \frac{Z_{0oi}}{\sqrt{1 - \left\{ \cos \left(\dfrac{\pi}{2} \sqrt{\dfrac{\epsilon_{effo}}{\epsilon_{effe}}} \right) \right\}^2}} \qquad (6.81)$$

indicating that Z_{0oa} must in general be greater than Z_{0oi}. This in turn implies that the conductors must be set narrower than initially calculated (before compensation) and the separating gap must be increased. Both requirements are completely acceptable in practice.

Dydyk also shows that at the coupler's centre frequency the $ABCD$ matrices reduce to

$$\begin{bmatrix} A & B \\ C & D \end{bmatrix}_e = \begin{bmatrix} 0 & j\,Z_{0e} \\ j\,Y_{0e} & 0 \end{bmatrix} \qquad (6.82)$$

and

$$\begin{bmatrix} A & B \\ C & D \end{bmatrix}_o = \begin{bmatrix} 0 & j\,Z_{0oi} \\ j\,Y_{0oi} & 0 \end{bmatrix} \qquad (6.83)$$

which represents a situation identical to that of an ideal directional coupler having the following scattering matrix:

$$[S] = \begin{bmatrix} 0 & C & -j\sqrt{1-C^2} & 0 \\ C & 0 & 0 & -j\sqrt{1-C^2} \\ -j\sqrt{1-C^2} & 0 & 0 & C \\ 0 & -j\sqrt{1-C^2} & C & 0 \end{bmatrix} \qquad (6.84)$$

where C is related to Z_{0e}, Z_{0o} and Z_0 as defined earlier in this chapter.

From these results it should be clear that the isolation is ideally zero and the directivity infinite in this case. In practice, due to losses and with the frequency departed from the design centre, the actual directivity will degrade.

As an example of the application of this theory to a practical design of a coupled section on an alumina substrate it is found that we require a compensating capacitance of 0.047 pF, an actual odd mode impedance of 31.7 Ω and a coupled region length of just over 2.3 mm. Dydyk applies Eesof's 'Linecalc' software routine to both uncompensated and compensated coupled sections (both having nominally 7 dB coupling) to generate the data repeated here in Table 6.1.

Rickard[23] has shown that a coupling factor variation of only ±0.3 dB (on −10 dB) is achievable over a two-and-a-half-octave bandwidth: 8−20 GHz. The directivity was 20 dB over this band, compared with 5−13 dB uncompensated, and the VSWR was always better than 1.4. No information was reported on the phase-shift variation over the band.

This performance considerably exceeds that of a comparable coupler using a dielectric overlay. However, it may be difficult to implement this approach with fairly large coupling factors (e.g. −3 dB or greater) because of the already very narrow coupling gap. At low frequencies the capacitors might be formed either using a thin-film add-on technique as described in Chapter

Table 6.1. The data provided in this table are due to Dydyk.[24]

Parameter	Uncompensated	Compensated
Centre frequency (GHz)	12.0	12.0
Coupling C(dB)		
Desired	7.0	7.0
Achieved	7.25	7.0
Z_0 (Ω)	50	50
Z_{0e} (Ω)	80.85	80.85
Z_{0o} (Ω)	30.92	31.71
ϵ_{effe}	7.20	7.20
ϵ_{effo}	5.31	5.32
w (mm)	0.444	0.440
s (mm)	0.081	0.089
l (mm)	2.505	2.327
Directivity (dB)		
Desired	Infinite	Infinite
Achievable	13.25 dB	Essentially infinite

2 or conceivably as chip components. At high frequencies the capacitors could simply be interdigital structures at the coupler ends.

6.11.6 The effect of a dielectrically anisotropic substrate

Provided the crystal axes of a dielectrically anisotropic substrate are correctly aligned with respect to the microstrip then the net electric displacements are similar for both even and odd modes. Thus the phase velocities will approach equality, depending upon the actual values of the permittivity tensor members.

Szentkuti[25] has shown that the degree of anisotropy of sapphire (see Chapter 3) and of some other readily available substrate materials is not sufficient to produce a really significant improvement. *If* a material having permittivity values of 25 and 10 were available then Szentkuti shows that the velocities differ by less than 1 per cent over the range:

$$0.05 \leqslant \frac{w}{h} \leqslant 1.2$$

where $s/h = 0.05$.

Also, unfortunately, in the case of sapphire at least, the crystal axis alignment would mean that the wave would experience a very complicated permittivity environment, changing in the direction of propagation and dependent on the actual location on the substrate.

6.11.7 Microstrip multiplexers

Various planar or quasi-planar technologies have been adopted to facilitate the function of multiplexing—the separation of a broadband input signal into a number of contiguous channels. Technological approaches to this problem at microwave frequencies include both microstrip and also suspended-substrate stripline.

An example of a microstrip quadruplexer has been reported by Uysal et al.[26] Their device comprises four 3 dB directional couplers and four phase shifting networks, as indicated in Figure 6.16. The design is critically dependent upon differing electrical lengths for the microstrip lines connected to the two intermediate 3 dB couplers as shown in Figure 6.16. Uysal described an experimental quadruplexer designed to split a 2–18 GHz signal into four contiguous output channels each 4 GHz wide. The design was realized in MIC form using a standard 25 mm × 50 mm alumina substrate. Some empirical adjustment of the phase lengths is indicated and it appears clear that there is room for further advances in this design approach.

6.11.8 Multi-section couplers

Because they tend to occupy considerable substrate area ('real estate'), particularly in terms of length, multisection couplers have generally not found much favour amongst solid-state microwave designers. However, at least one novel approach to the design of this type of coupler has been reported by Nakajima and co-workers.[27] This team described a 2–18 GHz

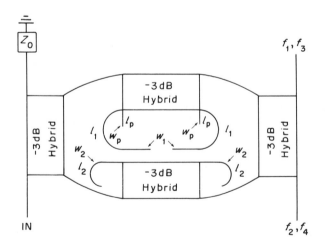

Fig. 6.16. Schematic diagram of a quadruplexer using ×4, 3 dB hybrids. (Reprinted, with permission, from Uysal, Aghrami and Mohamed.[26])

multisection coupler providing 6 dB of coupling using five semi-re-entrant tight-coupling sections.

A single section of such a coupler structure is shown in plan-schematic form in Figure 6.17 and in cross-section in Figure 6.18. A plan-schematic diagram of the entire five-section coupler is shown in Figure 6.19. Nakajima

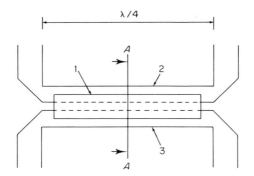

Fig. 6.17. A plan schematic view of a single section semi-re-entrant coupler. (© 1990 IEEE. Reprinted, with permission, from Nakajima, Yamashita and Asa.[27])

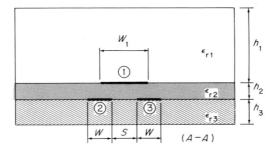

Fig. 6.18. A cross-sectional view of a semi-re-entrant coupled section. (© 1990 IEEE. Reprinted, with permission, from Nakajima, Yamashita and Asa.[27])

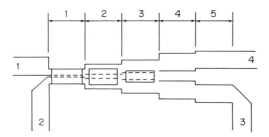

Fig. 6.19. The plan schematic view of a 5-section directional coupler. (© 1990 IEEE. Reprinted, with permission, from Nakajima, Yamashita and Asa.[27])

et al. use a capacitance matrix approach and provide a table showing all the design parameters for the three semi-re-entrant coupled sections and the two parallel (edge) coupled sections. All the physical dimensions are fully realizable—the first and narrowest coupling gap is 0.08 mm and the narrowest width is 0.23 mm for section three.

The coupler was fabricated on a Duroid 5880 substrate of nominally 0.508 mm thickness. Relative through-arm power transfer remained about 2 dB whilst the coupled arm delivered power around 6 dB down on the incident power (with approximately 2 dB of ripple).

6.11.9 Re-entrant mode couplers

The types of fine-line geometries and interdigital structures described in earlier sections of this chapter lead to design and fabrication difficulties where tight coupling is required. One method of overcoming some of these difficulties is to consider combline approaches (dealt with in Section 6.13) but very fine interdigital geometries, with their attendant production difficulties, remain.

The re-entrant concept is not new since Seymour Cohn originally described it for realisation in coaxial form and later Lavendol showed how it could be adapted for stripline technology. More recently Pavio and Sutton[28] have described a quadrature coupler using this approach. Pavio and Sutton's structure, shown in cross-section in Figure 6.20, comprises a subsidiary parallel-coupled line arrangement with a 'floating under-plane' of restricted width w_1. This subsidiary arrangement is layered on top of a conventional alumina or gallium arsenide substrate with a ground plane.

Several types of circuits are fabricated by Pavio and Sutton using this technology. In one case a 3 dB coupler was fabricated on a 0.381 mm thick alumina substrate and a polyamide film, 0.0075 mm thick and of permittivity 3.7, was used for the subsidiary arrangement. Capacitive compensation was incorporated, as described in Section 6.11.5 of this chapter, and was realized in the form of interdigitated capacitors. Although transmission loss was similar to that of a conventional coupler, the VSWR was superior. The phase response varied in the range 82°–87° over the 5–19 GHz band, indicating that further work is required to more closely phase-linearize these types of devices. These workers also described a 180° balun designed according to these principles (baluns are described in Chapter 8).

6.11.10 Patch couplers

At the higher millimeter-wave frequencies (high tens into hundreds of GHz) none of the foregoing coupler technologies are particularly practical on account of the extremely small overall dimensions. Many designers therefore adopt technologies other than microwave (e.g. fin-line, image line).

Fig. 6.20. Microstrip re-entrant mode coupler cross-section. (© 1990 IEEE. Reprinted, with permission, from Pavio and Sutton.[28])

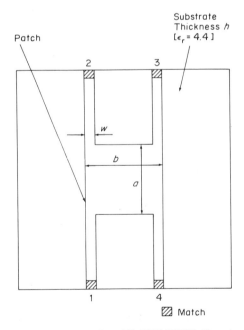

Fig. 6.21. Geometry of a patch coupler. (© 1990 IEEE. Reprinted, with permission, from Fusco and Merugu.[29])

It is, however, possible to consider special forms of microstrip patch coupler at these high frequencies and Fusco and Meruga[29] have described such a device, shown in schematic plan view in Figure 6.21. These workers describe a three-dimensional wave analysis of the coupler using a TLM numerical simulation to develop amplitude and phase responses over the frequency range 80–94 GHz. With this technique field distribution and radiation characteristics can also be identified.

6.12 THICKNESS EFFECTS, POWER LOSSES, AND FABRICATION TOLERANCES

6.12.1 Thickness effects

So far in this chapter we have neglected the effects due to the finite thickness of the microstrip lines.

Thickness effects play only a small part in the accurate design of coupled lines, provided the lines are of thick- or thin-film manufacture and the coupling is around or less than -10 dB. Rizzoli[30] has found that thick-film strips taper at the edges, thus allowing 'thin-strip' calculations to be effective.

The nominal spacing for a -3 dB coupler on alumina is about 10 μm. Therefore the effect of, say, 5 μm thickness 'side walls' must be very significant in such a case, especially where the side walls approach a vertical disposition. This is now a considerably complicated structure to analyse theoretically.

For design corrections it seems sensible to introduce modifications taking into account strip thickness and enabling the expressions of Section 6.5 to be used. With the thickness t we require a modified form of eqn (6.16) for $C_f(t)$:

$$C_f(t) = 0.5 \left\{ \frac{\sqrt{\epsilon_{\text{eff}}(t)}}{cZ_0(t)} - C_p \right\} \tag{6.85}$$

where $\epsilon_{\text{eff}}(t)$ and $Z_0(t)$ are for single microstrip lines and are given in Chapter 3 (eqns 3.19 to 3.25, with eqn 3.54 and also eqns 3.52a to 3.53b).

An extra term is also required in eqn (6.14) to account for the increased gap capacitance, which is entirely due to electric field in air. This is

$$C_{gt} = 2\epsilon_0 \frac{t}{s} \tag{6.86}$$

These expressions, eqns (6.85) and (6.86), were originally given by Gupta et al.[31]

It is found that, for 'thin' microstrip on alumina, the odd-mode characteristic impedance decreases by 2 or 3 per cent due to the thickness effect. The change is negligible for the even mode.

6.12.2 Power losses

In coupled microstrip lines there are two principal sources of power loss: one due to the conductors and the other due to the dielectric. Garg and Bahl[2] have reported expressions for both the conductor and dielectric losses and these expressions are now given here, together with extra information which enables the effects of surface roughness to be determined (for conductor loss, see Chapter 4 for *single*-line details):

$$\alpha_{c,i} = \left[\frac{8.686 R_s}{120 \pi Z_{0i} hc\, \{C_{ai}(t)\}^2} \right]$$

$$\times \left\{ \frac{\partial C_{ai}(t)}{\partial(w/h)} \left(1 + \frac{w}{h}\right) - \frac{\partial C_{ai}(t)}{\partial(s/h)} \left(1 - \frac{s}{h}\right) + \frac{\partial C_{ai}(t)}{\partial(t/h)} \left(1 + \frac{t}{h}\right) \right\} \quad (6.87)$$

$$\text{dB/unit length}$$

where this expression holds for the strips and ground plane combined, and

$C_{ai}(t)$ = odd- or even-mode line capacitance with air as the dielectric
R_s = sheet resistivity (ohms per square metre) of the metallization
i = o,e—for the odd or even mode

The attenuation α_{co} due to the odd mode is always greater than that due to the even mode, and the sensitivity of α_{co} to changes in the spacing s is also greatest. This is consistent with a physical interpretation where it must be remembered that interruptions are then (for the odd mode) being forced upon *opposing-edge* current components.

The loss for each mode is first calculated separately; then the total loss is determined as the average of these two results.

It should be noted that eqn (6.87) yields a rather low estimate of the conductor loss in practical situations. Compare, for example, the information supplied by Jansen.[32] With most plastics the discrepancy is very significant (several decibels) and there is a noticeable difference even when a polished sapphire substrate is used. Surface roughness is the dominant effect, although the precise nature of the conductor material is also very important. We previously considered these effects in Chapter 4, and a modifying expression (eqn 4.56) was given to improve the prediction of conductor loss. Exactly the same expression, with identical precautions, should be used to modify eqn (6.87) for coupled-line conductor losses.

Dielectric loss is almost always much less than conductor loss, due to the low-loss substrates that are generally used. The dielectric loss is given by[31]

$$\alpha_{d,i} = 27.3 \frac{\epsilon_r(\epsilon_{effi} - 1) \tan \delta}{\sqrt{\epsilon_{effi}}(\epsilon_r - 1) \lambda_0} \text{ dB/unit length} \quad (6.88)$$

where $\tan \delta$ is the loss tangent of the dielectric material and λ_0 is the free-space wavelength.

In X-band (8–12 GHz) we typically achieve $\alpha_d \simeq 1.2$ dB/m using an alumina substrate with a dielectric loss tangent of 0.0005. On the other hand, the *conductor* loss is approximately 8 dB/m for moderate line spacings, increasing to 20 dB/m or more for close spacing.

6.12.3 Effects of fabrication tolerances

This has been investigated by several workers, including Shamasundra and Gupta.[33] We are primarily interested in the maximum change in the coupling factor

$$\frac{|\Delta C|_{max}}{C} = \left|\frac{\Delta w}{w}S_{C(w)}\right| + \left|\frac{\Delta h}{h}S_{C(h)}\right| + \left|\frac{\Delta s}{s}S_{C(s)}\right| + \left|\frac{\Delta \epsilon_r}{\epsilon_r}S_{C(\epsilon_r)}\right| \qquad (6.89)$$

where C has been previously defined as

$$C = \frac{Z_{0e} - Z_{0o}}{Z_{0e} + Z_{0o}}$$

The sensitivity terms are defined by

$$S_{C(p)} = \frac{p}{C}\frac{\partial C}{\partial p} \qquad (6.90)$$

where C is the quantity whose sensitivity is to be determined, due to a change in a controlling parameter p.

A complete calculation of any sensitivity is actually quite a lengthy procedure. Complete functions giving the coupling factor have to be written down, as far as possible, using expressions given in Section 6.5. The derivative has then to be determined, as required, for eqn (6.90). For coupled microstrip lines on alumina a 4 per cent change in coupling factor will generally result from the following tolerance excursions:

$$\Delta s = \pm 2.5 \ \mu m$$
$$\Delta w = \pm 2.5 \ \mu m$$
$$\Delta h = \pm 25 \ \mu m \ (\text{on } h = 0.65 \ mm)$$
$$\Delta \epsilon_r = \pm 0.25 \ (\text{on } \epsilon_r = 9.7)$$

Comparatively large variations tend to occur in the substrate thickness h, and this usually influences the coupling factor C to a greater extent than other parameter variations.

For relatively tight coupling, certainly with $s/h < 0.02$, tolerance excursions in s influence C to an increasing extent until these excursions begin to actually dominate.

It is useful and interesting to note some figures which apply to a nominally -6 dB coupler manufactured on alumina. A 2.54 per cent variation in coupling has been quoted due to tolerances on various parameters.[33] The

principal contributions to this variation are Δh (1.39 per cent) and Δs (1.01 per cent). This weighting remains approximately the same for somewhat lower coupling factors, i.e. as s is increased, but the *total* percentage change rises to just over 6 per cent.[33]

Compared with the coupling factor, the VSWR is very much less sensitive to variations in the physical parameters; the main controlling parameters are ϵ_r and h. Over a wide range of coupling factors the VSWR remains close to a constant value, which generally lies between 1.2 and 1.4. The sensitivity of the VSWR only increases significantly when the coupling becomes very tight—around -3 dB or more. As previously discussed, different types of coupling structure are usually employed for such values.

6.13 PLANAR COMBLINE DIRECTIONAL COUPLERS

All of the coupling arrangements *so far* described here have three identifiable characteristics:

(a) They are 'backward-wave' devices, i.e. the coupled wave travels in the opposite direction to the input wave.
(b) They are all basically quadrature couplers, i.e. the coupled wave is 90° phase displaced, or close to that, with respect to the input wave.
(c) The maximum achievable coupling factor is in the region of -1.5 dB. Even this requires special multiconductor configurations. It is imposed by close tolerance requirements leading to poor reproducibility.

These backward-wave microstrip couplers suit a wide variety of circuit applications ranging from parallel-coupled filters to balanced transistor amplifiers. However, there are some applications, such as mixer circuits, where a 180° coupler phase shift is required.

Although not strictly a microstrip arrangement, planar combline couplers[34] are briefly considered here because they do yield 180° phase shifts and possess especially tight coupling. Planar combine directional couplers offer *forward* coupling of very large amounts; in fact approaching 0 dB can be achieved at X-band frequencies and above without any serious tolerance problems. A diagram of such a coupler is shown in Figure 6.22.

This arrangement may be regarded as a system of two interpenetrating comb transmission lines. As a coupler it derives its properties from the predominantly capacitive coupling between the two lines which arises from the interpenetration of the transverse 'finger' elements. For fuller details of the device properties, design, and performance the references given should be consulted.

Rigg and Carroll[35] have given a detailed design procedure for a general n-line 'herringbone' coupler, the fundamental structure of which is similar to the combline. They applied the design procedure to successfully realize

Fig. 6.22. The general layout of a planar combline directional coupler. (Reproduced from D. J. Gunton, *Electronic Letters,* **11**, 24, 607–608, by permission of the Controller, HMSO.)

a three-line equal power-splitting balun with a 3 dB bandwidth of three octaves centred on 1.8 GHz, a 180° phase relationship, and a linear phase variation of 100°/GHz across the band.

An analysis and design technique for combline couplers has been reported by Islam.[36]

6.14 CROSSTALK AND SIGNAL DISTORTION BETWEEN MICROSTRIP LINES USED IN DIGITAL SYSTEMS

As remarked in Chapter 3, microstrip lines are being used increasingly in electronic logic circuitry such as that involved in moderate- to high-speed digital systems. High-density packaging and short rise-time pulses are common in such systems and thus the problem of crosstalk 'noise' can be considerable.

Inevitably there will exist microstrip lines which are intended to operate, ideally, as if they were isolated from other microstrips on the same substrate. In practice, significant coupling ('crosstalk') can exist even when the lines are separated by several strip widths. We are therefore still concerned with a structure having the same general geometry shown in Figure 6.1, except that the separation s is now usually much greater than we have so far encountered. All the relationships developed for this structure still hold here. The nomenclature is the same as that used in Appendix A and the lines are terminated in the characteristic impedance: $Z_0 = \sqrt{Z_{0e} Z_{0o}}$.

It is necessary to evaluate the maximum fast crosstalk, as a percentage, in order to design effective digital systems using microstrip transmission lines. In most digital subsystems an epoxy-glass reinforced circuit board is used which will form the microstrip substrate; it has a permittivity of approximately 4.6. Some careful measurements have been performed by Catt[37] with adjacent microstrip lines on such a substrate, and a family of curves resulting from those measurements are given here as Figure 6.23. A few points calculated using eqn (A.33) are also given on the graph. Equation (A.33) is repeated here for convenience:

Fig. 6.23. Maximum fast crosstalk (per cent) as a function of separation s for a structure having the dimensions shown on the inset diagram. (The points indicate values computed using a program based upon the work in Ref. 1 where $\triangle \rightarrow w/h = 2.5$, $\square \rightarrow w/h = 0.625$.) (Reproduced by permission of Dr I. Catt,[37] Figure 25.)

$$C = \frac{Z_{0e} - Z_{0o}}{Z_{0e} + Z_{0o}}$$

All the points fit very closely to the measured curve. Thus, at least on a 'thin' epoxy-glass substrate, this expression for the coupling factor C may be reliably used to calculate the maximum fast crosstalk between microstrip lines. There is one difficulty in the calculations here: data for Z_{0e} and Z_{0o} are difficult to obtain. Where $s/h < 2.0$ these impedances may be determined using the analysis formulae given in Section 6.5. However, in most crosstalk problems we have $s/h > 2.0$ (this corresponds, for example, to a separation of 2.0 mm on a 1.0 mm thick substrate). In cases where approximate values are acceptable the available curves may be used and extrapolated. If greater accuracy is required then Z_{0e} and Z_{0o} must be computed accurately because, as s/h increases, Z_{0e} approaches Z_{0o}, and eqn (A.33) involves the subtraction of two quantities of similar magnitudes. The full computer analysis due to Bryant and Weiss, outlined in Section 6.3, may well have to be used.

It should be mentioned that several workers have advocated the use of an expression for crosstalk which involves the square roots of the characteristic impedances.[38] When compared with the well-established equation, (A.33), and measurements, this expression yields results that are approximately 50 per cent too low.

Coekin[39] has pointed out that the crosstalk can be halved if a grounded conductor can possibly be interposed between the two signal-carrying runs of microstrip. Also, if this interposed conductor can actually be connected through to the ground plane via close-spaced plated-through holes, then a

further halving of the crosstalk can be achieved. Thus, if we started with 2 per cent crosstalk between adjacent microstrips we could reduce this to 0.5 per cent by this method—if sufficient space is physically available between the lines.

For ultra-high density integration, with flip-chip logic ICs, alumina would almost certainly be used as substrate material. The temptation to implement close line spacings must be resisted because of the severe crosstalk problem that can arise.

An important transient analysis of distortion and coupling in lossy parallel-coupled microstrips is reported by Gilb and Balanis.[40] These workers use a spectral domain approach together with complex permittivity to characterize the frequency dependent complex propagation coefficient over a wide frequency range. Original (undistorted) Gaussian-shaped pulses approximately 50 ps in duration are considered and it is shown that both the signal and, more noticeably, the sense (coupled) line responses are significantly distorted. The principal reason for this distortion is the presence of dispersion. An example of the predicted sense line distortion is given in Figure 6.24.

If the microstrip lines amount to interconnects on a densely packed substrate then these results should be used as a guide for the designer. Gilb and Balanis only report results for a line separation of 1 mm on one type of substrate (permittivity 12.2). It is otherwise presumed that the distortion

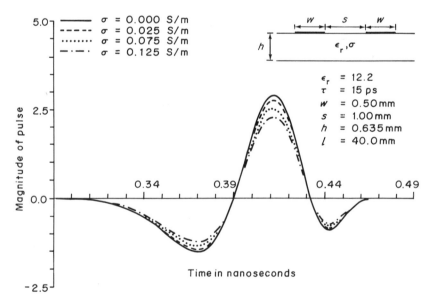

Fig. 6.24. Pulse distortion on coupled, lossy microstrip lines, sense line response. (© 1990 IEEE. Reprinted, with permission, from Gilb and Balanis.[39])

decreases as the separation increases, but quantification for a variety of substrates and geometries would be useful.

Another important report concerns a study of the step response of multiparallel-coupled lines. This work was performed by Orhanovic, Tripathi and Wang[41] who used a numerical solution to a set of partial differential equations describing the structures.

These workers provide results applying to as many as three coupled lines of unequal widths. Most of the published results refer to structures fabricated on 0.1 mm thick alumina substrates and they show the step responses as terminal voltage waveforms.

Discontinuities associated with coupled lines have been investigated by Sabban and Gupta.[42] They use a planar-lumped model to evaluate performance on the basis of equivalent planar waveguides and they apply the analysis to a coupled microstrip section with chamfered right-angled bends connecting to feeding single lines.

Sabban and Gupta report results up to 18 GHz which show that their model provides an accurate prediction of performance, based on a 0.1 mm thick GaAs substrate.

Amongst other microstrip structures (including meander lines) Wertgen and Jansen[43] report theoretical results, using a novel Green's function database technique, for a coupled right-angled bend fabricated on a 0.2 mm thick GaAs substrate. The results extend to a maximum frequency of 26 GHz.

6.15 CHOICE OF STRUCTURE AND DESIGN RECOMMENDATIONS

In this chapter most of the attention has been devoted to the basic parallel-edge-coupled microstrip line structure. This is because of the fundamental nature of that structure, with its importance in relatively simple couplers and its significance in circuit applications such as parallel-coupled band-pass filters. The design of this structure is therefore summarized first here, with reference to coupler realization.

6.15.1 Design procedure for coupled microstrips, where the mid-band coupling factor $C' \lesssim -3$ dB

(a) Provided the coupling is loose enough for the relationship $Z_0^2 \simeq Z_{0e}Z_{0o}$ to hold with sufficient accuracy, then eqns (6.8) and (6.9) give the required characteristic impedance:

Even mode: $Z_{0e} = Z_0 \sqrt{\dfrac{H}{Q}}$

Odd mode: $$Z_{0o} = Z_0 \sqrt{\frac{Q}{H}}$$

where $H = 1 + 10^{C'/20}$ and $Q = 1 - 10^{C'/20}$, and *the coupling factor C' is directly substituted as '$-C'$dB'*. We note also that this is the *mid-band* coupling factor.

(b) An *approximate* synthesis is then performed using the method due to Akhtarzad et al. (see Section 6.6). A simple CAD algorithm may be devised using the formulae provided. It is likely that discrepancies in the region of 10 per cent will arise in this process, which yields initial values of the 'shape ratios' w/h and s/h.

(c) These initial values can then be used to determine the corresponding impedances, Z_{0e} and Z_{0o}, employing the quite accurate (within 3 per cent) analysis formulae given in Section 6.5. If higher accuracy is required, and the increased time and cost are acceptable, the much more accurate computational technique developed by Bryant and Weiss may be used. This was explained in Section 6.3. Some adjustment for finite conductor thickness may sometimes be necessary.

(d) These new (stage c) values for Z_{0e} and Z_{0o} are compared with the original requirements, determined at stage (a), and discrepancies are noted. Shape ratios w/h and s/h can now be adjusted slightly to correct the impedances, and hence the coupling factor C', to the desired values. It will usually be best if s alone is slightly altered, since the resulting mismatch to the feed lines will be small.

For guidance, we observe that s affects the odd mode more than the even mode. Increasing s means that more field is within the (air) coupling gap and Z_{0o} increases. As a rough approximation x per cent change in s results in $x/10$ per cent change in Z_{0o}. For example, a 5 per cent increase in s yields about 0.5 per cent increase in Z_{0o}.

Finally the value of the mid-band coupling factor must be checked, which is given by eqn (6.6):

$$C' = 20 \log \left| \frac{Z_{0e} - Z_{0o}}{Z_{0e} + Z_{0o}} \right|$$

6.15.2 Relatively large coupling factors (typically $C \gtrsim -3$ dB)

The problems of unequal even- and odd-mode wave velocities worsen rapidly as the coupling increases, e.g. the approximation $Z_0^2 \simeq Z_{0e}Z_{0o}$ becomes poor when $C' > -3$ dB. An exact relationship for Z_0 was given as eqn (6.12), but this involves the even- and odd-mode electrical lengths θ_e and θ_o. In order to find these lengths we need, at the outset, to know the geometry (i.e. the shape ratios) for the structure, which is just what we were trying to find in the first place. One approach could be:

(a) Carry through a *rough*, first-order synthesis using the approximation $Z_0^2 \simeq Z_{0e}Z_{0o}$, and hence an initial w/h, s/h.

(b) Calculate approximate values of θ_e and θ_o using the values obtained at step (a). The analysis procedure given in Section 6.5 is well suited to this calculation, eqns (6.23) and (6.24) yielding the effective microstrip permittivities. The electrical lengths are

$$\theta_{e,o} = \frac{2\pi l}{\lambda_g} \sqrt{\epsilon_{effe,o}}$$

An approximate value of Z_0 is then obtained using eqn (6.12). If Z_0 differs from that of the feed lines by an unacceptable amount (VSWR specification) the width w of the microstrip lines should be altered to compensate. The discussion of Section 3.12.3 contains guidance on this type of adjustment. Altering w will alter Z_{0e}, Z_{0o}, θ_e, and θ_o; thus the procedure must be iterated to find the optimum value of w.

(c) The evaluation of w must also ensure that the required mid-band coupling factor is obtained, from eqn (6.6):

$$C' = 20 \log \left| \frac{Z_{0e} - Z_{0o}}{Z_{0e} + Z_{0o}} \right| \quad \text{dB}$$

(d) When this procedure has been used and the parameter values have been met as closely as possible the final values of Z_{0e} and Z_{0o} are used to synthesize, as described in Section 6.15.1.

(e) For close coupling, which is the case here, the effect of finite microstrip thickness is often significant and the analysis procedure should account for this by incorporating eqns (6.85) and (6.86) of Section 6.12.

6.15.3 Length of the coupled region

In a directional coupler the coupled region extends for one-quarter of a wavelength at mid-band and the designer requires the physical length of this region.

Since, in microstrip, we are faced with the problem of different phase velocities applicable to the even and odd modes we must also consider the different wavelengths associated with these modes. It is necessary to approximate the design to suit the *average* value of wavelength. This problem was treated in detail in Section 6.8 and the expressions derived (eqns 6.36 and 6.37) are repeated here:

$$\lambda_{ge} \simeq \frac{300}{F} \frac{Z_{0e}}{Z_{01e}} \quad \text{mm}$$

and

$$\lambda_{go} \simeq \frac{300}{F} \frac{Z_{0o}}{Z_{01o}} \quad \text{mm}$$

where F is directly substituted in gigahertz, $Z_{0e,o}$ are the even- and odd-mode characteristic impedances for the coupled microstrips with a substrate, and $Z_{01e,o}$ are the same impedances, but for the air-spaced microstrips. The average value

$$\lambda_{gm} \simeq \frac{1}{2}(\lambda_{ge} + \lambda_{go})$$

is taken for approximate design purposes. For further improved accuracy a weighted-average calculation has been suggested.[44]

6.15.4 Frequency response

There are two fundamental effects governing the frequency-dependent behaviour of coupled microstrips:

(a) the basic frequency response of *any* pair of parallel-coupled lines (neglecting any dispersion)
(b) the effects of microstrip dispersion.

These features, (a) and (b), were discussed in detail in Section 6.9.

Great care is needed when using the fairly complicated expressions involved (equations of the form of eqn (6.40) incorporated into eqn (6.41)). Although it appears that the even-mode electrical length θ_e may be evaluated 'fairly accurately' using eqn (6.40) and $\theta_e = 2\pi l/\lambda_{ge}$, there is evidence that the odd-mode calculation overestimates changes in $\epsilon_{effo}(f)$.[6]

For greater accuracy, particularly for frequencies into the millimetre-wave ranges, the expressions due to Kirschning and Jansen[9] should be used (see Section 6.9.2).

6.15.5 Coupled structures with improved performance

Although several alternative methods have been reported, the technique invented by Lange[10] still yields an electrical performance which is generally superior to that of its 'closest rivals'. The main disadvantage is the requirement for short wire connections. A design procedure now exists for the 'Lange coupler',[11-15] and combinations of Lange couplers have been built which cover the wide frequency range 2–18 GHz. The design repeatability of such couplers is generally ± 0.05 dB, on the coupling factor, that is ± 2.5 μm excursions in coupling gaps result in ± 0.1 dB changes in coupling factor.

Of the other techniques which have been developed the end-connected compensating lumped capacitors (Section 6.11.5) is possibly the next best—electrically. This is a particularly elegant technique because it can usually be realized as an *entirely planar structure*. (We recollect that the Lange coupler requires arrays of bond wires in its implementation.) However,

when tight or fairly tight coupling is required ($C' \gtrsim -3$ dB) the already narrow coupling gap makes it difficult to incorporate the lumped capacitive structures in this capacitor-compensated coupler.

The two equal capacitances predominantly affect the odd-mode and eqn (6.78) gave the required value:

$$C_L = \frac{\cos \theta_o}{4 \pi f Z_{0oi}}$$

Typically, for a -10 dB coupler, ± 0.3 dB was achieved, with a directivity of 20 dB and VSWR < 1.4 over the frequency range $8 \le f \le 20$ GHz.

Closely shielded couplers may be used in some instances but, to achieve good compensation, the close proximity of the shield to the circuit can cause difficulties. Dielectric overlays are a further possible alternative but, like close shielding, it can be difficult to design and control under production conditions.

6.15.6 Effects of conductor thickness, power losses and production tolerances

It has already been remarked that the principal effect of finite conductor thickness is to reduce Z_{0o} slightly. This reduction amounts to approximately 2–3 per cent for a 5 μm conductor on alumina, where the coupling gap is in the approximate range $10 \le s \le 500$ μm. The main power loss mechanism is conductor loss and eqn (6.48) is recommended for the calculation of this quantity.

Equations (6.89) and (6.90) enable one to calculate the sensitivity of coupling factor to fabrication tolerances. The expressions will not be reproduced here, but the following tolerance excursions typically give a worst-case change of 4 per cent in coupling factor (for alumina substrates):

$$\Delta s = \pm 2.5 \ \mu m$$
$$\Delta w = \pm 2.5 \ \mu m$$
$$\Delta h = \pm 25 \ \mu m \ (\text{on } h = 0.65 \ \text{mm})$$
$$\Delta \epsilon_r = \pm 0.25 \ (\text{on } \epsilon_r = 9.7)$$

6.15.7 Crosstalk between microstrip lines used in digital systems

This has been discussed in Section 6.14. With the provisos given in the main discussion the following expression should be used to evaluate the crosstalk:

$$C = \frac{Z_{0e} - Z_{0o}}{Z_{0e} + Z_{0o}}$$

(which is identical to that used for the coupling factor). Now, because the spacing is usually very substantial, and often $s \gg 2w$ here, Z_{0e} is very close

to the value of Z_{0o} and we have to evaluate a *small difference*. Thus, accurate computer-based techniques are recommended, or at least empirically curve-fitted formulae from curves using such computer-based techniques.

6.15.8 Post-manufacture circuit adjustment

After testing it may well be found that a completed coupled-line circuit requires adjustment. Either the actual circuit may be etched, laser trimmed, etc., or the artwork used in preparing the circuit may be judiciously modified to correct errors.

Changing s or w slightly will alter the impedances and hence the coupling factor. It will generally be necessary to alter *both* w and s at the same time so that the correct impedance relationships are preserved. For example, if s is increased by 1 per cent then Z_{0o} (mainly) will also increase, by approximately 0.1 per cent—see Section 6.15.1. Therefore w/h should be increased very slightly, by less than 1 per cent to reduce the impedance again.

The foregoing is only an outline of the procedure: in practice the precise trends for a particular case would soon become clear.

REFERENCES

1. Bryant, T. G., and Weiss, J. A., 'Parameters of microstrip trnasmission lines and of coupled pairs of microstrip lines', *IEEE Trans.*, **MTT-16**, No. 12, December 1968, 1021–1027.
2. Garg, R., and Bahl, I. J., 'Characteristics of coupled microstriplines', *IEEE Trans.*, **MTT-27**, No. 7, July 1979, 700–705.
3. Garg, Ramesh, Private communication concerning coupled-line formulae.
4. Akhtarzad, S., Rowbotham, T. R., and Jones, P. B., 'The design of coupled microstrip lines', *IEEE Trans.*, **MTT-23**, No. 6, June 1975, 486–492.
5. Getsinger, W. J., 'Dispersion of parallel coupled microstrip', *IEEE Trans.*, **MTT-21**, 1973, 145–146.
6. Easter, B., and Gupta, K. C., 'More accurate model of the coupled microstripline section', *IEE Journal on Microwaves, Optics and Acoustics,* **3**, No. 3, May 1973, 99–103.
7. Krage, M. K., and Haddad, G. I., 'Frequency dependent characteristics of microstrip transmission lines', *IEEE Trans.*, **MTT-20**, 1972, 678–688.
8. Richings, J. C., and Easter, B., 'Measured odd and even mode dispersion of coupled microstrip lines', *IEEE Trans.*, **MTT-23**, 1975, 826–828.
9. Kirschning, M., and Jansen, R. H., 'Accurate wide-range design equations for the frequency-dependent characteristic of parallel coupled microstrip lines', *IEEE Trans.*, **MTT-32**, No. 1, January 1984, 83–89.
10. Napoli, L. S., and Hughes, J. J., 'Characteristics of coupled microstrip lines', *RCA Rev.*, **31**, September 1970, 479–498.
11. Lange, J., 'Interdigitated strip-line quadrature hybrid', *1969 G-MTT Symposium, Digest of Papers,* pp. 10–13. (See also the paper by Lange, J., in *IEEE Trans.*, **MTT-20**, November 1972, 777–779).

12. Presser, A., 'Interdigitated microstrip coupler design', *IEEE Trans.*, **MTT-26**, No. 10, October 1978, 801–805.
13. Ou, W. P., 'Design equations for interdigitated directional coupler', *IEEE Trans.*, **MTT-19**, No. 2, February 1975, 253–255.
14. Osmani, R. M., 'Synthesis of Lange couplers', *IEEE Trans.*, **MTT-29**, No. 2, February 1981, 168–170.
15. Fusco, V. F., *Microwave Circuits*, Prentice-Hall, Reading, Massachusetts, 1987.
16. Waugh, R. W., 'Sensitivity analysis of the Lange coupler', *Microwave Journal*, November 1989, 121–129.
17. Tajima, Y., and Kamihashi, S., 'Multiconductor couplers', *IEEE Trans.*, **MTT-26**, No. 10, October 1978, 795–801.
18. Private discussions.
19. Waugh, R., and LaCombe, D., 'Unfolding the Lange coupler', *IEEE Trans.*, **MTT-20**, November 1972, 777–779.
20. Weiss, J. A., 'Microwave propagation on coupled pairs of microstrip transmission lines', in *Advances in Microwaves* (Eds L. Young and H. Sobol), Vol. 8, Academic Press Inc., New York, 1974, pp. 295–320.
21. Kouki, A. B., Khebir, A., and Mittra, R., 'Finite element modeling of two-dimensional transmission line structures using a new asymptotic boundary condition', *IEEE MTT-S International Microwave Symposium*, 1990, Digest of Papers, 717–720.
22. Shelad, B., and Spielman, B. E., 'Broadband (7–18 GHz) 10 dB overlay coupler for MIC application', *Electron Lett.*, **11**, No. 8, 17 April 1975, 175–176.
23. Rickard, D. C., 'Thick-film MIC components in the range 10–18 GHz', *European Microwave Conference, Digest of Papers,* Rome, 1976, 687–691.
24. Dydyk, M., 'Accurate design of microstrip directional couplers with capacitive compensation', *IEEE MTT-S International Microwave Symposium*, 1990, Digest of Papers, 581–584.
25. Szentkuti, B. T., 'Simple analysis of anisotripic microstrip lines by a transform method', *Electron. Lett.*, **12**, 1976, 672–673.
26. Uysal, S., Aghvami, A. H., and Mohamed, S. A., 'Microstrip quadruplexer using −3dB hybrids', *Proceedings of the 19th European Microwave Conference,* London, September 1989, 905–910.
27. Nakajima, M., Yamashita, E., and Asa, M., 'New broad-band 5-section microstrip-line directional coupler', *IEEE MTT-S International Microwave Symposium*, 1990, Digest of Papers, 383–386.
28. Pavio, A. M., and Sutton, S. K., 'A microstrip re-entrant mode quadrature coupler for hybrid and monolithic circuit applications', *IEEE MTT-S International Microwave Symposium*, 1990, Digest of Papers, 573–576.
29. Fusco, V. F., and Merugu, L. N., 'Full wave analysis of 94 GHz patch coupler', *IEEE MTT-S International Microwave Symposium*, 1990, Digest of Papers, 649–652.
30. Rizzoli, V., 'Highly efficient calculation of shielded microstrip structures in the presence of undercutting', *IEEE Trans.*, **MTT-17**, No. 2, February 1979, 150–157.
31. Gupta, K. C., Garg, Ramesh and Bahl, I. J., *Microstrip Lines and Slotlines,* Artech House, Inc., Dedham, Massachusetts, 1979.
32. Jansen, R. H., 'High-speed computation of single and coupled microstrip parameters including dispersion, higher-order modes, loss and finite strip thickness', *IEEE Trans.*, **MTT-26**, No. 2, February 1978, 75–82.
33. Shamasundra, S. D., and Gupta, K. C., 'Sensitivity analysis of coupled microstrip directional couplers', *IEEE Trans.*, **MTT-26**, No. 10, October 1978, 788–794.
34. Gunton, D. J., and Paige, E. G. S., 'Directional coupler for gigahertz frequencies,

based on the coupling properties of two planar comb transmission lines', *Electron. Lett.*, **11**, No. 17, 21 August 1975, 406–409.

35. Rigg, P. R., and Carroll, J. E., 'The canonical coupler: a new approach to microwave coupler design', *European Microwave Conference, Digest of Papers,* Brighton, 1979, pp. 490–494.

36. Islam, S., 'An analysis and a design technique for microstrip comblines', *Microwave Journal*, November 1989, 79–91.

37. Catt, T., 'Crosstalk (noise) in digital systems', *IEEE Trans. on Dig. Comp.*, **EC-16**, No. 6, December 1967, 743–763.

38. Okugawa, S., and Hagiwara, H., 'Analysis and computation of crosstalk noise between microstrip transmission lines', *Electronics and Communications in Japan*, **53-C**, No. 7, 1970, 128–138.

39. Coekin, J. A., *High-speed pulse techniques,* Pergamon Press, Oxford, 1975, pp. 203–205.

40. Gilb, J. P., and Balanis, C. A., 'Transient analysis of distortion and coupling in lossy coupled microstrips', *IEEE **MTT-S** International Microwave Symposium,* 1990, Digest of Papers, 641–644.

41. Orhanovic, N., Tripathi, V. K., and Wang, P., 'Time domain simulation of uniform and nonuniform multiconductor lossy lines by the method of characteristics', *IEEE **MTT-S** International Microwave Symposium*, 1990, Digest of Papers, 1191–1194.

42. Sabban, A., and Gupta, K. C., 'A planar-lumped model for coupled microstrip line discontinuities', *IEEE **MTT-S** International Microwave Symposium*, 1990, Digest of Papers, 247–254.

43. Wertgen, W., and Jansen, R. H., 'Novel Green's function database technique for the efficient full-wave analysis of complex irregular (M)MIC-structures', *Proceedings of the 19th European Microwave Conference*, London, September 1989, 199–204.

44. Kajfez, D., and Govind, S., 'Effects of difference in odd- and even-mode wavelengths on a parallel-coupled bandpass filter', *Electron. Lett.*, **11**, 5, 6 March 1975, 117–118.

7 Power Capabilities, Transitions, and Measurement Techniques

7.1 POWER-HANDLING CAPABILITIES

There are two main aspects to any consideration of power-handling in microstrip: continuously working (CW) and pulse operation. Under CW conditions the major problems and limitations are *thermal*, whereas under pulse conditions the principal limitation is *dielectric breakdown*.

7.1.1 Maximum average power P_{ma} under CW conditions

This is controlled by heat flow and the associated temperature rise due to conductor and dielectric losses (principally conductor losses in most cases). The problem is very well treated in detail by Gupta et al.,[1] and only the salient features will be presented here. Gupta calculates the power absorbed by a unit length of microstrip line, due to conductor losses. The density of heat flow, resulting from both conductor and dielectric losses, is then determined and the consequent temperature rise is found. The final expression for the temperature rise above ambient is

$$\Delta T = \frac{0.2303h}{K}\left[\frac{\alpha_c}{w_{\text{eff}}} + \frac{\alpha_d}{2w_{\text{eff}}(f)}\right] °\text{C/W} \tag{7.1}$$

where α_c and α_d are the conductor and dielectric losses respectively, in decibels per metre (length), w_{eff} and $w_{\text{eff}}(f)$ are effective microstrip widths (note that $w_{\text{eff}}(f)$ rises slowly with increasing frequency, see Section 4.8.2), and K is the thermal conductivity of the substrate. Thus, as expected, the substrate dominates the temperature rise due to continuous power transmitted along the line.

Apart from beryllium oxide (BeO), alumina and sapphire give the best (lowest) temperature rise characteristics when compared with other common substrates. For a wide range of microwave frequencies, 4–20 GHz, 50 Ω

lines on alumina and sapphire substrates yield a figure of approximately 0.02°C/W. Progressively worse materials are, for the same conditions:

silicon	0.04°C/W
gallium arsenide (GaAs)	0.06°C/W
quartz	0.2°C/W
polystyrene	1.0°C/W

If we consider a temperature rise (θ) of 75°C above ambient, a 50 Ω microstrip line on an alumina substrate could then carry a maximum average power of

$$P_{\text{ma}} = \frac{\theta}{\Delta T}$$

$$= \frac{75}{0.02}$$

that is

$$P_{\text{ma}} = 3750 \text{ W}$$

In practice this is somewhat (not greatly) frequency dependent, but such microstrip lines can be expected to reliably carry a few kilowatts of average power. In contrast, lines on polystyrene or other plastic substrates are restricted to only around 100 W. Beryllium oxide provides a thermally excellent substrate, with $\Delta T \simeq 1 \times 10^{-3}$°C/W and an average power of about 75 kW can be handled by a 50 Ω line on beryllium oxide at 10 GHz. However, beryllium oxide is rarely used as a substrate because its dust is toxic. This ceramic is encountered as a pressing attached firmly to power semiconductor devices; e.g. the collector of a transistor, the drain of a power FET, or the anode of an IMPATT. In microstrip, this portion of the device would be bonded to the ground plane using high thermal-conductivity materials. The power limitations on the actual microstrip then remain as previously outlined.

In most cases, therefore, alumina or sapphire substrates are likely to be encountered. Some improvement can be obtained when wider microstrip lines are used, that is lines of lower characteristic impedances. Since both the effective widths w_{eff} and $w_{\text{eff}}(f)$ in eqn (7.1) are strongly dependent upon the physical width w, then ΔT will decrease as the width increases. If the previously calculated 50 Ω line on alumina is, instead, 20 Ω, then the power capacity will be increased by a factor of approximately four and up to about 15 kW may be transmitted. Unfortunately most practical circuits will require narrow lines, of relatively high impedance, and the power capacity will therefore be *decreased*. Referring to the same example as

before, a 100 Ω line will be restricted to approximately 470 W maximum CW power.

7.1.2 Peak (pulse) power-handling capability

In a microwave system employing MICs the coaxial connectors, rather than the microstrip, are likely to set the ultimate limit to the peak power. To maintain a low VSWR, i.e. to achieve a good transition, the final portion of the coaxial connector, which actually contacts the microstrip, is almost invariably of a 3 mm subminiature form. This is usually the case even where N-type or APC7 coaxial lines are used in the main system external to the MIC. The final portion of the coaxial connector sets a limit because of air breakdown. With any transmission line having a characteristic impedance Z_0 and maximum (breakdown) voltage V_{mb} the peak power allowable is given by

$$P_{mp} = \frac{V_{mb}^2}{2Z_0} \qquad (7.2)$$

Now the breakdown electric field strength of dry air is 3 MV/m, and the internal difference in radii of the 50 Ω connector is approximately 1.5 mm— as can be seen from Figure 7.1 shown in the next section. Thus the breakdown voltage is

$$3 \times 10^6 \times 1.5 \times 10^{-3} = 4.5 \text{ kV}$$

and the peak power is, from eqn (7.2):

$$P_{mp} \simeq 200 \text{ kW}$$

In practice it would probably be unwise to greatly exceed 100 kW. Voltage antinodes due to internal mismatches reduce the allowable power still further.

Gupta et al.[1] mentions two useful improvements which can be made to the microstrip itself in order to improve peak power handling:

(a) Since sharp edges intensify the electric field and thus reduce the allowable breakdown voltage, thick, rounded conductors are preferred.
(b) To avoid air breakdown near the edges, the strip conductor can be painted with a dielectric paint having a permittivity identical to that of the substrate.

A careful technique of preferential etching should enable (a) to be achieved and (b) is relatively straightforward to implement. However, the introduction of such dielectric paint along the edges of the microstrip conductor will significantly alter the line parameters (ϵ_{eff} and Z_0) and it is not clear how one might calculate appropriate design adjustments. An empirical cut-and-try approach thus seems inevitable here, including tests for the final

microstrip-limited peak power when this, and not the transitions, forms the limit.

7.2 COAXIAL-TO-MICROSTRIP TRANSITIONS

These are available in a variety of forms, the choice mainly depending on the overall electrical quality required. For non-critical applications commercially produced transitions, operating from semi-rigid (3.58 mm diameter) coaxial lines, are available from a number of companies. The final contact to the microstrip is variously formed as a wedge or a tab. In most applications the wedge will probably be preferred as it is more mechanically rugged. Some connections to a microstrip circuit can often be made with comparatively low-grade (and therefore relatively cheap) 'transitions'—for d.c. bias ports inexpensive BNC connectors can simply be used. Where only temporary connection is required a pressure contact is quite satisfactory. A permanent transition will usually be completed with some form of epoxy bond surrounding the tab-microstrip metals.

It will be appreciated that the transition region, from the 'cylindrical' coaxial structure to the planar rectangular microstrip, presents quite a considerable discontinuity in the system. When used in conjunction with a microstrip line having 50 Ω characteristic impedance the commercial transitions typically yield a VSWR equal to 1.06 at moderate (e.g. X-band) frequencies. This is adequate for a wide variety of operating circuit and system applications.

In critical situations such as measurement arrangements a far superior VSWR performance is often desired and various compensation techniques have been developed. For a microwave transistor characterization jig a form of screw compensation has been developed.[2] A metal screw is introduced through a hole in the jig beneath the ground plane, in the vicinity of the transition. There is a further hole in line with this and actually in the ground plane, thus exposing a small region of the substrate underside surface (this arrangement can be seen in Figure 7.1b). The exact field situation is complex, but it seems clear that the screw mainly serves to reduce the microstrip line capacitance very locally so as to compensate (reduce) the transition capacitance. Extremely careful manufacture and time-domain reflectometry measurements have resulted in the transition shown in Figure 7.1.[3] The dimensions shown apply to a transition launching into a 50 Ω microstrip line on a 0.5 mm thick sapphire substrate (C axis perpendicular to the ground plane), and a reflection coefficient less than 0.01 was obtained. With some extra dimensional and shape refinements, described in the paper, the achievement amounts to a reflection coefficient less than 0.005 (VSWR < 1.01) over the range $1 \leq f \leq 18$ GHz. Over the same frequency range the insertion loss was less than 0.3 dB.

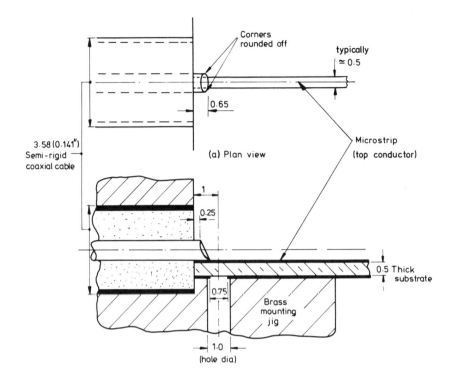

Fig. 7.1. Compensated coaxial-microstrip transition (all dimensions in millimetres). (© 1976 IEEE. Reprinted, with permission, from England.[3])

The use of alumina substrates demands some modifications. For a similar performance a 1.0 mm diameter ground-plane hole was found to be necessary, where the microstrip characteristic impedance is approximately 50 Ω. For further details the reference should be consulted.

In the case of a different substrate (thickness and/or permittivity) it is necessary to repeat these careful measurements with a cut-and-try approach all over again.

If it could be developed, a general design procedure for such transitions would be most useful.

7.3 WAVEGUIDE-TO-MICROSTRIP TRANSITIONS

The principal transmission medium for microwaves at millimetric wavelengths consists of rectangular waveguide. Where microstrip is used at such

wavelengths transitions from waveguide to microstrip are therefore required. Since the waveguide itself is relatively narrow-band the transitions are also only required to operate over typically 10 to 50 per cent bandwidths.

One essentially single-frequency solution is to first introduce a waveguide-to-coaxial line transition (the T-transformer, for example) and then to follow this with a coaxial line-to-microstrip transition as already described. This may be convenient in some restricted cases.

Transitions of improved bandwidth and other advantages are, however, conceivable. The desirable features of *any* waveguide-to-microstrip transitions are:

(a) Low return loss. This means that inherent reflections will be low and reflection measurements may be made on the microstrip circuit.
(b) Low insertion loss.
(c) Sufficient bandwidth for the application.
(d) Fairly easy and well-reproducible re-connection to the microstrip.
(e) An in-line design for ease of connecting test equipment.
(f) Straightforward mechanical reproduction.

Several different solutions to the problem have been presented in the literature and some contrasting techniques will be presented here.

7.3.1 Ridgeline transformer insert

In the design approach of Schneider *et al.*[4] the waveguide impedance is transformed to the microstrip impedance by use of a fairly broad-band stepped ridgeline transformer which is mechanically connected to the microstrip by a tab and a single pressure screw. A four-step ridgeline transformer designed to work over the 27.5–31.3 GHz band gave more than 30 dB return loss (reflection coefficient <0.03) and less than 0.1 dB insertion loss over the band. A quartz substrate of 0.76 mm thickness was used for the microstrip circuit and the arrangement is shown in Figure 7.2.

For a required bandwidth and return loss, the necessary impedance at each step in the transformer can be calculated using Cohn's design method.[5] The stepped single ridgeline selected is quite compatible with the asymmetrical microstrip line. It is also easily machined and connected to the waveguide. The mechanical dimensions required of ridgeline structures, to realize a specified impedance, can be calculated from information given by Hopfer.[6] So that the substrate will stop against ridgeline, the last (deepest) step of this ridgeline is made sufficiently deep—actually 2.84 mm in this case. By tapering the edge of the transformer at the junction the local reactance there is decreased, hence improving the performance.

Good contact is maintained both to the microstrip and its ground plane by applying pressure via a single insulated screw operating on the contact

Fig. 7.2. Rectangular waveguide-to-microstrip transition using a ridgeline transformer for the 27.5–31.3 GHz band: (a) complete mechanical structure and (b) ridgeline transformer dimensions (all in millimetres). (Reprinted with permission from *The Bell System Technical Journal*, © 1969, the American Telephone and Telegraph Company, from Schneider *et al.*[4] Figure 5.)

tab (this screw is not shown in Figure 7.2). For further details the paper by Schneider *et al.*[4] should be consulted.

More recently Menzel and Klassen[7] have reported developments of the ridged waveguide to microstrip transition, including a 30 GHz design based on alumina-substrated microstrip and a 60 GHz design in which waveguide 15 was connected to microstrip on a GaAs substrate. These workers use mode matching techniques based upon an LSE and LSM field analysis. The structures are shown in Figure 7.3. Using an unloaded ridged waveguide in transition with 0.056 mm wide microstrip on a 0.2 mm thick GaAs substrate the minimum reflection coefficient S_{11} was found to be about 0.15 for an optimum ridge width of 0.5 mm.

7.3.2 Mode changer and balun

A different approach was adopted by van Heuven.[8] This transition begins, in the waveguide, with a form of mode-changing section which rotates and concentrates the fields until they principally exist in a parallel line. This symmetrical line is then matched to the asymmetrical microstrip line by the

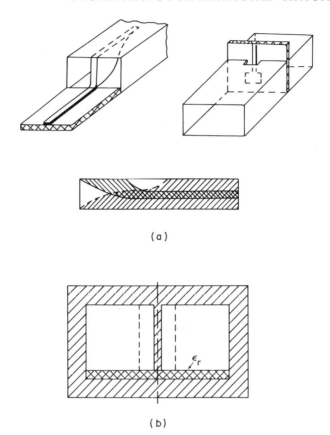

(a)

(b)

Fig. 7.3. (a) Transitions from metal waveguide to microstrip. (b) Waveguide-to-microstrip transitions using ridged waveguide. (Reprinted, with permission, from Menzel and Klassen.[7])

use of a 'balun' portion. The arrangement is shown in some detail in Figure 7.4; the substrate is usually quartz.

The conductors on the substrate are electrically connected to the top and bottom walls of the waveguide, in the region marked (A) in Figure 7.4(b). Each conductor is then gradually tapered until, towards the end of the region marked (B), a balanced-conductor or 'symmetrical' line is formed. The characteristic impedance of this symmetrical line should be approximately the same as that of the final microstrip. In the region marked (C) a balun is formed; notice that one side of this has to be the microstrip ground plane and therefore this side is electrically connected back to the (ground) top and bottom waveguide walls.

Serrated chokes may be used to produce good r.f. connections to the waveguide. Since the ground plane divides the waveguide into two separated

(a) WAVEGUIDE - END VIEW

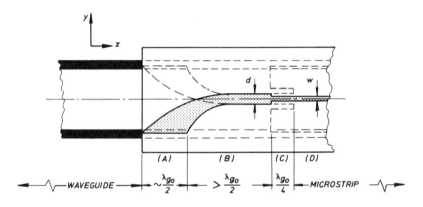

(A) WAVEGUIDE CONNECTIONS AND FIRST TAPER
(B) TAPER TO BALANCED-CONDUCTORS LINE
(C) BALUN
(D) MICROSTRIP

(b) LONGITUDINAL VIEW

Fig. 7.4. Waveguide-to-microstrip transition using tapers and a balun. (© 1976 IEEE. Reprinted, with permission, from van Heuven.[8])

guides each having a cut-off frequency well above the band in use further waveguide-mode propagation beyond the transition is prevented. The height of the guide is increased above normal to compensate for the loading due to the quartz substrate.

Van Heuven's transition gave the following performance:

bandwidth	18–26 GHz, and, over this,
VSWR	1.11
insertion loss	0.2 dB

Over a significant, but narrower, band (17.7–19.7 GHz) the VSWR remained less than 1.1. This corresponds to a reflection coefficient less than 0.05, which is only slightly worse than for Schneider's transition[4]—although this was operating at a higher frequency.

Further details, including mechanical dimensions, can be found in van Heuven's paper.[8] Final design adjustments are largely empirically determined.

7.3.3 A waveguide-to-microstrip power splitter

A type of waveguide-to-microstrip power splitter which is suitable for at least the lower microwave frequencies has been developed by Wu et al.[9] This device uses a T-bar waveguide to microstrip transition, the theory for which is based upon current distributions and the dyadic Green's function techniques—treating the T-bar as an internal antenna. From this analysis an expression for the T-bar radiation resistance is obtained. The physical structure and simplified model are shown in Figure 7.5.

The design goal is to select a T-bar structure whose radiation resistance is matched to 50 Ω. The design described by these workers exhibits an output resistance which is within 2 Ω (4%) of the requisite 50 Ω value. A return loss below 20 dB is obtained over the 3.3–4.6 GHz band and the power difference between output ports is below 0.1 dB.

7.4 TRANSITIONS BETWEEN OTHER MEDIA AND MICROSTRIP

As discussed in Chapter 2, a variety of other media exists which serve some useful functions in certain special applications and operate effectively at approximately the same frequencies as microstrip. Techniques have been developed which provide transitions between some of these structures and microstrip.

The first example we take is the triplicate-to-microstrip transition. Although, at the time of writing, this has not yet been fully analysed (as an uncompensated transition) it has been simply treated by Hall and James[10] who also report some measurements of power loss due to radiation. On a plastic ($\epsilon_r = 2.32$) substrate this loss was found to be approximately 11 per cent. Such an uncompensated transition may well yield an acceptable performance when the microstrip substrate is of plastic, but worse results are likely with alumina and some form of compensation would be desirable. This might be of the type previously described in Section 7.2 for coaxial-to-microstrip transitions.

A microstrip-to-slotline transition was first described by Robinson and Allen[11] and the details have been thoroughly discussed by Gupta et al.[12] The simple basic structure yields a VSWR < 1.2 over the frequency range

Fig. 7.5. Waveguide-to-microstrip power splitter. (a) Structure of the waveguide to microstrip power divider. The side view illustrates that the microwave signal is coupled to two T antennas on an alumina substrate. The front view of the power divider consists of two T-shaped coupling antennas. (b) Simplified model using symmetry of the waveguide to microstrip power divider. The plane of symmetry in (a) is an electric wall where a metal plate can be inserted without changing the field distribution. Thus the power divider problem is changed to the waveguide-to-microstrip transition design. (© 1990. Reprinted, with permission, from Wu, Schneider and Trambarulo.[9])

2–4 GHz. With some care in design, and the microstrip orientated at right angles to the slotline, a VSWR < 1.1 was achieved over a higher range of frequencies 8–10 GHz. It is apparent that a six-port junction will provide further improvements, and the structure is useful up to a frequency of at least 12 GHz.[13]

The design of dielectric-loaded slotline to microstrip transitions has been considered and reported by Qin and Wang.[14] In their design approach they have the microstrip coupled transversely across the slotline which is constructed on a relatively high permittivity substrate (permittivity greater than 9.6) with a lower permittivity (2.5) polymer dielectric overlay. Qin and Wang provide design equations covering the 3–4 GHz band.

A fairly simple fin-line to microstrip transition, designed as one component in a single-sideband mixer operating over the 26–38 GHz frequency range, is described by Climer.[15] This component, shown in outline in Figure 7.6, is built on a quartz substrate and the entire mixer design was based on microstrip. The approach of applying the local oscillator input to the fin-line port ensures that this oscillator is isolated from both the signal input and its output as well as being rejected by the diplexer. Disadvantages include higher signal losses and greater difficulty in designing the signal filter.

Horng et al.[16] have presented a full-wave analysis of shielded microstrip line-to-line transitions (edge-coupled and overlay structures). The derivation of their results depends upon a method of moments, the use of dyadic Green's function and a Galerkin procedure. They provide a check on results using a power conservation approach and show that all results are within 1 per cent accuracy. Data is presented for high-grade alumina-type substrates operated over the frequency range 0.5–11 GHz.

Transitions between microstrip and coplanar waveguide are briefly considered by Gupta et al.[1] and more extensively by Viitanen and Kuismann[17] who report an analysis of this type of transition at frequencies up to 20 GHz.

7.5 INSTRUMENTATION SYSTEMS FOR MICROSTRIP MEASUREMENTS

In general we have to choose between *frequency-domain* and *time-domain* measurements. The great majority of microwave measurements are best

Fig. 7.6. Microstrip-to-fin-line transition (Reprinted, with permission, from Climber.[15])

performed in the frequency domain because, with present systems technology, a high acuracy and a large amount of information are then potentially available. However, there exist some requirements where time-domain measurements are valuable and high-resolution time-domain reflectometer (TDR) systems should then be used. Pulse-rise times less than 28 ps are common with such systems; more details will be given in a later section.

In this section we are concerned with frequency-domain instrumentation. There are several questions that should be considered, for example:

(a) Are single-frequency techniques suitable?
(b) Are swept-frequency techniques suitable?
(c) Are single-port or two-port (or more) measurements desired?
(d) What information is actually required about the circuit?
(e) How important is the speed and ease of measurement?
(f) What accuracy is required?

Single-frequency techniques were used in early microwave measurements but these have been largely superseded by swept techniques nowadays. In some instances standing-wave indicators (SWI) are useful, and these are generally operated in conjunction with a swept-frequency test signal.

At the most sophisticated level computer-augmented fully automatic measurement systems are available; the cost is naturally relatively high and can only occasionally be justified. Such systems employ sweep (frequency) generators, frequency synthesizers, down-converters, detectors, signal-processing electronics, and a computer to control the system. Full amplitude and phase information is determined for any circuit under study. One-port or two-port circuits are readily handled, and systems have been developed to cope with multiport studies.[11]

For many microstrip measurements such comprehensive systems are not necessary and, indeed, often only the *magnitude of a one-port network reflection coefficient* is required. A suitable specific system is indicated in Figure 7.7. This arrangement may be used for substrate measurements and many measurements on microstrip resonators including dispersion, discontinuities, and various Q-factors.

The microstrip circuit under test is connected to the test port of a dual directional coupler via a coaxial-to-microstrip transition. The dual directional coupler has high directivity and good, uniform coupling throughout the band. It also exhibits an effective source impedance close to 50 Ω, i.e. a good match throughout the band. This directional coupler enables the ratio:

$$\frac{\text{Reflected signal (from the microstrip circuit)}}{\text{Incident signal (direct from the swept-signal generator)}}$$

to be determined and displayed on an adapted CRO. There remain three other very important items:

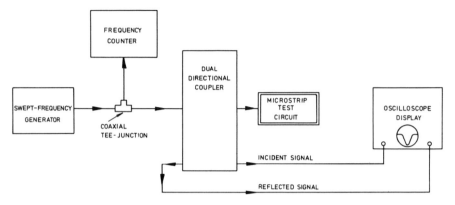

Fig. 7.7. Example of a one-port instrumentation system (typically having all-coaxial interconnections)

 (a) a swept-signal generator (SSG)
 (b) a frequency counter
 (c) a special oscilloscope-display indicator ('CRO').

Item (a), the SSG, continuously varies or 'sweeps' the r.f. test signal, generated by a suitable source, so that a frequency is obtained that varies linearly with time. A variety of sweep rates are available ranging from some megahertz per second to several gigahertz per millisecond. The slow rates are particularly useful when driving an 'X–Y' pen recorder read-out, and the faster rates are generally used when displaying information on the 'CRO' screen. All solid-state sources cover, individually, approximate octave bands from around 0.5 to 18 GHz. Typical r.f. output power is 50–100 mW and the entire system will operate continuously over the very wide frequency range. (The directional coupler may have to be changed for lower frequencies, in the hundreds of megahertz bands.) With some adaptation, the same system configuration may be used, in coaxial technology, up to about 34 GHz or 50 GHz in special cases. Beyond this some waveguide systems are available to well over 100 GHz. At these higher frequencies the r.f. sources are usually backward-wave oscillator tubes and the available output power is often lower than that obtainable when the frequency is below 18 GHz. For some measurements, such as Q-factor, the flatness of the r.f. power as a function of frequency is important and the 'raw' generators cannot directly provide a sufficiently flat output. In such cases an external feedback 'levelling loop' must be connected. This consists of a directional coupler and a detector which feeds a correcting d.c. signal to levelling PIN diodes *that are usually already provided within the SSG* (but this should always be checked).

Item (b), the frequency counter, ensures that any required single frequency can be accurately recorded directly. Most commercially available counters

operate up to at least 18 GHz on direct input, the first stage of frequency down-conversion is often implemented by means of a phase-locked converter. Higher frequencies can be accommodated with extra conversion and suitable scale calibration. Accuracies of ±2 MHz on 10 GHz are generally achieved.

Item (c), the 'CRO' indicator, displays the *relative* signal amplitude as a function of frequency. The ratio of the reflected signal amplitude to that of the incident signal (reference) is continuously displayed. These signals are actually introduced to the main CRO signal-processing electronics after down-conversion to lower frequencies, and the down-converter has been omitted from Figure 7.7 since it can be incorporated within the CRO unit. Amplifiers and precision attenuators enable accurate calibration of the vertical scale to be achieved and typical scalings are 0.25 dB/div., 1 dB/div., 5 dB/div., and 10 dB/div.

Good synchronization is essential between the sweep-generating circuits of the SSG and the horizontal sweep within the 'CRO'. In practice careful adjustment of the SSG sweep rate is necessary. Slow *manual* sweep of the SSG output is most useful for finally selecting identified points on the characteristic; the frequency and relative amplitude of these points can then be recorded.

7.6 MEASUREMENT OF SUBSTRATE PROPERTIES

Unlike many coaxial lines and waveguides microstrip depends inherently on the properties of its dielectric substrate. Suitable substrates were described in Chapter 2. The most significant property is the permittivity and its variations.

In the early stages of preparing microwave integrated circuits, substrates are easily obtained that are metallized all over their broad faces. This structure, shown in Figure 7.8(a), can be treated as a parallel-plate resonator to which an r.f. signal may be coupled (Figure 7.9). The permittivity of the substrate material is then deduced from the expression

$$\epsilon_{\mathrm{r}} = \left(\frac{c}{2lf_{n,m}}\right)^2 (n^2 + m^2) \tag{7.3}$$

where ϵ_{r} is strictly the value in a direction normal to the broad faces (important—especially in the case of anisotropic substrates), l is the length of the substrate sides, and n, m are mode numbers.

It is possible to form an almost totally enclosed rectangular resonator by metallizing not only the broad faces but also the narrow edges of a substrate. An aperture is then needed for coupling; the arrangement is illustrated in Figure 7.10(a) with extra details of the field in the coupling region given in Figure 7.10(b). Errors involved with such resonator measurements have

(a) PARALLEL-PLATE RESONATOR

(b) DIELECTRIC-FILLED, ALMOST WHOLLY-METALLIZED RESONATOR

Fig. 7.8. Metallized substrates forming rectangular microwave resonators

(a) OVERALL VIEW

(b) COUPLING DETAILS
(only E-Field shown)

Fig. 7.9. Coaxial line coupling to a parallel-plate resonator. (© 1973 IEEE. Reprinted, with permission, from Ladbrooke et al.[19])

been described by various workers. Howell[18] concentrated on the effects of losses and fringing, Ladbrooke et al.[19] gave details of coupling errors, and Owens et al.[20] pursued a reexamination of the results given by both Howell and Ladbrooke. The deductions due to Owens et al.[20] are significant because

Fig. 7.10. Coupling via an aperture to an almost wholly metallized resonator. (© 1973 IEEE. Reprinted, with permission, from Ladbrooke *et al.*[19])

the permittivities of sapphire were under study, and these values have already been accurately determined by independent methods. Although it is very difficult to account accurately for a correction due to edge fringing, some adjustment must be made for this in order to achieve accurate results.

7.7 MICROSTRIP RESONATOR METHODS

Resonator methods are generally preferred for microstrip measurements because other conceivable methods suffer from significant difficulties which will first be briefly described.

Microstrip short-circuit terminations which have definable characteristics are very hard to implement (see Chapter 5) and matched loads having good quality with low VSWR are difficult to achieve. Any *transmission* measurements in microstrip therefore have to be made through transitions to other lines (usually coaxial) in which matched loads, etc., are available. Errors are then introduced due to the microstrip-to-coaxial transitions, and the influence of these errors on required parameters can be very difficult to establish, although substantial improvements in transition design have recently been achieved. Standing-wave measurement techniques have also been developed, to a limited extent, but the precision obtainable does not yet approach coaxial or waveguide standards.

If a microstrip line is formed as some closed loop on a substrate, or left open- or short-circuited at both ends, it represents a resonator at certain microwave frequencies. By rather loosely coupling an r.f. test signal to such a resonator, so that frequency-pulling effects are avoided, the reflected energy can be studied and resonator characteristics may then be deduced.

This section deals with a variety of resonator configurations useful for determining quantities such as effective microstrip permittivity (ϵ_{eff}) and various discontinuity parameters. Q-factor measurements are described in Section 7.8 and some measurement techniques for parallel-coupled microstrips are outlined in Section 7.9.

7.7.1 The ring resonator

This arrangement seems to be due to Troughton,[21,22] principally for dispersion measurements, and results obtained using ring resonators have been quoted in support (or otherwise) of certain dispersion theories and other measured results. Although other shapes have been tried, the most popular design is the simple circular ring shown in Figure 7.11.

Since the ring has no open ends it is almost free of radiative losses and may be used to determine line losses alone. For rings in which $w/h < 1$ and l (mean circumference) $\gg w$ we have, at resonance,

$$l = \frac{n\lambda_{\text{g}}}{2} \tag{7.4}$$

where n is the integer order of resonance. If dispersion measurements are to be performed, then $\epsilon_{\text{eff}}(f)$ is required and, since $c/f\sqrt{\epsilon_{\text{eff}}(f)} = \lambda_{\text{g}}$, eqn (7.4) becomes

$$\frac{nc}{2f\sqrt{\epsilon_{\text{eff}}(f)}} = l \tag{7.5}$$

With f in gigahertz and l in millimetres, this gives

$$\epsilon_{\text{eff}}(f) = \left(\frac{150n}{fl}\right)^2 \tag{7.6}$$

Fig. 7.11. Ring resonator

For n resonances with a particular ring, n values of $\epsilon_{eff}(f)$ are therefore obtainable.

The main advantage usually claimed for ring resonators is the freedom from open-end effects. However, while this appears to be an important advantage, it by no means signifies that the electrical length of the ring can be accurately established, for the following drawbacks apply to ring resonators:

(a) curvature effects, which result in the mean physical circumference of the ring not being the most accurate value to use in calculating resonance wavelengths
(b) local radial distortion of the field in the vicinity of the coupling gap
(c) field interactions across the ring
(d) the possibility of resonance splitting due to width non-uniformities around the ring.

The problem of curvature effects, (a), appears to have been satisfactorily resolved by Owens[23] who demonstrated that a curvature correction can be applied to greatly improve the accuracy of ring measurements of dispersion. Both the measurements and the computations were very carefully performed and were compared independently. The cited paper should be consulted for further details.

It is clear that the curvature problem is greatest for wide ($w/h \gg 1$) rings of relatively small diameter. The drawback of field interactions, (c), is also greater for $w/h \gg 1$ and for rings of relatively small radii. Therefore, if rings of sufficiently large radius are used, problems (a) and (c) are diminished and the overall effect of (b) is reduced since the coupling region occupies a smaller fraction of the total ring length. But it must be remembered that a larger ring increases the chance of variability not only in width but also in the substrate permittivity and thickness under the line, and the possibility of resonance splitting, (d), still remains.

7.7.2 The side-coupled, open-circuit-terminated, straight resonator[21,22]

The microstrip configuration for this structure is shown in Figure 7.12. This microstrip resonator, in common with several other structures which are used, has abruptly open-circuited ends. The behaviour of this 'discontinuity' is such that a considerable proportion of the field fringes beyond the physical ends of the line, and the effect of this is best accounted for by considering the line to be somewhat longer electrically.

This situation has already been dealt with (see Section 5.2.1) and the same notation will be used here. There are two open-end-effect extensions to be considered, as indicated in Figure 7.13. It is apparent that a total *effective* length of $(l + 2l_{eo})$ exists, where l_{eo} is a hypothetical extension of

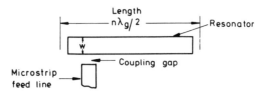

Fig. 7.12. Side-coupled open-ended straight resonator

Fig. 7.13. Illustrating the effective length of an open-ended straight resonator

the line to make up for the end-fringing. Also, the electrical length θ can be written as

$$\theta = \beta(l + 2l_{eo}) \tag{7.7}$$

With similar considerations to those used for the ring resonator, the dispersion expression for this straight resonator becomes

$$\epsilon_{eff}(f) = \left\{ \frac{150n}{f(l + 2l_{eo})} \right\}^2 \tag{7.8}$$

This particular straight resonator technique may be occasionally useful, but it possesses the following disadvantages:

(a) There is the local transverse distortion of the field in the region of the side coupling gap. This is an identical problem to (b) mentioned under ring resonators, and it may be expected to lead to even larger errors with straight resonators because the coupling region can occupy a relatively greater fraction of the total line length.
(b) The position of the feed line, relative to either open end of the resonator, is fairly important. A symmetrical location at the centre of the resonant line may often be best. In any event, other positions lead to limitations over the number of resonances which may be excited.

Positioning the feed line at right angles to the resonator and in the vicinity of the open-circuit end complicates considerably the fringing fields there and is thus undesirable.

7.7.3 Series gap coupling of microstrips

The properties and parameters of a symmetrical series gap were covered in Section 5.3. A considerable amount of success has been enjoyed in several areas of microstrip measurements and in *some* filter designs by the use of end-to-end series gap coupling (these types of filters are considered in Chapter 8).

There are, in general, two possibilities for feeding with series gap structures where the feed line may have a characteristic impedance Z_{02} differing considerably from that of the feed line Z_{01}. The types of geometry which are possible are indicated in Figure 7.14.

Coupling configurations (a) and (c) ('unsymmetrical longitudinally') are often disadvantageous for at least two reasons:

(1) It is hard to predict the degree of coupling, let alone assess the effects of adjusted gaps.
(2) These configurations have not, as yet, been analysed theoretically. Equivalent circuits and parameters are therefore unavailable.

Coupling configurations (b) and (d) overcome these objections.

Practical coupling gap designs for microstrip measurement resonators are constrained by a prime requirement that the degree of coupling be kept as loose as possible while still offering a tolerable signal-to-noise ratio for measurement. Even slight loading of the resonator due to the external circuit will affect the resonant frequency. Furthermore, all the techniques

(a) UNEQUAL WIDTHS AT GAP (b) EQUAL WIDTHS AT GAP

CASE WHERE $Z_{01} < Z_{02}$

(c) UNEQUAL WIDTHS AT GAP (d) EQUAL WIDTHS AT GAP

CASE WHERE $Z_{01} > Z_{02}$

Fig. 7.14. Practical microstrip series-gap-coupled arrangements

about to be outlined employ *two* coupling gaps. These must be manufactured as nearly as possible identically in each case, because otherwise the gap end-effect lengths must definitely differ due to the differing capacitance values. Errors will then result since the expressions applicable to each measurement method assume equal coupling gaps and hence identical end-lengths. Different end-effect lengths at the gaps result in too many unknowns, thus prohibiting accurate solution of the resonance expressions.

These circuits do not have the disadvantages which apply to the arrangements discussed in the previous paragraphs. They do, however, all involve the end-effect lengths both at the gaps and at the abruptly open-circuited ends.

Richings[24] performed experiments to ascertain suitable coupling gap separations (g) for resonators on alumina, and the present writer has carried out other investigations where the substrate material is sapphire. For resonators on 0.5 mm thick monocrystalline sapphire the coupling gap separation may be calculated approximately from the empirical expression

$$g = \frac{8,000\log Z_0}{Z_0} - 148 \ \mu m \tag{7.9}$$

where Z_0 is the characteristic impedance of the microstrip line forming the resonator.

Use of this formula ensures that fairly loose coupling is maintained. When resonators are to be constructed on alumina the gap needs to be somewhat greater for similar conditions to apply. In the case of 0.65 mm thick alumina substrates, with permittivity approximately 10.0, gap separations are approximately 30 per cent greater than those calculated for sapphire. This is because the thicker material and lower permittivity both serve to increase the influence of the series capacitance and hence to increase the coupling.

7.7.4 Series-gap-coupled straight resonator pairs[25]

This arrangement is particularly useful for dispersion measurements and Figure 7.15 shows both the physical layout and effective lengths involved. In this method the *combined* end-effect length l_e is eliminated for the purpose of calculating the effective microstrip permittivity $\epsilon_{eff}(f)$ using measured resonant frequencies and lengths of each resonator. For this technique to operate successfully we make $l_2 \simeq 2l_1$. (Incidentally, this means that l_2 resonates at many more frequencies than l_1, which has the extra advantage of providing further measurement data.) Refer to Figure 7.15 and assume that:

(a) l_{eo} and l_{eg} are both independent of resonator length so that $l_{eo1} = l_{eo2} = l_{eo}$.

(b) $l_e = l_{eo} + l_{eg}$ (combined end-effect length). Then

(a) Practical arrangement

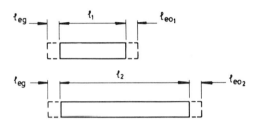

(b) Effective lengths of the resonators

Fig. 7.15. Open-circuited pairs of straight resonators

$$l_1 + l_e = \frac{n\lambda_{g1}}{2} \tag{7.10}$$

and

$$l_2 + l_e = n\lambda_{g2} \tag{7.11}$$

where the unknowns are λ_{g1}, λ_{g2}, and l_e, and n is the (integral) order of resonance.

From these expressions, eqns (7.10) and (7.11), eliminating l_e:

$$l_2 - l_1 = n\lambda_{g2} - \frac{n\lambda_{g1}}{2} \tag{7.12}$$

Now, there is the general relationship between $\epsilon_{\text{eff}}(f)$ and wavelengths:

$$\epsilon_{\text{eff}}(f) = \left(\frac{\lambda_0}{\lambda_g}\right)^2 \tag{7.13}$$

Therefore we may write

$$\epsilon_{\text{eff}}(f) = \left(\frac{c}{\lambda_{g1}f_1}\right)^2 = \left(\frac{c}{\lambda_{g2}f_2}\right)^2 \tag{7.14}$$

where f_1 applies to λ_{g1} and f_2 applies to λ_{g2}. We may allow the same $\epsilon_{\text{eff}}(f)$ value for each result since f_1 and f_2 have deliberately been forced to be close values in designing the resonators, and $\epsilon_{\text{eff}}(f)$ changes slowly with frequency.

Equation (7.12) therefore becomes

$$l_2 - l_1 = \frac{nc}{f_2\sqrt{\epsilon_{\text{eff}}(f)}} - \frac{nc}{2f_1\sqrt{\epsilon_{\text{eff}}(f)}} \tag{7.15}$$

and solving this for $\epsilon_{\text{eff}}(f)$ gives

$$\epsilon_{\text{eff}}(f) = \left\{ \frac{nc(2f_1 - f_2)}{2f_1 f_2(l_2 - l_1)} \right\}^2 \tag{7.16}$$

This expression may be used to obtain dispersion results in cases where both resonators, l_1 and l_2, are resonant at close frequencies.

It is also necessary to evaluate the combined end-effect length l_e. Equations (7.10) and (7.11) can readily be solved for l_e. Initially dividing the equations eliminates several common quantities (again assuming $\epsilon_{\text{eff}}(f)$ to be constant):

$$\frac{l_1 + l_e}{l_2 + l_e} = \frac{\lambda_{g1}/2}{\lambda_{g2}} = \frac{c/2f_1\sqrt{\epsilon_{\text{eff}}(f)}}{c/f_2\sqrt{\epsilon_{\text{eff}}(f)}} = \frac{f_2}{2f_1}$$

and after a little manipulation this yields the required end-effect length:

$$l_e = \frac{f_2 l_2 - 2f_1 l_1}{2f_1 - f_2} \tag{7.17}$$

An approximate *a priori* knowledge of l_e is required so that $l_2 = (2l_1 + l_e)$, which ensures that f_1 and f_2 are close frequencies.

The special requirements arising from this method are that, for each set of results, a pair of resonators must be made with identical widths and coupling gaps. Both resonators may lie on the same substrate, or each may occupy one of a pair of substrates with very closely matched thicknesses, in order to accommodate a long resonator across a diagonal or to avoid resonator interaction.

Additional values of $\epsilon_{\text{eff}}(f)$ can be obtained from the odd-mode resonances of the longer resonators, where

$$l_2 + l_e = (2n - 1)\frac{\lambda_{g2}}{2}$$

$$= (2n - 1)\frac{c}{2f_2\sqrt{\epsilon_{\text{eff}}(f)}} \cdot \tag{7.18}$$

or

$$\epsilon_{\text{eff}}(f) = \left\{ \frac{(2n - 1)c}{2f_2(l_2 + l_e)} \right\}^2 \tag{7.19}$$

where the appropriate value of l_e is interpolated between values calculated from the adjacent frequencies, using eqn (7.17).

7.7.5 The resonant technique due to Richings and Easter[24,29]

These appear to have been the first to devise an experimental technique enabling separation of the end-effects due to the series gap and the foreshortened open circuit. The new feature which is introduced is a second series-gap-coupled quarter-wavelength line resonator. As in the previous, paired resonators, method at least a very approximate *a priori* knowledge of the end-effect lengths is required. The physical structure is shown in Figure 7.16.

The quarter-wavelength line, of physical length l_2, transforms the open circuit at the foreshortened end of l_2 to an effective short circuit in the vicinity of the gap g_2. This is the crucial feature of the method, because it means that with two gaps the initial circuit can be solved (with length l_1) involving gap end-effect extensions. For 'stage 2' of the measurements the quarter-wavelength line l_2 is removed by chemical etching. If we assume that the foreshortened open-circuit end-effect lengths are all identical and equal to l_{eo} and that the gap end-effect lengths are equal to l_{eg}, we may adopt the same notation as in Section 7.7.4 (preceding eqn 7.10). Thus we obtain the following expressions for 'stage 1' shown in Figure 7.16, resonator l_1:

$$n\lambda_g'/2 = l_1 + 2l_{eg} \tag{7.20}$$

and, with l_1 removed,

$$n\frac{\lambda_g''}{2} = l_1 + l_{eg} + l_{eo} \tag{7.21}$$

(a) Stage 1

(b) Stage 2 (l_2 Removed)

Fig. 7.16. The two-stage resonator technique originally devised by Richings and Easter

where n is the (integer) order of resonance and the frequencies must differ slightly, and hence the different wavelengths λ'_g and λ''_g.

These wavelengths are evaluated by measuring the two resonant frequencies f' and f'', knowing $\epsilon_{eff}(f)$, and applying the procedure previously described following eqn (7.13) in Section 7.7.4. The effective microstrip permittivity $\epsilon_{eff}(f)$ can either be determined independently (Section 7.7.4) or by constructing a further circuit under this method and eliminating l_{eo} and l_{eg} from the resonance expressions.

We can now evaluate l_{eg} and l_{eo}. Setting the experimentally determined quantities $n\lambda'_g/2 = x$ and $n\lambda''_g/2 = y$ we obtain, from eqn (7.20),

$$l_{eg} = \frac{x - l_1}{2} \tag{7.22}$$

and, from eqn (7.21),

$$l_{eo} = y - l_{eg} - l_1 \tag{7.23}$$

Using this technique, and also sets of varied-length resonators to accommodate different frequencies, Richings[24] obtained several useful and consistent results.

For adequate accuracy in measurement, the gaps must be made identical to within a few micrometres for resonators manufactured on alumina-type substrates.

7.7.6 The symmetrical straight resonator[26]

All previously described straight resonator measurement methods have suffered from the drawback of requiring some *a priori* knowledge of the end-effect equivalent lengths. The 'symmetrical resonator' method does not require this rather uncertain piece of design information. This arrangement also avoids the critically placed quarter-wave transforming line associated with Richings' method. The basic structure is shown in Figure 7.17.

Electrically, taking a plane situated in the physical centre of the microstrip resonator, a very nearly identical circuit is seen whether one looks towards the source (coupled in through g_1) or towards the matched load. In the network analyser measuring system the source impedance is generally very close to 50 Ω, so a matched 50 Ω load is also used. The gaps g_1 and g_2 are kept very nearly equal and set for loose coupling.

For 'stage 1', referring to Figure 7.17(a) and using the notation previously employed, the resonance equation is

$$\tfrac{1}{2}n\lambda_{g1} = l + 2l_{eg} \tag{7.24}$$

where the microstrip wavelength λ_{g1} is known from the effective microstrip permittivity $\epsilon_{eff}(f)$ obtained by another method such as the paired resonators. The resonant frequency f_1 is measured and l_{eg} becomes

(a) Stage 1.

Coaxial line with
matched load

(b) Stage 2.
(Load and load-section removed)

Fig. 7.17. Symmetrically loaded resonator technique

$$l_{eg} = \frac{nc}{4f_1\sqrt{\epsilon_{eff}(f)}} - l \qquad (7.25)$$

With the coaxial-matched load removed and the second section of line chemically etched away from the substrate we have 'stage 2' of this method. The resonance equation in this case is, referring to Figure 7.17(b) and using the previously employed notation,

$$\tfrac{1}{2}n\lambda_{g2} = l + l_{eg} + l_{eo} \qquad (7.26)$$

whence

$$l_{eo} = \frac{nc}{2f_2\sqrt{\epsilon_{eff}(f)}} - l - l_{eg} \qquad (7.27)$$

where all quantities on the right-hand side are known, including the new measured resonant frequency f_2.

This general type of resonator layout has also been used to calibrate through-reflect-lines (TRL) for MMIC on-wafer microstrip lines.[27] Calibrated versus resonator-derived loss measurements were reported over the frequency range 1–22 GHz for magnitude and from 1 to 26 GHz for phase. Both reflection coefficient S_{11} and forward transmission S_{21} were measured for microstrip and also coplanar lines. It was concluded that microstrip provides a superior TRL capability, given the relatively good agreement between calibrated and independent measurements.

7.7.7 Resonance methods for the determination of discontinuities other than open circuits[28]

In order to measure quantities *other than* equivalent discontinuity capacitance it is necessary to set up resonator arrangements which ensure effective

cancellation of the capacitive effect in the vicinity of the discontinuity. For the determination of inductance, a voltage node must be placed in the discontinuity region and use is made of the fact that such a node will occur $\lambda_g/4$ away from a voltage maximum (at a true open-circuit plane). Thus, it is important to very accurately know wavelength and open-end effect for the line concerned. The arrangement shown in Figure 7.18(a)[29] indicates how through-arm inductances L_1 may be measured. The measurement of L_3, and of capacitance C, is indicated in Figure 7.18(b) and (c) respectively.

In all cases the parameters are first determined as equivalent extra lengths of the resonator, and then finally as lumped equivalent elements. For the measurement of the impedance ratio n^2 it is necessary that the transformed impedance (in the vicinity of the junction) be finite and also well defined. To obtain this the intermediate eighth-wavelength sections of line are used since these yield input reactances equal to $-jZ_0$ Ω.

In order to complete the measurements, the secondary arm of the T-junction has to be removed by etching-off, and the circuits have to be re-measured. The demands on the accuracy of initial circuit fabrication and the etching-off operation are generally high. For further details the cited reference should be consulted.

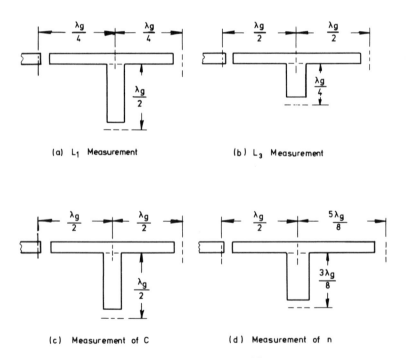

(a) L_1 Measurement (b) L_3 Measurement

(c) Measurement of C (d) Measurement of n

Fig. 7.18. T-junction measurement arrangements. (© 1975 IEEE. Reprinted, with permission, from Easter[29])

7.8 *Q*-FACTOR MEASUREMENTS

Reliable techniques for the determination of the *Q*-factor of a microwave resonator are well known. Fairly recently network-analyser-compatible swept-frequency techniques for measuring the *Q*-factor have been well described in a comprehensive paper by Aitken.[30] These techniques are well suited to microstrip resonator measurements where, for example, the instrumentation system shown in Figure 7.7 is in use.

Both the *Q*-factor and the coupling coefficient are obtained using the techniques described by Aitken. It is indicated that errors in measured *Q*-factors can be kept below 10 per cent provided the coupling coefficient η (Aitken's β) is within the range

$$0.15 \leq \eta \leq 0.75$$

Below 0.15 the signal-to-noise ratio becomes very poor and the measurement accuracy suffers, although the accuracy can be improved if signal-averaging techniques are used. For straight microstrip resonators the coupling gap separation given by eqn (7.9) should ensure that η lies within the range given, but some adjustment may occasionally be required. Coupling gaps for ring resonators require experimental optimization.

Measurements of the *Q*-factors of ring resonators yield the losses due, predominantly, to conductor resistivity and substrate dielectric losses. There is very little loss due to radiation.

On the other hand, measurements conducted using open-circuit-ended straight resonators are influenced by the 'end-effects'. The *Q*-factor due to these end-effects alone may be found in terms of the reflection coefficients ρ at each end. This is useful since, once *Q* is measured, ρ is known and the parameters which describe the behaviour of the open circuit can be obtained. To simplify the analysis we start by assuming that the microstrip resonator is isolated,[31] i.e. the coupling does not influence the *Q*-factor significantly. The situation is shown in Figure 7.19.

It is shown in Appendix B that the external *Q*-factor applicable to this type of resonator can be expressed in terms of the end reflection coefficient ρ as:

Fig. 7.19. Equivalent equal loads $(G + jB)$ at both ends of a straight, symmetrical, open-ended resonator

$$Q_e = \frac{\pi l \rho}{1 - \rho^2} \qquad (7.28)$$

(This is given in Appendix B as eqn (B.23).)

By simple substitution in eqn (7.28) we can obtain the results for end reflection coefficients, given measured external Q-factors.

For example, an external Q-factor of 29.8 leads to an end reflection coefficient of 0.9. This would be a poor Q-factor if it applied to a resonator for most circuit design purposes and should more likely be the attribute of some resonant structure designed for a microstrip antenna (see Chapter 10). In contrast a measured external Q-factor of 312 (fairly typical of microstrip in circuits) corresponds to an end reflection coefficient of 0.99.

It should be noted that the actual measured Q-factor will also include both conductor and dielectric losses which are accounted for by an unloaded Q-factor, Q_u. Therefore the total measured Q for the resonator is given by

$$(1/Q) = (1/Q_u) + (1/Q_e) \qquad (7.29)$$

Since Q_u can casily be measured, using the simple arrangements described earlier in this chapter (sections 7.5 and 7.7.1), Q_e becomes the only unknown in eqn (7.29) and can readily be found.

7.9 MEASUREMENTS ON PARALLEL-COUPLED MICROSTRIPS

The basic principles of parallel-coupled microstrip lines, and design procedures using them, were discussed in Chapter 6.

In order to check any theory and hence develop sound designs, it is highly desirable that adequate measurements can be performed to support the theoretically calculated results. Gupta et al.[1] have quite fully described some test arrangements for determining coupled microstrip line parameters. Particular parameters required are:

even- and odd-mode characteristic impedances
even- and odd-mode phase velocities
coupling factor (principally the mid-band, i.e. maximum value).

Gupta describes the measurement technique, originally given by Napoli and Hughes,[32] in which the even and odd modes are separately excited and outputs from the coupled lines are fed into the 'reference' and 'test' channels of a microwave network analyser. Resonant, coupled, half-wavelength lines are used and the wavelengths applicable to each mode are found from the resonant frequencies. It is suggested that a slotted line be inserted in the test channel input to determine the VSWR values, and hence the characteristic impedances from

$$Z_{0e} = Z_0\sqrt{r_e} \qquad (7.30a)$$

and

$$Z_{0o} = \frac{Z_0}{\sqrt{r_o}} \qquad (7.30b)$$

where r_e and r_o are VSWR values for the even and odd excitations respectively.

However, it would seem that the VSWR could equally well be obtained from reflection coefficient measurements using the network analyser. (The required interrelationship is given as eqn 1.17 in Chapter 1.) Since, especially with 'loose' coupling, Z_{0e} and Z_{0o} do not differ greatly the measurements of r_e and r_o must be accurate. The value of Z_0 must also be known with confidence.

Richings[24] points out that Napoli and Hughes' measurement technique is fairly complex, requiring four identical coupling gaps and four transitions. Richings has given details of a single-port measurement technique, the basic circuit for which is shown in Figure 7.20. This technique has been used for open microstrip circuits and also for shielded circuits. Careful measurements account for the end-effects, and the resonance conditions are

$$\frac{n\lambda_{ge}}{2} = l + l_{ege} + l_{eoe} \qquad (7.31a)$$

and

$$\frac{n\lambda_{go}}{2} = l + l_{ego} + l_{eoo} \qquad (7.31b)$$

where the subscripts, and the equivalent end-effect lengths, have the meanings previously given. Two distinct resonances are observable, leading

Fig. 7.20. Resonant method for determining the phase velocities in two parallel-coupled microstrips

to the measurement of two frequencies f_{oe} and f_{oo}. Effective permittivities and phase velocities can then be evaluated from the expressions

$$\epsilon_{effe} = \left(\frac{c}{f_{oe}\lambda_{ge}}\right)^2 \qquad (7.32a)$$

$$\epsilon_{effo} = \left(\frac{c}{f_{oo}\lambda_{go}}\right)^2 \qquad (7.32b)$$

$$v_{pe} = \frac{c}{\sqrt{\epsilon_{effe}}} \qquad (7.33a)$$

and

$$v_{po} = \frac{c}{\sqrt{\epsilon_{effo}}} \qquad (7.33b)$$

The accuracy of this technique is subject to the limitations discussed for the measurement of single line parameters (Section 7.7.5). Further details are given in Richings' thesis,[24] where power-loss measurements are also described.

Richings has given a different circuit for the determination of the coupling factor (Figure 7.21). As described in Appendix A (the comments following eqn A.37) the maximum coupling occurs when the coupled region extends for a quarter-wavelength. This is provided by the overlapping length, $l' = \lambda_{gm}/4$ in Figure 7.21. The resonance condition for this circuit is

$$\frac{Z_0}{\tan\theta} + \frac{1}{2}\left(\frac{Z_{0e}}{\tan\theta_e} + \frac{Z_{0o}}{\tan\theta_o}\right) \pm \frac{1}{2}\left(\frac{Z_{0e}}{\sin\theta_e} - \frac{Z_{0o}}{\sin\theta_o}\right) = 0 \qquad (7.34)$$

The appropriate values of Z_{0e} and Z_{0o} are obtained by theoretical calculation and the electrical lengths ($\theta_{e,o}$) are determined by an iterative

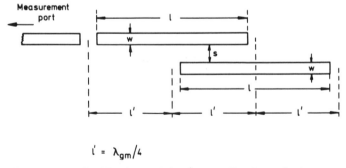

$$l' = \lambda_{gm}/4$$

Fig. 7.21. Resonant method for determining the coupling factor between two parallel-coupled microstrips

method using a computer. Thus the coupling factor can also be evaluated.

Rizolli[33] has demonstrated a method using properly terminated parallel-coupled microstrip to ensure separate excitation of even and odd modes. Rizolli's method requires short circuits to ground, which can be effectively implemented in microstrip at high microwave frequencies (see Section 5.4), but which become difficult at millimeter-wave frequencies.

Babu and Banmali[34] have described a technique for measuring coupled microstrip line parameters without the use of a network analyser. Only a signal generator, a conventional standing-wave indicator, a variable short-circuit stub, and two coaxial-to-microstrip transitions are required. These workers show how scattering-parameter measurements can be performed, with corrections for the transitions, to obtain the characteristic impedances, phase velocities, and losses. Circle diagram techniques are employed and the cited reference should be read for further details.

It is difficult to assess the accuracy of this method from the given information because so few final results are actually quoted.

7.10 STANDING-WAVE INDICATORS IN MICROSTRIP

All the measurements described so far in this chapter have required transitions from the measuring system transmission medium (often coaxial line) to the microstrip circuit under study. Even the best, carefully compensated transitions must interfere to some extent with reflections from the microstrip circuit. This fact prompted the development of some standing-wave indicators (SWI) in microstrip.

Possibly the most thoroughly designed microstrip SWI is the one reported by Ladbrooke.[35] In this indicator the circuit, whose characteristics are to be studied, is made to move in relation to a fixed probe. Instead of a slot, as used with conventional coaxial or waveguide SWI, a thin dielectric leaf has the circuit pattern printed on it and this leaf is added to the basic structure. Ladbrooke employed a mica slice for the thin leaf and pressed this down uniformly on to the main substrate by using a backing of expanded polystyrene ($\epsilon_r = 1.04$). The reference given is recommended for further details.

It is probably unlikely that a microstrip SWI would be useful in *routine* microstrip measurements. This is because computer-augmented automatic correction is readily achievable, effectively removing the parameters of the transitions from the 'raw' measurements made in coaxial line or waveguide systems. Microcomputers of ever-increasing power-to-cost ratios tend to make this approach very attractive.

However, a particular virtue of a microstrip SWI is the facility for probing the fields around, and even substantially away from, the circuit structure.

In his paper Ladbrooke,[35] for example, describes the measurement of a wave launched from the foreshortened open-circuit end of a microstrip line. This is predominantly due to the surface-wave fields discussed in Section 4.10.

7.11 TIME-DOMAIN REFLECTOMETRY (TDR) TECHNIQUES

The basic principles and operating practices associated with TDR techniques, in which a pulse is transmitted through the circuit, are well known and have been published elsewhere. Although *frequency-domain* studies generally yield accurate information regarding the performance of microwave circuits and systems, TDR possesses the following possible advantages:

(a) immediate recognition of the nature of various elements, by waveform examination

(b) approximate location of specific elements or discontinuities in the circuit or system

(c) fairly simple instrumentation.

Since the operation of TDR depends inherently on pulse reflections from the elements under study we return to the reflection coefficient Γ caused by any load Z_L

$$\Gamma = \frac{Z_L - Z_0}{Z_L + Z_0} \tag{7.35}$$

In several simple but significant cases this is easily interpreted, e.g.

Matched load:

$$Z_L = Z_0$$

and hence $\Gamma = 0$

There is no reflection and an incident pulse transmitted will not suffer any distortion due to this load.

Short circuit:

$$Z_L = 0$$

and hence $\Gamma = -1$

This means that an incident pulse transmitted will suffer complete cancellation in the round-trip time required for the pulse to reach and return from the short-circuit location.

Thus, TDR can be effectively used for checking (broad-band) matched circuits, short circuits, and a wide variety of other elements and

discontinuities—provided that the system resolution is sufficiently precise. The *time* resolution is particularly critical.

We recall the delay time calculated for 50 Ω microstrip on an alumina substrate which was given in Section 3.11. This was 8.6 ns/m, or 8.6 ps/mm. Using this value we find that a TDR with a system pulse-rise time of about 30 ps can only resolve distance separations of approximately 2 mm, at best, along 50 Ω microstrip lines. In practice the effects of multiple reflections and noise complicate the displayed waveform considerably and limit the resolution.

Piller[36] has presented a very interesting extension of both TDR and TDT (time-domain transmission) measurement techniques. Substantial information is given and Piller claims that time constants down to approximately 0.1 ps are within the capabilities of his methods. The principal features of this technique are the careful interpretation of maximum pulse amplitude and the separate employment of a ramp waveform. Using a 30 ps TDR system, specially adapted, where the minimum voltage is 2 mV and the maximum voltage is 1 V, it is claimed that the following minimum circuit parameter values can be resolved:

With TDR: $C_g \simeq 15$ fF $L_1 \simeq 35$ pH

With TDT: $Cs \simeq 1$ fF $L_g \simeq 1$ pH

where

subscript 'g' means 'to ground'
subscript 'l' means 'lead'
subscript 's' means 'series'

Measurement precautions mentioned by Piller include:

(a) a calibration measurement with a very narrow gap (so that $C_g \rightarrow 0$)
(b) the use of a good reference point
(c) high noise rejection and jitter rejection
(d) expansion of the waveform plots or traces
(e) use of an averaging interpretation of the measurements.

Piller's paper[36] should be consulted for further details.

REFERENCES

1. Gupta, K. C., Garg, Ramesh, and Bahl, I. J., *Microstrip Lines and Slotlines*, Artech House, Inc., Dedham, Massachusetts, 1979, pp. 71–78.
2. Operating manual for the Hewlett-Packard 11608A microwave transistor measurement fixture.
3. England, E. H., 'A coaxial to microstrip transition', *IEEE Trans.*, **MTT-26**, January 1976, 47–48.

4. Schneider, M. V., Glance, B., and Bodtmann, W. F., 'Microwave and millimetre wave hybrid integrated circuits for radio systems', *Bell System Tech. J.*, July/August 1969, 1703–1726.
5. Cohn, S. B., 'Optimum design of stepped transmission-line transformers', *IRE Trans*, **MTT-3**, No. 3, April 1955, 16–21.
6. Hopfer, S., 'The design of ridged waveguides', *IRE Trans.*, **MTT-3**, October 1955, 20–29.
7. Menzel, W. and Klaassen, A., 'On the transition from ridged waveguide to microstrip', *Proceedings of the 19th European Microwave Conference*, London, Sept. 1989, 1265–1269.
8. Van Heuven, J. H. C., 'A new integrated waveguide-microstrip transition', *IEEE Trans.*, **MTT-24**, March 1976, 144–147.
9. Wu, Y.-S., Schneider, M. V. and Trambarulo, R., 'Waveguide-to-Microstrip power splitter', *IEEE MTT-S International Microwave Symposium* 1990, Digest of Papers, 475–478.
10. Hall, P. S., and James, J. R., 'Analysis and design of triplate corporate feeds at high frequencies', *European Microwave Conference, Digest of Papers*, Brighton, 1979, pp. 106–110.
11. Robinson, G. H., and Allen, J. L., 'Slotline application to miniature ferrite devices', *IEEE Trans.*, **MTT-17**, 1969, 1097–1101.
12. Gupta, K. C., *et al.*, *Microstrip lines and Slotlines*, Artech House, 1979, pp. 234–244.
13. Cullen, A. L., 'The six-port and the microprocessor in microwave measurements', *European Microwave Conference, Digest of Papers*, Brighton, 1979, pp. 74–82.
14. Qin, T.-K., and Wang, L.-F., 'Design of dielectric loaded slot-line-microstrip transitions', *Proceedings of 2nd International Symposium on recent Advances in Microwave Technology*, Sept 4–8 1989, 332–335.
15. Climer, B. J. 'A broadband single-sideband mixer with low spurious levels', *Proceedings of the 19th European Microwave Conference*, London, Sept. 1989, 737–742.
16. Horng, T. S., Yang, H. Y., and Alexopoulos, N. G., 'A full-wave analysis of shielded microstrip line-to-line transitions', *IEEE MTT-S International Microwave Symposium* 1990, Digest of Papers, 251–254.
17. Viitanen, A. J., and Kuismann, P. A., 'Analysing the asymmetric coupled microstrip line and the microstrip-coplanar guide transition'. *Proceedings of 2nd International Symposium on recent Advances in Microwave Technology*, Sept 4–8 1989, 887–892.
18. Howell, J. Q., 'A quick accurate method to measure the dielectric constant of microwave integrated-circuit substrates', *IEEE Trans.*, **MTT-21**, March 1973, 142–143.
19. Ladbrooke, P. H., Potok, M. H. N., England, E. H., 'Coupling errors in cavity-resonance measurements on MIC dielectrics', *IEEE Trans.*, **MTT-21**, August 1973, 560–562.
20. Owens, R. P., Aitken, J. E., and Edwards, T. C., 'Quasi-static characteristics of microstrip on an anisotropic sapphire substrate', *IEEE Trans.*, **MTT-24**, No 8, August 1976, 499–505.
21. Troughton, P., 'High *Q*-factor resonators in microstrip', *Electron. Lett.*, **4**, No. 24, 29 Nov. 1968, 520–522.
22. Troughton, P. 'Measurement techniques in microstrip, *Electron. Lett.*, **5**, 1969, 25–26.
23. Owens, R. P., 'Curvature effect in microstrip ring resonators', *Electron. Lett.*, **12**, No. 14, 8 July 1976, 356–357.

24. Richings, J. G., 'The realization of filter networks in microstrip', *PhD Thesis*, University College of North Wales, Caerns, UK. June 1973.
25. Deutsch, J., and Jung, H. J., 'Measurement of the effective dielectric constant of microstrip lines in the frequency range from 2 GHz to 12 GHz', *Nachrichtentech. Z.*, **12**, 1970, 620–624.
26. Edwards, T. C., 'Microwave design parameters for microstrip transmission lines on sapphire substrates', *M. Phil. Thesis*, Royal Military College of Science, Shrivenham, March 1976.
27. Woodin, C. and Goff, M., 'Verification of MMIC on-wafer microstrip TRL calibration', *IEEE MTT-S International Microwave Symposium* 1990, Digest of Papers, 1029–1032.
28. Easter, B., Gopinath, A., and Stephenson, I. M., 'Theoretical and experimental methods for evaluating discontinuities in microstrip', *Radio and Electronic Engineer*, **48**, No. 1/2, January/February 1978, 73–84.
29. Easter, B., 'The equivalent circuit of some microstrip discontinuities', *IEEE Trans.*, **MTT-23**, No. 8, August 1975, 655–660.
30. Aitken, J. E., 'Swept-frequency microwave *Q*-factor measurement', *Proc. IEE*, **123**, No. 9, September 1976, 855–862.
31. Owens, R. P., Private communication.
32. Napoli, L. S., and Hughes, J. J., 'Characteristics of coupled microstrip lines', *RCA Review*, **31**, 1970, 479–498.
33. Rizzoli, V., 'Resonance measurement of even and odd mode propagation constants in coupled microstrips', *Proc. International Microwave Symposium*, Rickey's Hyatt House, Palo Alto, CA, 12–14 May 1975.
34. Babu, G. R., and Banmali, 'Measurement of coupled microstrip line parameters', *Int. J. Electronics*, **45**, No. 5, 1978, 551–550.
35. Ladbrooke, P. H., 'A novel standing-wave indicator in microstrip', *Radio and Electronic Engineer*, **44**, No. 5, May 1974, 273–280.
36. Piller, U., 'Time-domain immittance measurements', *European Microwave Conference, Conference Proceedings*, Montreaux, 1974, pp. 61–65.

8 Passive MICs

8.1 TERMINATIONS AND ATTENUATORS

In waveguides or coaxial lines it is usually preferable for short circuits to be realized rather than open circuits since relatively high-quality short circuits can be manufactured. In microstrip the opposite situation applies, open circuits being in far greater use than short circuits, although neither form of termination can be made with coaxial or waveguide quality due to the hybrid fields associated with microstrip. These terminations are discussed in greater detail in Sections 5.2 and 5.4. The application of open-circuit terminations will be seen in conjunction with filters and other passive circuits, which are dealt with later on in this chapter, active circuits such as oscillators which are considered in Chapter 9 and also antennas, the subject of Chapter 10.

With some care in fairly simple circuit designs quite broad-band matched terminations can be realized. For frequencies up to about 12 GHz a simple rectangular resistor deposition is suitable and may be expected to yield a VSWR below 1.1. The actual deposition may be thick or thin film. The resistor is kept in-line with the microstrip and the far end is shorted to ground via 'wrap-around' metallization.

At frequencies above 12 GHz series-inductive reactance becomes an increasingly severe problem and the resistor element has to be viewed as a series LR network—at least to a first approximation. Lacombe[1] devised a fairly simple broad-band matching arrangement for a 50 Ω resistor to form a 50 Ω microstrip termination with VSWR < 1.5 over the range $1 \leq f \leq 18$ GHz. The details will not be outlined here, because considerably improved designs for matched terminations have more recently been developed. These improved designs rely upon attenuator structures, which are therefore considered first.

Rickard[2] has described a π-type microwave attenuator, manufactured in thick-film microstrip. The electrical circuit configuration and the microstrip layout are both indicated in Figure 8.1. Resistance values are $R_s = 17$ Ω

(a) Circuit topology (b) Microstrip circuit

Fig. 8.1. Network and microstrip circuit layout for a π-type attenuator

and $R_p = 290 \, \Omega$, both ± 15 per cent tolerance. To maintain lumped behaviour the resistances are made less than an eighth-wavelength long at the highest frequency (i.e. less than 1 mm). The interconnecting resistor, R_s, is formed as a 'sideways S' shape within the width of the microstrip. The resistors R_p are effectively grounded by using quarter-wave open-circuit transforming stubs, each having a characteristic impedance of 30 Ω. This attenuator had a final performance, over the frequency range $8 \le f \le 20$ GHz, as follows:

<div align="center">

Attenuation 3 dB \pm 0.2 dB

VSWR < 1.1

</div>

Such attenuators have several important applications. For example, they are used in broad-band power-levelling loops where the microwave output power is maintained constant over a wide range of frequencies.

Finlay et al.[3] have described a precision microstrip matched load using two 3 dB attenuator pads of a slightly different basic design to that described here (actually 'pseudo-tee'). These two pads are followed by a 50 Ω resistor which is grounded at its far end. The length of microstrip line between the pads is approximately $\lambda_g/4$ at 15 GHz and between the final pad and the resistor is approximately $\lambda_g/16$ at 15 GHz. An excellent performance is obtained, load VSWR being less than 1.1 for all frequencies up to 18 GHz. This is somewhat degraded by the SMA transition to the microstrip line, which results in a VSWR below 1.2 up to 13 GHz but also a peak of 1.22 at about 15 GHz. Such performance is compatible with most requirements likely to be encountered in microstrip circuits.

8.2 BRANCH-TYPE COUPLERS AND POWER DIVIDERS

Branch-type couplers differ most significantly, compared with other couplers considered earlier, in having d.c. as well as microwave coupling. Although

their inherent bandwidths are not great (typically up to about 50 per cent) they can be used for comparatively high microwave power transmission.

A branch-line directional coupler is shown in Figure 8.2(a), with the 'even'- and 'odd'-mode equivalent 'circuits' given in Figure 8.2(b). The terms 'even' mode and 'odd' mode refer to the overall effects of having voltage nodes or voltage antinodes present at the branch- or main-line junctions respectively. In Figure 8.2(a), λ_{gm} is the mid-band wavelength in the microstrip line.

Particularly significant properties of these couplers are:

(a) Special-purpose chokes or filters may be connected into the joining branches, especially if these are made (say) $3\lambda_{gm}/4$ in length rather than only $\lambda_{gm}/4$.
(b) High microwave power may be handled with very little danger of breakover.
(c) They are well suited to tight coupling requirements: 0 dB is feasible over approximately a 50 per cent. bandwidth for a six-branch coupler, provided that a VSWR of about 1.2 is tolerable. (A fourteen-branch arrangement gives a VSWR of 1.05, but naturally occupies much more substrate area.)
(d) They may be used for the purpose of impedance transformation.[4]

8.2.1 'Standard' designs

Design techniques appropriate for stripline have been described by Howe.[5] These may be translated into the microstrip form by following the procedures given in Chapters 3 and 4, and Sections 5.10 and 5.11, for the T-junctions.

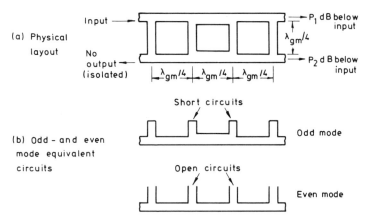

Fig. 8.2. Branch-line directional coupler. (Reproduced by permission of Wiley Eastern Limited, New Delhi, from K. C. Gupta and A. Singh, *Microwave Integrated Circuits*.)

When only four branching sections are required this branch-line coupler may be re-formed into a ring shape; this is shown in Figure 8.3. One important application for this arrangement is offered by the balanced mixer (and also phase detector) illustrated in Figure 8.4. It can be readily verified, by counting up the total number of quarter-wavelength phase shifts in each path, that the signal and local oscillator waves arrive at each Schottky-barrier diode in relative antiphase. This is a design requirement for this type of mixer. Unwanted waves are filtered out by the various low-pass filters shown on the circuit. The design of such filters is described in Section 8.3. The use of this 3 dB ring coupler enables both the signal and the local

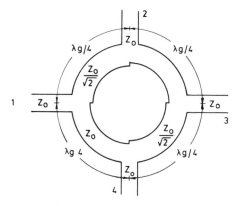

Fig. 8.3. Ring form of 3-dB branch-line directional coupler. (Reproduced by permission of Wiley Eastern Limited, New Delhi, from K. C. Gupta and A. Singh, *Microwave Integrated Circuits*.)

(a) Schematic layout

(b) Microstrip circuit

Fig. 8.4. Balanced mixer circuit based upon the structure of Figure 8.3. (Reprinted by permission of Wiley Eastern Limited, New Dehli, from K. C. Gupta and A. Singh, *Microwave Integrated Circuits*.)

oscillator powers to be coupled to the mixer diodes with very little loss, resulting in high sensitivity.

It is also worth considering the so-called 'rat-race' circuit, or 'hybrid-ring', which is shown in Figure 8.5. Output signals from ports 2 and 4 differ in phase by 180°. (In contrast to the branch-line coupler where the phase difference is 90°.)

An interesting and important design feature arises by considering the quarter-wave transformer action of this coupler. Only ports 2 and 4 exhibit this action because port 3 is half-wave separated from the input feeding port 1. Thus, the *net effective load* on the inner ring lines feeding ports 2 and 4 amounts to $2Z_0$ (two Z_0 loads appearing, equivalently, in series). Now, the characteristic impedance Z_0 of *any* quarter-wave-transforming line between two impedances Z_{01} and Z_{02} is known to equal $\sqrt{Z_{01}Z_{02}}$; this result was developed in Section 1.7. In this case, the two impedances are Z_0 and $2Z_0$ respectively, so the impedance of the intervening quarter-wave line (i.e. the ring) must be

$$Z_0' = \sqrt{Z_0 \cdot 2Z_0}$$

or

$$Z_0' = \sqrt{2}\, Z_0 \tag{8.1}$$

Thus the characteristic impedance of the line forming the ring itself must be $\sqrt{2}$ times that of the feeder line impedances. Where the impedance of all feeders is 50 Ω, the ring characteristic impedance is therefore 70.7 Ω.

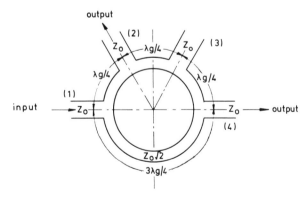

Fig. 8.5. 'Rat-race' or 'hybrid-ring' coupler. (Reproduced by permission of Wiley Eastern Limited, New Dehli, from K. C. Gupta and A. Singh, *Microwave Integrated Circuits*.)

8.2.2 Broadband and millimeter-wave branchline couplers

Considerable effort continues to be expended towards improving the design flexibility, expanding the bandwidth and extending the upper frequency capability of branchline and similar couplers. For example, Wright and Judah[6] describe a particularly broadband hybrid ring with flat coupling and Ashworth[7] develops design expressions for a two-branch broadband hybrid ring (or a two-branch guide) coupler of any coupling coefficient.

Mayer and Knöchel[8] describe branchline couplers with improved design flexibility and broad bandwidth. They show that, for a simple four-port branchline coupler, the impedances of the 'inner coupler' are given by:

$$Z_x/Z_o = \{1 + S_{11}(f_o)\}/\{1 - S_{11}(f_o)\} \tag{8.2}$$

$$Z_2/Z_o = \sqrt{-\frac{3}{2}\left(1 - \frac{Z_I}{Z_o}\right) + \sqrt{\frac{9}{4}\left(1 - \frac{Z_I}{Z_o}\right)^2 + \frac{Z_I}{Z_o}}} \tag{8.3}$$

$$Z_1/Z_o = Z_2/(Z_o\sqrt{2}) \tag{8.4}$$

These impedances refer to a branchline coupler of similar layout to one section of the device shown in Figure 8.2(a). The inner coupler is that bound by the first series and parallel arms respectively and this is also shown in Figure 8.6. Mayer and Knöchel also describe a two-branch coupler modified by the addition of half-wavelength open-circuited stubs centrally located on each inner arm as shown in Figure 8.7. The way this approach serves to broadband the coupler amounts to relocating, each side of the original centre frequency, two frequencies at which S_{11} is a maximum. Further details are described in the reference.

Mayer and Knöchel show that it is possible to increase the bandwidth from 10 per cent with an unmodified coupler to more than 40 per cent

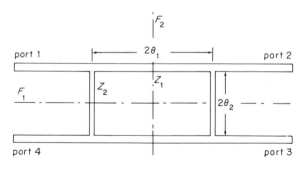

Fig. 8.6. Two-branch coupler with two-fold symmetry (θ_i = line lengths, Z_1 = characteristic impedances). (© 1990 IEEE. Reprinted, with permission, from Mayer and Knochel.[8])

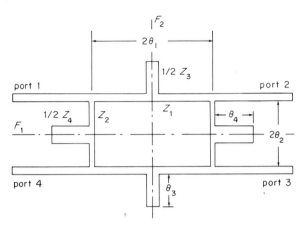

Fig. 8.7. Extended two-branch coupler with open stubs at the symmetry-planes. (© 1990 IEEE. Reprinted, with permission, from Mayer and Knochel.[8])

with the stub-modified device whilst maintaining very flat amplitude characteristics.

These workers also describe some more sophisticated broadband couplers and high-power in-phase power dividers.[9] An example of a broadbanded 0°/180° coupler is shown in Figure 8.8, where the stubs are clearly visible. A folded stub is used (Z_6) for one of the inner arms of this design. The high-power dividers described use the all-reactive broadband principle already outlined and therefore avoid the problem of the resistive element required in the Wilkinson divider and the difficulty of restricted bandwidth associated with the Gysel circuit. Bandwidths around 50 per cent are readily achievable. Other workers have also described various types of hybrid power dividers, but usually with resistive elements.

The branchline coupler principle is by no means restricted to microwave (below 30 GHz) applications and, for example, Meaney[10] has described a novel design for millimeter-wave systems. He (Meaney) points out that there are some problems with the direct use of microstrip in these types of application. In particular, there is the restricted range of realizable characteristic impedances, mentioned several times earlier in this text. There is also the fact that as the frequency increases so the aspect ratios of quarter-wavelength lines decreases, to the point where the physical lengths become shorter than the widths. This effect causes degradation of input match, bandwidth and isolation. There are also problems associated with mutual coupling and discontinuities.

By using curved inner series lines Meaney's design enables all input ports to be mutually orthogonally disposed and the discontinuities are also mostly eliminated. This coupler is based upon a three-branch structure to take advantage of the broadbanding effect of this topology and it uses three-

Fig. 8.8. Broadbanded 0°/180°-coupler. (© 1990 IEEE. Reprinted, with permission, from Knochel and Mayer.[9])

quarter-wavelength series lines together with quarter-wavelength parallel lines. Since the relatively long series lines are those with the lowest characteristic impedance and therefore are the widest, so their aspect ratios are maintained as reasonable values. The coupler provides a nominal 3 dB insertion loss (variations between 2.7 and 4.7 dB) with good return loss and isolation performance over the 26–40 GHz band.

When using curved microstrip lines it is important to account accurately for the effect of line width. For example Roy *et al.*[11] have shown that the design of rat-race hybrids is influenced markedly by the line width and these workers develop an expression for a suitable modification of this width allowing accurate design.

Branchline couplers and other similar circuit structures are readily amenable to CAD and a fairly simple design, on a workstation screen using Eesof's Touchstone™ software, is shown in Figure 8.9.

Other recent developments centre on MMIC realisation of couplers and power dividers (the topic of MMICs is considered in Chapter 9).

Tokumitsu *et al.*[12] announced a novel MMIC realisation approach using three micron/three layer dielectric films. They describe layered microstrip (thin film microstrip, or TFMS), inverted microstrip and also both normal and inverted 'quasi-coplanar' transmission structures. These workers designed and built 12 GHz four-port Wilkinson power divider circuits using vertical connections, 15 GHz branchline couplers and 20 GHz port-interchanged rat-race hybrids (using a line crossover)—all based on this technology. A more detailed exposition of the port-interchanged rat-race hybrid, shown with

Fig. 8.9. Example of CAD applied to a branchline coupler design (photo courtesy of Eesof, Inc., 1990)

comparison with the conventional topology in Figure 8.10, is presented in a separate paper by this group[13] where it is pointed out that the new version provides greater MMIC design flexibility. In addition, Tokumitsu *et al.* describe a 6-stage distributed amplifier using TFMS (distributed amplifiers are described in Chapter 9).

8.2.3 Microstrip baluns

The balun, a balanced-to-unbalanced transmission converter, was invented by Marchand in 1944 and has been available commercially in (mainly) coaxial form for several decades. MIC realizations of this type of circuit element have only emerged in comparatively recent times. Interest is stimulated primarily by the specific circuit requirements of mixers, multipliers, Class B push–pull amplifiers and antennas where balanced transmission line structures are needed to feed the circuit whereas unbalanced lines must then form the remainder of the circuit.

Baluns have been designed for both MIC and MMIC implementation. For example, Pavio and Kikel[14] report a compensated balun which provides

Fig. 8.10. Conventional and port-interchanged rat-race hybrids. (a) Circuit diagram of a conventional rat-race hybrid. (b) Circuit diagram of a port-interchanged rat-race hybrid. (Reprinted, with permission, from Nakamoto, Tokumitsu and Aidawa.[13])

an amplitude response that is flat within 1 dB over the 6–18 GHz bandwidth (both in hybrid and monolithic forms). Phase matching between the output arms is very good, being 180° ± 2°. The circuit construction is shown in Figure 8.11 in which it can be seen that suspended substrate technology is avoided and the transformation is achieved by means of a multilayer conductor structure. Tight coupling is obtained by using a very thin second dielectric layer t_n and, for the MMIC balun, this is realised with a 20 000 Å thick layer of silicon nitride on the GaAs surface.

Barber[15] has described enhanced coupled, even mode terminated, baluns and also mixers using this technology. Barber's approach starts with the creation of a planar balun circuit cell which comprises a pair of coupled transmission lines bridged at their outputs by a series-connected pair of capacitors. A final transmission line is connected from the centre point between the capacitors to a load resistor which is grounded. This overall structure forms one balun cell the circuit and physical cross-section of which is shown in Figure 8.12. To form a complete broadband balun typically eight of these cells are cascaded. One of the two input lines is fed whilst the opposite line is grounded (i.e. unbalanced) and the two outputs left floating in balanced format. The capacitors are formed in overlay technology and the appropriate line interconnections are made with vias.

These baluns are designed to operate over the 2–26 GHz band and Barber describes mixers (restricted to the 2–18 GHz band) using this design approach. For the mixers the IF port is formed by diplexing the local oscillator and IF frequencies using reactive filtering. A Schottky monolithic quad of diodes forms the mixing function and the mixers exhibit conversion loss varying between 7 and 10 dB over the band. The entire mixer, incorporating a total 16 balun cells, occupies a 6 mm square package.

(a)

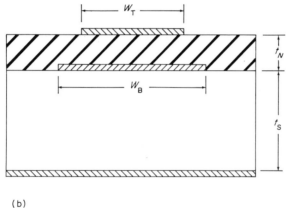

(b)

Fig. 8.11. Circuit construction and cross-sectional views of baluns. (a) Circuit construction of monolithic or hybrid balun. (b) Typical balun structure cross-section. (© 1990 IEEE. Reprinted, with permission, from Pavio and Kikel.[14])

8.3 A SUITABLE STRATEGY FOR MICROWAVE FILTER DESIGN

We shall first review the overall design strategy which may be followed for any microstrip filter. Table 8.1 shows the typical sequence which starts with a prototype specification (by the insertion-loss approach) and finishes with a practical, fully dimensioned, microstrip circuit.

(a)

1 SUBSTRATE	7 METAL VIA
2 ONE OF COUPLED LINES Tx	8 TRANSMISSION LINE NETWORK
3 ONE OF COUPLED LINES Tx'	9 RESISTOR
4 CAPACITOR Cx	10 GROUND
5 CAPACITOR Cx'	11 GROUND VIA
6 NETWORK INTERCONNECT	12 DIELECTRIC LAYER

(b)

Fig. 8.12. Planar baluns: circuit cell and cross-sections. (a) Planar balun circuit cell. (b) Cross section of a circuit cell. (© 1990 IEEE. Reprinted, with permission, from Barber.[15])

8.3.1 Low-pass filters formed with cascaded microstrips

For this design example we shall employ cascaded sections of microstrip as indicated against 3(1) of Table 8.1, although this technique is only really useful at frequencies up to several gigahertz. It is first necessary to analyse the direct substitution of lumped filter elements by such microstrip lengths.

Short ($< \lambda_g/4$) lengths of relatively high impedance line will behave predominantly as series inductances. Also, a very short ($\ll \lambda_g/4$) length of relatively low impedance line will act predominantly as a shunt capacitance.

Table 8.1

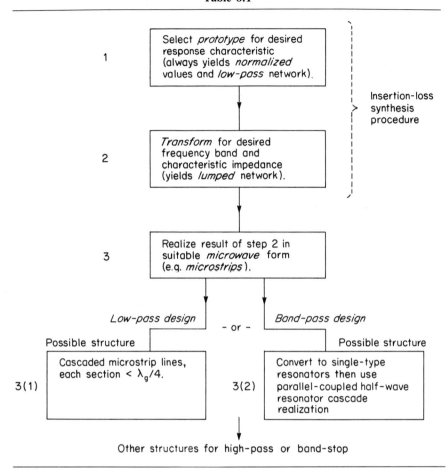

So a π-network of lumped elements can quite readily be realized with alternate sections of low- and high-impedance microstrip lines, *except that great care is necessary to fully calculate the equivalent line lengths.* In particular consider the π-equivalent arrangement applicable to a section of *any* transmission line. This was given previously, in Section 1.8 of Chapter 1, and is repeated here as Figure 8.13.

Let us first consider a predominantly inductive length of line. For short ($< \lambda_g/4$) and loss-free lines this reduces to the two equivalent circuits shown in Figure 8.14. Basic transmission line theory gives the input reactance of the line of length l:

$$X_L = Z_0 \sin\left(\frac{2\pi l}{\lambda_g}\right) \tag{8.5}$$

Fig. 8.13. Equivalent π-network of impedances representing a length l of transmission line having a propagation coefficient γ

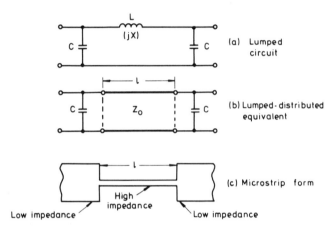

Fig. 8.14. Inductive length of line with adjacent capacitive lines

so that the length of this predominantly inductive line is

$$l = \frac{\lambda_g}{2\pi} \sin^{-1}\left(\frac{\omega L}{Z_0}\right) \tag{8.6}$$

Occasionally, the small-angle approximation can be used such that $2\pi l/\lambda_g \ll \pi/4$ and eqn (8.6) simplifies to

$$l \simeq \frac{f\lambda_g L}{Z_0} \tag{8.7}$$

But it must be emphasized that the full eqn, (8.6), will often be necessary in practice. The end susceptances B_L for the equivalent circuit, yielding values of C_L, are also given by

$$B_L = \frac{1}{Z_0}\tan\left(\frac{\pi l}{\lambda_g}\right) \tag{8.8}$$

Then, approximately, for short lengths:

$$C_{\mathrm{L}} \simeq \frac{l}{2fZ_0\lambda_{\mathrm{g}}} \tag{8.9}$$

The inductive length of line l will also be presented with adjacent lengths of low impedance and, therefore, wide microstrip lines. This situation results in a significant step discontinuity; the properties of such discontinuities were discussed in Section 5.8. The expressions for step capacitance, which were given in that section, should be used to evaluate the discontinuity capacitances here.

We have previously remarked that a short length of line having a relatively low characteristic impedance yields a capacitive element and this is shown, together with its equivalent circuit, in Figure 8.15.

The predominating shunt capacitance is determined by first considering the susceptance

$$B = \frac{1}{Z_0}\sin\left(\frac{2\pi l}{\lambda_{\mathrm{g}}}\right) \tag{8.10}$$

so that

$$l = \frac{\lambda_{\mathrm{g}}}{2\pi}\sin^{-1}(\omega C Z_0) \tag{8.11}$$

or, usefully, since these line lengths are almost always very short:

$$l \simeq f\lambda_{\mathrm{g}}Z_0 C \tag{8.12}$$

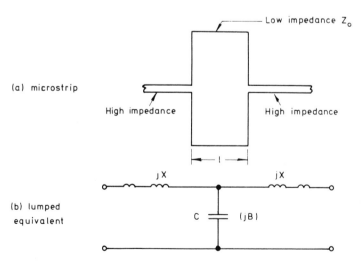

Fig. 8.15. Capacitive length of line with adjacent inductive lines

Since this low-impedance length of line has the equivalent T-network shown it must incorporate 'stray' series inductances. These inductance values will now be shown to be usually negligible. We have

$$X_L \simeq \frac{\pi l Z_0}{\lambda_g} \qquad (8.13)$$

But, *both* Z_0 and l are relatively small, so this complete reactance will be only a very small correction to the main inductive reactance of the adjacent inductive length of line.

The capacitance actually calculated for the capacitive length of line must in fact be less than the required total lumped value C_T because the inductive sections of line and the steps in width already contribute their capacitances, C_L and C_s respectively. Since these capacitances will generally differ due to the differing inductive lengths we finally have, for the capacitive line,

$$C_{line} = C_T - C_{L1} - C_{L2} - C_{s1} - C_{s2} \qquad (8.14)$$

where C_{L1} and C_{L2} are obtained from eqns (8.8) or (8.9), applied to each adjacent inductive line, and C_{s1} and C_{s2} are determined using the expressions given in Section 5.8 of Chapter 5. A more complete equivalent circuit for a 'C-L-C' portion of a filter is given in Figure 8.16. At frequencies where these structures are useful, generally only up to a few gigahertz, *inductive* effects due to the step discontinuities can usually be neglected.

A summary chart (Table 8.2) is presented showing the results applicable to inductive and capacitive elements realized in the form of microstrip lines.

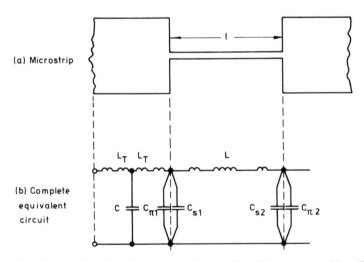

Fig. 8.16. Inductive length of microstrip line with adjacent capacitive lines and lumped equivalent circuit including discontinuity (step) capacitances and inductances

Table 8.2

Mainly *inductive* element	Mainly *capacitive* element

$$X_{\mathrm{I}} = Z_{0\mathrm{I}} \sin\!\left(\frac{2\pi l_{\mathrm{L}}}{\lambda_{\mathrm{gL}}}\right)$$

End-capacitances

$$C_{\mathrm{L}} \simeq \frac{l_{\mathrm{L}}}{2fZ_{0\mathrm{L}}\lambda_{\mathrm{gL}}} \text{(eqn 8.9)}$$

Main inductive length

$$l_{\mathrm{L}} = \frac{\lambda_{\mathrm{gL}}}{2\pi}\sin^{-1}\!\left(\frac{\omega L}{Z_{0\mathrm{L}}}\right) \quad \text{(eqn 8.6)}$$

For short-lengths *only*

$$l_{\mathrm{L}} \simeq \frac{f\lambda_{\mathrm{gL}}L}{Z_{0\mathrm{L}}}$$

$$B_{\mathrm{C}} = \frac{1}{Z_{0\mathrm{c}}}\sin\!\left(\frac{2\pi l_{\mathrm{c}}}{\lambda_{\mathrm{gc}}}\right)$$

End-inductances*

$$L_{\mathrm{C}} \simeq \frac{l_{\mathrm{c}}Z_{0\mathrm{c}}}{2f\lambda_{\mathrm{gc}}}$$

Main capacitive length

$$l_{\mathrm{c}} = \frac{\lambda_{\mathrm{gc}}}{2\pi}2\pi\sin^{-1}(\omega C Z_{0\mathrm{c}})$$

Since lengths are usually short we can often use

$$l_{\mathrm{c}} \simeq f\lambda_{\mathrm{gc}}Z_{0\mathrm{c}}C \quad \text{(eqn 8.12)}$$

*Usually negligible

8.3.2 Worked example of a low-pass filter design

We shall assume that an alumina substrate is to be used with permittivity 9.6 and thickness 0.635 mm. The specification includes the important data:

$$f_{\text{cut-off}} = 2 \text{ GHz}$$

Insertion loss = 30 dB at 4 GHz

Source and load impedances both 50 Ω

Let us further assume that a suitable prototype has been selected, to meet this specification, and that the transformed, lumped low-pass network has

been calculated (i.e. stage 2 of Table 8.1). Figure 8.17 shows the network topology, and the lumped element values are

$$C_1 = 2.55 \text{ pF} \quad L_2 = 6.25 \text{ nH} \quad C_3 = 2.1 \text{ pF}$$
$$L_4 = 3.66 \text{ nH} \quad C_5 = 0.49 \text{ pF}$$

Due to the fairly low microwave frequency of 2 GHz we can neglect the effect of dispersion in our calculations based upon this frequency.

To arrive at the general microstrip layout needed to realize this filter we just trace through, component by component, the circuit of Figure 8.17, replacing the lumped components by appropriate lengths of microstrip as we proceed. The result is illustrated in Figure 8.18.

Some *a priori* design information must be provided about the microstrip lines, because expressions for inductance and capacitance depend upon *both* characteristic impedance and length. It is reasonable to initially fix the characteristic impedances because their orders of magnitude are roughly prescribed according to

(a) Z_{0c} (capacitances) $< Z_0$ (50 Ω) $\ll Z_{0L}$ (inductances). Also, the resulting width, w_c, must not allow transverse resonances to occur at the highest frequencies of operation (Section 4.9).

Fig. 8.17. Network topology of the prototype for the 2-GHz low-pass filter in the example

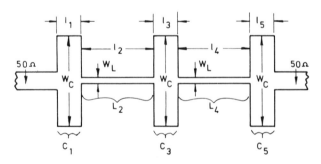

Fig. 8.18. General microstrip configuration for a five-section low-pass filter consisting of cascaded 'low-high-low' impedance lines (suitable for frequencies up to several gigahertz)

(b) Z_{0L} must not be so large that its manufacture becomes inordinately difficult—as a narrow line—or its current-carrying capability becomes a limitation.

In this case the values given below are reasonable starting values of impedance:

Capacitive lines: $Z_{0c} = 25 \, \Omega$ (this turns out not to be too wide)

Inductive lines: $Z_{0L} = 90 \, \Omega$ (this is sufficiently within the limits of manufacture for such narrow lines).

The lengths of the various sections of microstrip line are calculated for the critical cut-off frequency specified: 2 GHz. Since the same computer program based upon the methods described in Chapters 3 and 4 evaluates microstrip line widths as well as wavelengths, all these values are now given in Table 8.3. (The wavelength on a 50 Ω line is included automatically, but it will not be required in the following calculations.)

Because the parameters for the various lengths of line are interrelated we must perform initial calculations and then correct progressively via an iterative procedure as follows:

(a) Determine initial values for the lengths of microstrip lines (l_L) to replace inductances.
(b) Calculate the end capacitances associated with these lengths l_L.
(c) Calculate the step-discontinuity capacitances. (Use the expressions given in Section 5.8.)
(d) Reduce the originally specified lumped capacitance values by the appropriate amounts as calculated in (b) and (c), taking care to refer to the microstrip configuration so that the correct substractions are carried out.
(e) Calculate initial values for the lengths of microstrip lines (l_c) to replace these new capacitances.
(f) Determine the inductances of the T-sections associated with these capacitive lengths of line.
(g) Reduce the originally specified lumped inductance values by the appropriate amounts as calculated in (f).

Table 8.3

Characteristic impedance (Ω) (given data)	25	50	90 (Z_{0L})
Width (mm)	2.00	0.632	0.132
Wavelength (mm)	55.7 (λ_{gc})	64.7	65.0 (λ_{gL})

(h) *Repeat* the entire procedure, starting with the new values (from step g) in step (a) again.

When successive inductance values converge fairly closely (within about 10 per cent, say) average values can be taken to speed the procedure, with little final error. It is not necessary to show all the design calculations for this example, but some economical methods will be indicated. The lengths (a) to replace inductances are calculated from eqn (8.6):

$$l_{\rm L} = \frac{\lambda_{\rm gL}}{2\pi}\sin^{-1}\left(\frac{2\pi fL}{Z_{\rm OL}}\right) \tag{8.15}$$

If a pocket-type calculator is used 'sin^{-1}' is often in degrees and has to be converted into radians. With this modification and the actual values substituted, eqn (8.15) becomes

$$l_{\rm L} = \frac{65.0}{360}\sin^{-1}(0.1396L)\,({\rm mm}) \tag{8.16}$$

where L is nanohenries. A similar expression is obtained for capacitive line lengths, where C is in picofarads.

An accurate expression for the equivalent end capacitances due to these inductive lines can be obtained from eqn (8.8)

$$C_{\rm L} = \frac{1}{2\pi fZ_{\rm OL}}\tan\left(\frac{\pi l_{\rm L}}{\lambda_{\rm gL}}\right) \tag{8.17}$$

Again, working in degrees and substituting the known values we have

$$C_{\rm L} = 0.8842\tan(2.9032l_{\rm L}) \tag{8.18}$$

Finally, an accurate expression for the inductances associated with the capacitive sections of line can be derived from eqn (1.23a). After modifying this for degrees and making the appropriate substitutions we have

$$L_{\rm c} = 1.9894\tan(3.232l_{\rm c}) \tag{8.19}$$

This now completes the expressions required for the line sections. The step-discontinuity capacitances are also required. These may be calculated using eqn (5.22), although strictly the width ratio $w_1/w_2 = 15.2$ lies outside the range of eqn (5.22), thus incurring some error. This gives a value of 0.056 pF for the step capacitance. As an appropriate check the capacitance can also be calculated under the rather gross assumption that the 2.0 mm wide capacitive lines are almost open circuit (i.e. neglecting the existence of the 0.132 mm lines). Using eqn (5.1), given in Section 5.2, we then obtain 0.075 pF, which is not very greatly different to the previous value of 0.056 pF. It is probably reasonable to approximate the result to 0.06 pF.

To show the procedure here, the new value required to be realized for C_3 is

$$C_3' = C_3 - C_{L2} - C_{L4} - 2C_{STEP}$$
$$= 2.1 - 0.5182 - 0.2429 - 0.12$$
$$\simeq 1.21 \text{ pF}$$

When the lengths of the lines have all been fully calculated the final dimensions are obtained as outlined in steps (a) to (h) here; these dimensions are given on Figure 8.19.

In a fully implemented CAD routine the entire equivalent filter network can be completely redrawn and the new frequency response characteristics can be automatically plotted out. Adjustments are then made to the structure until the response is considered satisfactory. A practical trimming of the actual filter in order to achieve the desired performance can only rarely be avoided.

One final point must be mentioned regarding the choice of characteristic impedances. Perez[16] has emphasized that the ratio Z_L/Z_c affects the deviation of the cut-off frequency from the exact model figure. There is thus an optimum value for this ratio which seems to be around 200 to 9. Unfortunately these values cannot be realized using microstrip lines on an alumina substrate due to the width extremes which were discussed previously.

For low-pass filter design at higher microwave frequencies, up to at least 20 GHz, the 'high-low-high' cascaded sections are unsuitable due to transverse resonances, radiation, and a number of other undesirable features. High-impedance sections of line can still be used but the low-impedance sections must be discarded completely. Instead, the high-impedance lines may interconnect a series of distributed-resonant patch conductors that are usually disc shaped. Thus, circular disc resonators effectively 'replace' the low-impedance lines used at lower frequencies, but it is important to place the interconnecting lines at the correct locations for the resonant modes of these discs.[17]

An alternative approach to the design of low-pass filters has been described by Al-Charchafehi and Vassilikos.[18] These workers use 'delta stubs' (tapered microstrip stubs) and they report a filter with less than 1 dB insertion loss at frequencies below 2 GHz and 37 dB at 4 GHz.

Fig. 8.19. Complete 2-GHz microstrip low-pass filter (line widths as illustrated are disproportionately large here)

8.4 BAND-PASS FILTERS

This section deals mainly with an approach to band-pass microstrip filter design by the use of a cascade of half-wave resonators—each quarter-wave parallel coupled to its neighbours. This type of filter yields a narrow-to-moderate bandwidth; structures providing a broader bandwith are considered towards the end of the discussion.

Before proceeding with this we must study some preliminaries, beginning with the transformed (lumped-element) band-pass filter network shown in Figure 8.20. The designer would arrive at this at the end of step 2 calculations, as indicated in Table 8.1. Unfortunately, a serious difficulty is encountered with this network in its present form.

It is evident that direct realization of this filter requires the design and interconnection of *two distinctly different kinds* of resonant LC circuits (hereafter referred to as 'resonators'). Both series and parallel types of resonators would be required in any attempt to *directly* realize this filter. Any microstrip resonators which could be used and electromagnetically coupled together in some way would individually exhibit *one* type of resonance—series *or* parallel—but not both simultaneously from similar structures.

There are two principal approaches to planar coupling between adjacent resonators: *end* coupling or *edge* coupling.

8.4.1 End-coupled bandpass filters

The general layout of an end-coupled microstrip filter is shown in Figure 8.21. In practice the right-hand side of this filter would either connect to the load (the next stage) or would comprise further filter sections, dependent upon the performance requirements.

A microstrip version of this type of filter has been described by Cunningham et al.[19], to operate at a centre frequency of 28.76 GHz with a 2.78% bandwidth. The application is for a front-end filter in a down-converter for a millimeter-wave multipoint multichannel video distribution system.

Fig. 8.20. Transformed, lumped-element, band-pass filter network

Fig. 8.21. General microstrip layout for an end-coupled band-pass filter (series coupling gaps between cascaded straight resonator elements)

Further specification parameters for this filter included a local oscillator rejection of at least 10 dB (at 27.6 GHz) and an image rejection (at 26.6 GHz) of at least 35 dB. The filter's insertion loss was simply to be as low as possible. This design approach was chosen because end-coupled filters are known to be relatively narrow-band, which should suit the 2.78 per cent bandwidth demanded here. The technology was chosen to be microstrip throughout for reasons of low cost.

In all types of filters based upon transmission line resonators the coupling gaps between resonators are of prime importance and the resonator lengths obviously also represent a particularly significant design result. For these end-coupled resonator filters the coupling needs to be as tight as possible, implying gaps that are usually much smaller than the substrate height. The resonator lengths depend not only upon the microstrip wavelength, but also on coupling reactances and the effect of shunt capacitances due to the gaps. Cunningham shows that this length, excluding the effective gap lengths, is given by the following expression:

$$l_j = \frac{\lambda_g}{2\pi}(\pi - X_{j-1,j} - X_{j,j+1}) - b_{j-1,j} - b_{j,j+1} \qquad (8.20)$$

where $b_{j-1,j}$ are lengths due to gap shunt capacitances and

$$X_{j-1,j} = 0.5[\tan^{-1}(2B_{j-1,j}/Y_0)] \qquad (8.21)$$

A three-resonator filter was designed to meet the specification with a final bandwidth of 750 MHz and a center frequency of 28.78 GHz. The out-of-band response (critical in this application) was found to be superior to that of a parallel-coupled filter.

Other types of end-coupled filters have been developed, some only partially based upon microstrip but usually mainly involving microstrip-type elements. For example, Tzuang et al. have reported what they term a quasi-planar broadside filter of this general type.[20] This design, in contrast to conventional end-coupled filters, is relatively broadband (a 30 per cent bandwidth is specified). The filter structure includes a ground plane milled to provide a slot beneath the microstrip section to form suspended microstrip. The reader is referred to the paper for further details, but the amplitude response is shown in Figure 8.22.

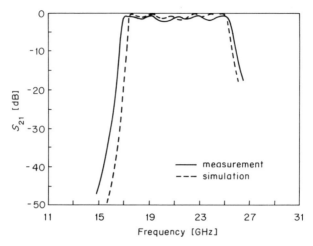

Fig. 8.22. Performance of quasi-planar broadside end-coupled bandpass filter. (© 1990 IEEE. Reprinted, with permission, from Tzuang, Chiang and Su.[20])

8.4.2 Parallel-coupled (or edge-coupled) bandpass filters

As discussed in Chapter 6 for microstrip couplers, maximum coupling is obtained between physically parallel microstrips when the length of the coupled region is $\lambda_g/4$, or some odd multiple thereof. To achieve resonance, each resonator element has to be $\lambda_g/2$ in length, or any multiple thereof. Therefore the microstrip circuit must have the general layout shown in Figure 8.23, where $l_1, \ldots, l_4 \simeq \lambda_g/4$.

We will assume, for the moment, that this fairly straightforward cascade of parallel-coupled (or 'edge-coupled') microstrip resonators can be designed on a basis of all-parallel resonator networks together with intervening circuits known as *inverters*.

Outlining the four main design steps:

(a) Determine the one-type resonator network, to realize the specification, from the original prototype.
(b) From the network parameters, evaluate the even- and odd-ordered characteristic impedances, Z_{0e} and Z_{0o}, applicable to the parallel-coupled microstrip. (This will be shown in detail presently.)
(c) Relate the values of Z_{0e} and Z_{0o} to microstrip widths and separations (w, s). Procedures for this step have already been given in detail in Chapter 6.
(d) Calculate the whole resonator length $2l'$, slightly less than $\lambda_g/2$, and therefore of the coupled-section length l', which is slightly less than $\lambda_g/4$.

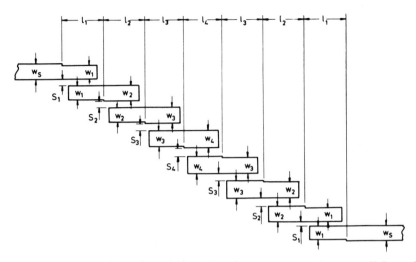

Fig. 8.23. General microstrip configuration for a seven-section, parallel-coupled band-pass filter

Here λ_g is the mid-band and average microstrip wavelength. Allowance must be made for the semi-open-circuit microstrip end-effects which exist for all elements in this circuit.

8.4.3 Impedance and admittance inverters; basic design principles

The simplest form of an impedance inverter is the quarter-wave transformer. It should be recollected from Section 1.7 that such circuit components transform any load impedance by the quantity Z_0^2, where Z_0 is the characteristic impedance of the quarter-wave transforming section of line. Characteristic impedance squared terms will arise for impedance and admittance inverters.

It has already been stated that a network consisting of admittance inverters, usually called J-inverters, and uniform-type resonant circuits is to be made equivalent to the original network consisting of two types of resonant circuit—series and parallel. While the impedance level will vary through the system, two features must be maintained for equivalence:

(a) The resonant frequency must remain constant: $\omega_0^2 = 1/LC$.
(b) Impedances at similar planes in each network must be equal.

The details of impedance or admittance inverter relationships will not be derived here; the excellent book by Matthaei et al.[21] is recommended in this respect.

In the following expressions the 'first coupling structure' is that formed by w_1, s_1, w_1 (Figure 8.23) and the 'final coupling structure' consists also of w_1, s_1, w_1 at the opposite end of the filter. The quantities 'g' refer to the *prototype* element values, for example $g_0 = 1$ (the normalized termination value) and $g_1 = 0.781$—from the table for a sixth-order Chebyshev prototype which is standard data available in many references.[21] The admittance inverter parameters (J) are then given by:

For the *first* coupling structure:

$$\frac{J_{01}}{Y_0} = \sqrt{\frac{\pi\delta}{2g_0 g_1}} \qquad (8.22)$$

For the *intermediate* coupling structures:

$$\left.\frac{J_{j,j+1}}{Y_0}\right|_{j=1 \text{ to } (n-1)} = \frac{\pi\delta}{2\omega'_c \sqrt{g_j g_{j+1}}} \qquad (8.23)$$

For the *final* coupling structure:

$$\frac{J_{n,n+1}}{Y_0} = \sqrt{\frac{\pi\delta}{2g_n g_{n+1}}} \qquad (8.24)$$

In these three equations, δ is the fractional bandwidth

$$\delta = \frac{f_2 - f_1}{f_0} \qquad (8.25)$$

The frequency transformation from the low-pass prototype filter to the band-pass microwave filter is then

$$\frac{\omega'_i}{\omega_c} = \frac{2}{\delta}\left(\frac{f_i - f_0}{f_0}\right) \qquad (8.26)$$

where ω'_c in eqn (8.23) is the prototype cut-off frequency (it equals 1.0) and ω'_i has to be defined in the filter specification. (*The full design procedure will presently be shown by means of a specific numerical example, where the meaning of ω_c and ω'_i, etc., will become clear.*)

To proceed with the microstrip design, we now only require the odd- and even-mode coupled-line impedances Z_{0o} and Z_{0e}. These are given by

$$(Z_{0e})_{j,j+1} = Z_0(1 + aZ_0 + a^2 Z_0^2) \qquad (8.27)$$

and

$$(Z_{0o})_{j,j+1} = Z_0(1 - aZ_0 + a^2 Z_0^2) \qquad (8.28)$$

where

$$a = J_{j,j+1}$$

and Z_0 is the 'system' characteristic impedance (i.e. that of the lines feeding the filter).

Having obtained the values of Z_{0e} and Z_{0o}, section by section, the microstrip widths and separations are determined. The synthesis technique due to Akhtarzad *et al.* (see Ref. 4 in Chapter 6) is recommended for this, with refinements as described in Chapter 6. Calculations and a typical design are best shown by a specific example. A 0.635 mm thick alumina substrate with a permittivity of 10.0 is assumed to be used.

8.4.4 A worked numerical example of a parallel-coupled band-pass filter

Assume that the specification shown in Figure 8.24 is to be realized, where L is the insertion loss. The pass-band insertion loss (exactly related to the 'tolerance') has not been specified on the characteristic; it would usually be expected to lie below about 0.1 or 0.2 dB.

Step 1

Using eqns (8.25) and (8.26) a fractional bandwidth of 0.1 and transformation ratio of -1.619 should be obtained.

Step 2

Examine filter prototype specifications which meet the insertion-loss requirements. These are available in many handbooks (see, for example,

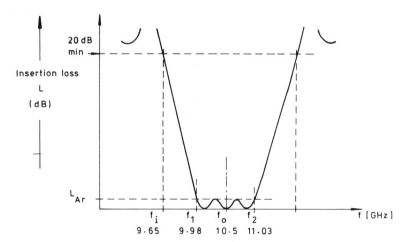

Fig. 8.24. Insertion-loss characteristic for the band-pass filter of the example

Ref. 21). The family of curves shown in Figure 8.25 apply to Chebyshev prototypes all giving a 0.01 dB pass-band ripple.

Notice that the independent variable has been adjusted for convenience to

$$\left|\frac{\omega'}{\omega_c}\right| - 1$$

Considerable care must be exercised regarding the value of ω_i'—it is the value *before* transformation and is *not* equal to ω_i. It must be obtained from eqn (8.26). In this case the value of $(|\omega_i'/\omega_c| - 1)$ is found to be 0.619, which is indicated on Figure 8.25.

A sixth order ($n = 6$) design therefore gives a 23-dB insertion loss at ω_i'. (This is equivalent to the 9.65 GHz point in the final requirement, which specified >20 dB. For ripple amplitudes >0.01 dB, only odd order designs are permissible, ensuring that $g_0 = g_n = 1.0$ accurately.) The table of values for a Chebyshev filter with $n = 6$ is given in Table 8.4.

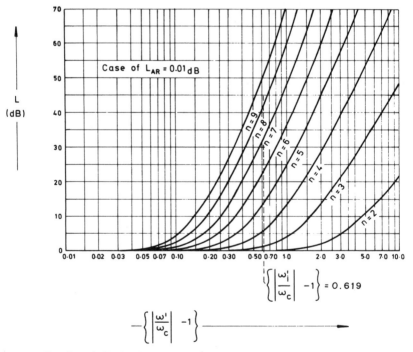

Fig. 8.25. Family of Chebyshev prototype filter insertion-loss characteristics for a pass-band ripple of 0.01 dB. (Reproduced by permission of McGraw-Hill Book Company from Matthaei, *et al.*,[21] Figure 4.03–4.)

Table 8.4

g_0	g_1	g_2	g_3	g_4	g_5	g_6	g_7
1.0	0.781	1.360	1.690	1.535	1.497	0.710	1.101

Table 8.5

j	$J_{i,j+1}/Y_0$	$(Z_{0e})_{j,j+1}(\Omega)$	$(Z_{0o})_{j,j+1}(\Omega)$
0	0.449	82.5	37.6
1	0.1529	58.8	43.5
2	0.1038	55.7	45.3
3	0.0976	55.4	45.6

Step 3

Calculate the inverter admittances (normalized for a 50 Ω system) and hence coupled-line impedances using eqns (8.22) to (8.24), and also (8.27) and (8.28). The results are shown in Table 8.5.

Step 4

The microstrip line dimensions are finally obtained as follows (substrate permittivity $\epsilon_r = 10$):

Using Akhtarzad *et al.* synthesis (see Ref. 4 in Chapter 6):

$$j = 0 \begin{cases} Z_{0se} = \tfrac{1}{2}Z_{0e} = 41.25 \ \Omega \\[1mm] Z_{0so} = \tfrac{1}{2}Z_{0o} = 18.8 \ \Omega \\[2mm] \text{Equivalent single-line synthesis yields:} \\[1mm] \left(\dfrac{w}{h}\right)_{se} = 1.5 \\[2mm] \left(\dfrac{w}{h}\right)_{so} = 4.8 \\[2mm] \text{Then coupled-line synthesis gives the required parameters:} \\[1mm] w/h = 0.65 \qquad s/h = 0.2 \end{cases}$$

$$j = 1 \begin{cases} Z_{0se} = 29.4 \ \Omega \quad Z_{0so} = 21.75 \ \Omega \\[6pt] \left(\dfrac{w}{h}\right)_{se} = 2.5 \qquad \left(\dfrac{w}{h}\right)_{so} = 4.0 \\[6pt] \dfrac{w}{h} = 1.0 \qquad \dfrac{s}{h} = 0.68 \end{cases}$$

$$j = 2 \begin{cases} Z_{0se} = 27.85 \ \Omega \quad Z_{0so} = 22.65 \ \Omega \\[6pt] \left(\dfrac{w}{h}\right)_{se} = 2.7; \qquad \left(\dfrac{w}{h}\right)_{so} = 3.7 \\[6pt] \dfrac{w}{h} \simeq 1.0 \qquad \dfrac{s}{h} \simeq 1.0 \end{cases}$$

$$j = 3 \begin{cases} \text{Since impedances are only 0.3 } \Omega \text{ different from the} \\ j = 2 \text{ case, approximately the same results are obtained.} \end{cases}$$

Figure 8.26 shows the final dimensions of this filter, assuming the substrate to have a thickness of 0.635 mm. 'Open' end-effects may be determined experimentally, as described in Section 7.9. Then the physical lengths of the resonators and the coupled regions are half-wavelengths and quarter-wavelengths respectively, with the equivalent end-effect lengths subtracted. (A wavelength is specified at the centre frequency f_0 and calculated as described in Sections 6.8 and 6.9.) The measured amplitude response of this filter is shown in Figure 8.27.

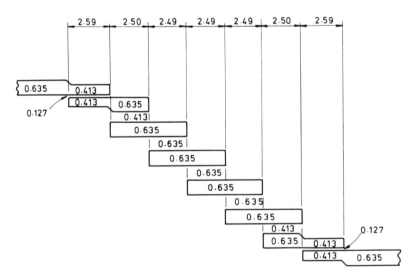

Fig. 8.26. Final design dimensions of the band-pass filter calculated in the example

Fig. 8.27. Measured (———) compared with specified (—.—.) insertion-loss characteristics for the 10–11 GHz parallel-coupled band-pass filter

This filter was manufactured using copper-conductor thin-film technology, with the dimensions given on Figure 8.26; *no* post-manufacture trimming was undertaken. We note that the following results apply:

	Desired (Figure 8.24)	Measured (Figure 8.27)
3-dB bandwidth	10.5%	12.5%
Centre frequency (GHz)	10.5	10.7

Judicious modification is necessary to adjust these values; e.g. making the resonators just 2 per cent longer should result in a closer centre frequency. The bandwidth is particularly sensitive to the separations of the first and last coupled sections, and these separations should be increased slightly so that the bandwidth is reduced.

8.4.5 CAD of parallel-coupled bandpass filters

A number of approaches have been developed for the computer-aided design of microstrip bandpass filters. The method described in outline follows largely the design of procedure developed above (i.e. a first-cut design, excluding losses).

The aim is to generate the artwork for a filter having the general layout shown in Figure 8.23, given an insertion loss specification with the format shown in Figure 8.28.

This indicates the following input requirements:

1. centre frequency f_o (GHz)
2. 'ripple' or 'band-edge' insertion loss L_R (dB)
3. insertion loss at the 'attenuation' points L_A (dB)
4. 'ripple' bandwidth B_R (GHz)
5. 'attenuation' bandwidth B_A (GHz).

Four other items of input information, not evident from the insertion loss specification, are also required:

6. the termination characteristic impedance Z_o (Ω)
7. substrate relative permittivity ϵ_r
8. substrate thickness h (mm)

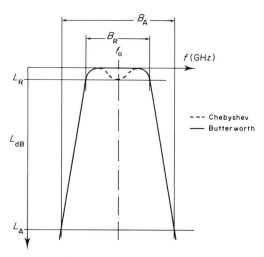

Fig. 8.28. General bandpass filter insertion-loss characteristic

9. Finally the appropriate function must be chosen by which the insertion loss is to be modelled. In this case a choice between Chebychev or Butterworth is given.

The main flow chart for the algorithm applicable to this CAD approach is shown in Figure 8.29 and the sequence of events now proceeds as follows:

1. Initially the required number of sections is calculated and for each section the even and odd mode characteristic impedances Z_{0e} and Z_{0o} (A) are determined, as given by Matthaei et al.[21]

2. These impedances (A) are applicable to any TEM-type parallel-coupled band-pass filter and the next step is for the algorithm to follow either 'triplate' (stripline) or microstrip technology design routes, as directed.

3. The 'triplate' design procedure is relatively straightforward and simple, being direct, and it will not be described here.

4. For microstrip design, the most challenging process is the accurate computation of the final widths (w) and separations (s). Firstly auxiliary single microstrip parameters are calculated, leading to approximate, starting-point, values of the aspect ratios of w/h and s/h as determined by the synthesis technique due to Akhtarzad et al. and provided in Chapter 6.

5. These aspect ratios are then used in an accurate analysis routine, also shown in Chapter 6, to find new values for the section impedances (B).

6. The new impedances (B) are compared with the cardinal values (A) and adjustments are made to w/h and s/h in order to correct the new values.

7. The latest w/h and s/h quantities are only output when the impedance differences $(A - B)$ are acceptably small.

8. Microstrip lengths are now required. Modified dispersion formulae are used to calculate the even-and-odd-mode phase velocities v_e and v_o at the centre frequency f_o and the weighted-mean phase velocity v_n is then evaluated using

$$v_n = \frac{Z_{0e} + Z_{0o}}{(Z_{0e}/v_e) + (Z_{0o}/v_o)} \tag{8.29}$$

Then the microstrip wavelength λ_{gn} is given by

$$\lambda_{gn} = v_n/f_o \tag{8.30}$$

9. Open-end-effect lengths must be calculated as derived from the single-line model and v_n for the parallel-coupled lines. These are also a function of frequency $[l_e, \text{coupled } (f)]$.

10. Each corrected, physical, 'quarter-wave' length is then

$$l = (\lambda_{gn}/4) - [l_e, \text{coupled } (f)] \tag{8.31}$$

and this completes the design algorithm.

Fig. 8.29. Flow chart applicable to the CAD of bandpass filters

8.4.6 Improvement to the basic edge-coupled filter response

Parallel (or edge)-coupled microstrip bandpass filters of the types designed according to principles described above display somewhat unsymmetrical frequency responses in addition to a significant second passband unless compensated. These highly undesirable effects are due to the field characteristics of the odd mode, as described in Chapter 6.

Various techniques are available for compensation of this odd-mode effect and some of these were also outlined in Chapter 6. Bahl[22] has shown that, with capacitive compensation introduced at the ends of each quarter-wave section (see Section 6.11.5), the stopband performance of these types of filters is greatly improved. The 'sharpness' or selectivity characteristics of both skirts are improved most markedly in the case of the upper skirt where the improvement is typically 5–10 dB. Also, the second passband is then more than 40 dB below the level of the fundamental passband—and, to further advantage, slightly shifted in frequency. Compensating capacitances typically lie in the range 0.002–0.055 pF, depending upon the specific filter section and the substrate. Bahl gives examples of GaAs and quartz substrates.

8.4.7 Filter analysis and design including all losses

As mentioned earlier, the preceding filter CAD approaches neglect the effects of power losses through the structure. It is extremely difficult, if not impossible, to anticipate these in quantitative detail and then to actually synthesize a design including all losses. Instead, it is suggested that the design be 'completed' to a first-cut level and that the resulting filter structure be then simulated on the computer to determine its insertion loss response including power losses. At one extreme the design philosophy could then be aborted and a different, less lossy, approach adopted (e.g. hairpin line resonators might be used if straight resonators were implemented at first). Alternatively, controlled variations on the exact design could be studied for their impact on loss-driven aspects of their characteristics.

The insertion loss of each section of the filter is best characterized in terms of the forward transfer scattering parameter S_{21}. This, in turn, involves section and mode impedances as follows:[21]

$$S_{21zl} = \left. \frac{2Z_{12n}/Z_0}{\left(1 + \dfrac{Z_{11n}}{Z_0}\right)^2 - \left(\dfrac{Z_{12n}}{Z_0}\right)^2} \right|_{l(n \,=\, \text{mean})} \tag{8.32}$$

where

$$\left. \frac{Z_{11e,o}}{Z_0} \right|_{l} = \left. -j\left(\frac{Z_{0e} + Z_{0o}}{2Z_0}\right)\cot\theta_{e,o} \right|_{l} \tag{8.33}$$

and

$$\left. \frac{Z_{12e,o}}{Z_0} \right|_{l} = \left. -j\left(\frac{Z_{0e} - Z_{0o}}{2Z_0}\right)\csc\theta_{e,o} \right|_{l} \tag{8.34}$$

The appropriate phase angles for use in these expressions should be determined using the formulae given above for the CAD routine i.e. the

weighted-mean phase velocity and microstrip wavelength. Further forward
scattering parameter functions are readily available to account for the loaded
and unloaded Q-factors of the resonators (accounting for the conductor and
dielectric losses) and the radiation loss from the semi-open ends of these
resonators.

Dissipative losses:

For the nth resonator we can write the dissipative loss S-parameter as:

$$S_{21_{ln}} = Q_u/(Q_u - Q_L) \tag{8.35}$$

where the loaded-Q (Q_L) is

$$Q_L = (f_o/\Delta f)\sin\{(2n - 1)\pi/2N\} \tag{8.36}$$

In this expression Δf is the 3 dB bandwidth of the filter which is determined
from the insertion loss function (i.e. Butterworth or Chebyshev) and N is
the total number of resonators.
 For each resonator Q_u is found from:

$$1/Q_u = 1/Q_c + 1/Q_d + 1/Q_r \tag{8.37}$$

The Q-factors due to conductor and dielectric losses are found by using
expressions given in Chapter 4 and the radiation conductance G_r is also
provided in Chapter 4. We require an expression for Q_r in terms of G_r and
this is the external Q-factor for a resonator that is equally end-loaded. Such
an expression is given in Appendix B and is:

$$Q_r = \frac{\pi\rho}{1 - \rho^2} \tag{8.38}$$

where

$$\rho = (G_rZ_o - 1)/(G_rZ_o + 1) \tag{8.39}$$

This expression is also equal to the input scattering parameter S_{11rad} and
therefore, using standard S-parameter theory, the contribution due to this
radiation mechanism is given by:

$$|\Delta S_{21}|^2 = 1 - |S_{11rad}|^2 \tag{8.40}$$

Once all these features are quantified it should be possible to fully simulate
the behaviour of the filter, having first converted the S-parameters to an
appropriate cascade family of parameters such as $ABCD$.

8.4.8 Band-pass filters with increased bandwidth (>15 per cent)

Where larger bandwidths are desired one of several alternative configurations
may be considered. It has long been appreciated that a series of stubs, all

$\lambda_g/4$ in length if open-circuited ($\lambda_g/2$ if short-circuited) and spaced $\lambda_g/4$ apart, will form a band-pass filter with a realizable bandwidth of the order of an octave. The stubs are placed alternately on each side of the microstrip. Kompa and Mehran[23] have reported a frequency-dependent design procedure for this type of filter in microstrip. Tzuang and Lo[24] have also described a tapped combline bandpass filter with a 4 GHz pass band.

Pang[25] has shown that the first and last coupling structures of the parallel-coupled filter, described in the previous section, can be replaced by continuous strip designs involving a quarter-wave transformer. In *triplate* Pang gives a 1.6 GHz filter with a 41 per cent 3 dB bandwidth based upon this approach. Apart from the bandwidth improvement there is an advantage regarding tolerancing, since the regions which originally involved the narrowest coupling gaps are now avoided altogether. The characteristic impedance of the transforming line length is not particularly straightforward to evaluate—it requires the use of the sineplane approach in the design. Microstrip realizations should be tried.

Mara and Schappacher[26] have given details of an interesting parallel-coupled filter design involving multiline couplings instead of the more usual two lines. Several filters are described. In one case a 6 GHz octave-bandwidth band-pass filter is described, based upon a 0.1 dB ripple five-section Chebyshev prototype. The ends of all the resonators were wire-bonded together (like the Lange coupler described in Section 6.7). Unfortunately the 6 GHz filter exhibits a spurious pass-band beginning at about 11.6 GHz, preceded by a characteristic in which the insertion loss does not improve beyond 18 dB—after the main pass-band. Doubtless some judicious tuning of the filter would overcome these problems.

8.4.9 Further developments in band-pass filter design

With the extensive knowledge available for microwave filter design, including the approaches described above, it has been proposed that an 'expert system' could be developed for the rapid and reliable availability of such design processes. Nakamura and Hibino[27] have reported such an expert system operating on Hewlett Packard 9000-series computers. Starting with the specification their procedure uses a microwave data base, a knowledge data base, a microwave circuit simulator and what they term 'inference' (the effect of interactions between the different requirements).

The knowledge data base includes sufficient information to allow for design margins and to choose the appropriate substrate on the basis of frequency of operation, temperature, etc. It also has the ability to ensure that the filter will fit the available substrate space (the real estate).

The microwave circuit simulator is designed to call up one of several commercially available CAEE software packages such as SuperCompact or

Eesof's Touchstone™. This is arranged to be available automatically, by operating the system on an Ethernet bus.

Final MIC/MMIC pattern drawing is facilitated by an appropriate subsidiary package that again operates on the Ethernet bus. Practically any orientation can be provided, for example the in-line type of layout shown in Figure 8.30.

Parallel-coupled bandpass filters may be provided as 'stand-alone' components or may be realized as part of a more complex circuit. An example of a stand-alone version of such a filter is shown in Figure 8.31. This filter is specified to provide a smooth and symmetrical bandpass characteristic over the frequency range 9.0–9.5 GHz.

As mentioned earlier in this text, work is also progressing in microwave applications of the important field concerning high-temperature superconductors. Although not strictly high-temperature, one example of work regarding the design of superconductive microstrip filters is that reported by Bonetti and Williams.[28] These workers use substrate materials such as lanthanum aluminate and lanthanum galate. Due to the relatively new nature of these materials in this type of application Bonetti and Williams report substrate measurements at about 4 GHz. The loss tangents are very low and the permittivities vary between 23.1 and 24.3.

The resulting bandpass filters, operated at 77 K, exhibit symmetrical insertion loss characteristics and a bandwidth of less than 4 per cent (150 MHz). Hairpin line resonators are employed throughout the filters.

8.5 SPUR-LINE BAND-STOP FILTERS

Bates and Pearson[29,30] evolved a simple and clever technique for designing a band-stop filter using a resonant structure (Figure 8.32) fabricated *within the width* of a microstrip line. It was demonstrated that this filter is almost completely non-dispersive, and the structure is so configured that radiation loss is very low compared with coupled-line or shunt-stub filters. The filter

Fig. 8.30. Example of bandpass filter pattern layout designed by expert system (output from 'ME-10'). (© 1990 IEEE. Reprinted, with permission, from Nakamura and Hibuno et al.[27])

Fig. 8.31. A proprietary bandpass filter module (photo courtesy of Miteq, Inc., 1990)

Fig. 8.32. One section of a spur-line band-stop filter. (Reproduced by permission of the Institution of Electrical Engineers from Bates.[29])

structure was shown to be equivalent to an open-circuit shunt stub of characteristic impedance Z_1 followed by a continuous section of microstrip having a characteristic impedance Z_{12}. These impedances are given by

$$Z_1 = \frac{Z_{0e}}{Z_{0o}}\left(\frac{Z_{0o} + Z_{0e}}{2}\right) \qquad (8.41)$$

and

$$Z_{12} = \frac{Z_{0o} + Z_{0e}}{2} \qquad (8.42)$$

where Z_{0o} and Z_{0e} have the meanings defined previously for coupled lines.

This filter can be designed according to the procedure described by Matthaei, Young, and Jones[21] and computer-aided optimization can then be employed. Bates[29] gives graphs of Z_1 and Z_{12} as functions of the parameters s/h and w/h, although a designer with access to computer facilities would probably need to develop suitable expressions for these impedances. The length a (see Figure 8.32) is electrically $\lambda_g/4$ at mid-band (f_0), but the gap equivalent length l_{eg} will affect the physical value of a such that

$$a = \frac{3 \times 10^8}{4f_0\sqrt{\epsilon_{effo}}} - l_{eg} \qquad (8.43)$$

where ϵ_{effo} is the odd-mode effective microstrip permittivity at the centre frequency (see Section 6.9) and l_{eg} is the gap end-effect extension (see Section 5.3).

The value of the series gap separation b has to be determined empirically and a value of approximately 50 μm appears to be suitable for designs on alumina, and similar substrates.[29] Once b is chosen, the end-effect length may be found by using

$$l_{eg} = C_{odd}v_{po}Z_{0o} \qquad (8.44)$$

C_{odd} is calculated from expressions and data given by Benedek and Silvester,[31] and Bates gives results which indicate that a very good agreement between calculation and experiment can be achieved for gap separations of about 100 μm or less.

The filter design procedure is thus:

(a) From the specified response desired use the technique given by Matthaei *et al.*[21] to find Z_1 and Z_{12}.
(b) Calculate Z_{0e} and Z_{0o} from Z_1 and Z_{12} using eqn (8.24) and 8.25).
(c) Employ computer-aided optimization to ensure that the network so far described meets the desired specification within acceptable limits.
(d) From Z_{0e} and Z_{0o} synthesize the desired w/h and s/h.
(e) Evaluate the length a, approximately.
(f) Choose the series gap separation b (\approx 50 μm, on alumina) and hence also determine l_{eg}.
(g) Iterate (e) and (f) until a is found more accurately and the centre frequency f_0 is close to that desired.

The band-stop filter performance is particularly good when two of these spurline structures are arranged back to back in cascade, with a separation slightly less than a quarter-wavelength. For example, where a = 3.2 mm the distance separating the structures should be about 3.0 mm.[30] With such a filter a rejection (insertion loss) exceeding 35 dB was achieved over the band $9.3 \leq f \leq 9.7$ GHz.

Several variations of the structure are conceivable. A 'double-sided' arrangement, i.e. two spur-line structures side by side across the microstrip

width, has been manufactured for a band centred on 30.5 GHz. The rejection exceeded 29 dB over a 2 GHz band.[30] This general concept, leading to unusual and very compact filters, must be capable of extension into many new designs.

Microstrip lends itself to entirely new structural shapes to perform circuit/system functions which were previously realized using triplate line, or coaxial, or waveguide technology.

8.6 PASSIVE MICs WITH SWITCHING ELEMENTS

MICs having switching functions, normally implemented with PIN diodes but increasingly using transistors, have been realizable for many years. The basic design concepts are simple, utilizing the fact that specific line lengths (e.g. quarter wave) are switched into or out of the circuit depending upon whether the PIN or transistor switch is activated or not.

Although strictly a form of active circuit, we mention this application here since it is not accommodated by any of the analog active functions covered in Chapter 9. Broadband cross point switch matrices have been reported, for example by Powell *et al.*,[32] using GaAsFETs (in some cases dual-gate GaAsFETs) and microstrip crossovers. Such switch matrices are readily capable of being implemented in MMIC form and, as 2 × 2 matrices operating over the 4–10 GHz band, can achieve over 40 dB isolation with 4 dB insertion loss. The circuits can be cascaded to form higher order matrices.

8.7 ISOLATORS AND CIRCULATORS

Various configurations have been studied to exploit ferrites in MICs and hence to realise isolators and circulators. The characteristics of ferrite in this context are described in Chapters 3 and 4.

Kane and Wong[33] report an isolator design based upon edge-guide mode principles in conjunction with a transverse slot or slit discontinuity (the transverse slit is described in Chapter 5). This configuration is shown in Figure 8.33 where the entire substrate is ferrite, the isolator section occupies an area approximately 10 mm × 8 mm and the slot measures approximately 6 mm × 0.8 mm.

At the centre frequency of 11 GHz the forward insertion loss is about 2 dB whereas the isolation is better than 18 dB.

A circulator designed for operation over the 31–37 GHz band has been described by Pan *et al.*[34] Their approach uses a new nickel–zinc ferrite substrate material, which has superior performance to previous materials and which is available in 0.127 and 0.254 mm thicknesses. The adhesion

Fig. 8.33. An edge-guide mode microstrip isolator with transverse slot discontinuity. (© 1990 IEEE. Reprinted, with permission, from Kane and Wong.[33])

between the ferrite and the metallization layer was poor in previous designs and this is improved in the new design by sputtering a seed layer of titanium tungsten (TiW) on to the ferrite surface prior to the final metallization. The thicknesses of the TiW and gold or copper layers are 500 and 2000 Å respectively.

The overall layout of the stub-tuned circulator is shown in Figure 8.34. It is possible to make circulators of particularly small dimensions using this technology and Pan *et al.* have demonstrated one one having a diameter of 3.8 mm and thickness 0.127 mm. In the case of a 7.6 mm diameter circulator a calibrated insertion loss of 0.6 dB and a return loss of better than 17 dB were measured.

This type of circulator is designed primarily for deep space communications but the design principles should be extendable to other applications.

Fig. 8.34. Wideband stub tuned microstrip circulator. (© 1990 IEEE. Reprinted, with permission, from Pan, Shih and Riley.[34])

REFERENCES

1. Lacombe, D., 'A multi-octave microstrip 50 Ω termination', *IEEE Trans.*, **MTT-20**, No. 4, April 1972, 290–292.
2. Rickard, D. C., 'Thick-film MIC components in the range 10–18 GHz', *European Microwave Conference, Digest of Papers*, Rome, 1976, pp. 687–91.
3. Finlay, H. J., Hopkins, L. G. T. and Ozamiz, J. M., 'Design and applications of precision microstrip multi-octave attenuators and loads', *European Microwave Conference, Digest of Papers*, Rome, 1976, pp. 692–696.
4. Ho, Chen Y., 'Transform impedance with a branch-line coupler', *MicroWaves*, May **1976**, 47–52.
5. Howe, H., *Stripline Circuit Design*, Artech House, Inc., Dedham, Massachusetts, 1973.
6. Wright, A. S. and Judah, S. K., 'Very broadband flat coupling hybrid ring' *Electron. Lett.*, **23**, 1987, 47–49.
7. Ashforth, J.V., 'Design equations to realise a broadband hybrid ring or a two-branch guide coupler of any coupling coefficient'. *Electron. Lett.*, **24**, 1988, 1276–1277.
8. Mayer, B., and Kl, R., 'Branchline-couplers with improved design flexibility and broad bandwidth' *IEEE MTT-S International Microwave Symposium* 1990, Digest of Papers, 391–398.
9. Knochel, R. and Mayer, B. 'Broadband printed circuit 0°/180° couplers and high power inphase power dividers' *IEEE MTT-S International Microwave Symposium* 1990, Digest of Papers, 471–478.

10. Meaney, P., 'A novel branch-line coupler design for millimeter-wave applications' *IEEE MTT-S International Microwave Symposium* 1990, Digest of Papers, 585–588.
11. Roy, J. S. *et al.*, 'Analysis of curved microstrip transmission lines and its applications' *Proceedings of 2nd International Symposium on Recent Advances in Microwave Technology (ISRAMT '89)* Bejing, People's Republic of China, Sept 4–8, 1989.
12. Tokumitsu, T. *et al.*, 'Multilayer MMIC using a 3μm × 3-layer dielectric film structure' *IEEE MTT-S International Microwave Symposium* 1990, Digest of Papers, 831–834.
13. Nakamoto, H., Tokumitsu, and Aikawa, M., 'A monolithic, port-interchanged rat-race hybrid using a thin film microstrip line crossover' *Proceedings of the 19th European Microwave Conference*, London, Sept. 1989, 311–316.
14. Pavlo, M., and Kikel, A., 'A Monolithic or Hybrid broadband compensated balun' *IEEE MTT-S International Microwave Symposium* 1990, Digest of Papers, 483–486.
15. Barber, R. G., 'Enhanced coupled, even mode terminated baluns and mixers constructed therefrom' *IEEE MTT-S International Microwave Symposium* 1990, Digest of Papers, 495–498.
16. Perez, R., 'Avoiding doubts in low-pass filters', *Microwave Systems News,* June **1979**, 58–65.
17. D'Inzeo, G., *et al.*, 'Design of circular planar networks for bias filter elements in microwave integrated circuits', *Alta Frequenza,* **48**, No. 7, July 1979, 425–431.
18. Al-Charchafchi, S. H. and Vassilikos, E, 'A microstrip filter using delta stubs' *SMBO International Microwave Symposium*, Brazil, July 24–27 1990.
19. Cunningham, G. J., Blenkinsop, P. A. and Palmer, J. H., 'Microstrip end-coupled filter design at MM-Wave frequencies' *Proceedings of the 19th European Microwave Conference,* London, Sept. 1989, 1210–1213.
20. Tzuang, C-K C. *et al.*, 'Design of a quasi-planar broadside end-coupled bandpass filter' *IEEE MTT-S International Microwave Symposium* 1990, Digest of Papers, 407–410.
21. Matthaei, G. L., Young, L., and Jones, E. M. T., *Microwave Filters, Impedance-Matching Networks and Coupling Structures*, McGraw-Hill, New York, 1965. (Back in print from October 1980: Artech House Inc., Dedham, Massachusetts.)
22. Bahl, I. J., 'Broadbanding microstrip filters using capacitive compensation' *Applied Microwave* 1989, 72–76.
23. Kompa, G., and Mehran, R., 'Microstrip filter analysis using a microstrip waveguide model', *Radio and Electronic Engineer,* **50**, No. 1/2, January/February 1980, 54–58.
24. Tzuang, Ching-Kuang C. and Lo, Wen-Teng, 'Printed-circuit realization of tapped combline bandpass filter' *IEEE MTT-S International Microwave Symposium* 1990, Digest of Papers, 131–134.
25. Pang, K. K., 'Design of microwave filters by sine-plane approach', *IEEE Trans.,* **MTT-21**, No. 10, October 1973, 607–611. (Also private communications.)
26. Mara, J. F., and Schappacher, J. B., 'Broadband microstrip parallel-coupled filters using multi-line sections', *Microwave Journal, April* **1979**, 97–99.
27. Nakamura, A. H. *et al.*, 'Expert system for microwave filter design' *IEEE MTT-S International Microwave Symposium* 1990, Digest of Papers, 1183–1186.
28. Bonetti, R. R. and Williams, A. E. 'Preliminary design steps for thin-film superconducting filters' *IEEE MTT-S International Microwave Symposium* 1990, Digest of Papers, 273–276.
29. Bates, R. N., 'Design of microstrip spur-line band-stop filters', *IEE Journal on Microwaves, Optics and Acoustics,* **1**, No. 6, November 1977, 209–214.

30. Bates, R. N., and Pearson, R. E., 'Designing bandstop filters for micro-wave frequencies', *Electronic Engineering,* April **1978**, 39–41.
31. Benedek, P., and Silvester, P., 'Equivalent capacitances for microstrip gaps and steps', *IEEE Trans.,* **MTT-20**, No. 11, November 1972, 729–733.
32. Powell, S., *et al.*, 'Broad band monolithic cross point switch matrices' *IEEE MTT-S International Microwave Symposium* 1990, Digest of Papers, 461–464.
33. Kane, R. C., and Wong, T., 'An edge-guide mode microstrip isolator with transverse slot discontinuity' *IEEE MTT-S International Microwave Symposium* 1990, Digest of Papers, 1007–1010.
34. Pan, J. J., Shih, M. and Riley, L., 'High performance millimeter-wave microstrip circulator for deep space communications' *IEEE MTT-S International Microwave Symposium* 1990, Digest of Papers, 1015–1017.

9 Active MICs and MMICs

9.1 INTRODUCTION

Microstrip networks are particularly easily adapted to circuit design using transistors of various types. In the case of an amplifier the circuit is appropriately optimized to provide low noise, moderate-to-high gain, or substantial output power. The specific types of transistor which can be implemented include:

bipolar transistors
GaAsFETs (gallium arsenide field effect transistors)
HEMTs (high electron mobility transistors)
HBTs (heterojunction bipolar transistors).

There are a few other types of transistor devices and it is likely that more will be developed during the lifetime of this text.

In the 1970s and early 1980s microwave amplifiers and oscillators were mainly implemented using various types of microwave active diodes since transistors with sufficient available gain at microwave frequencies, or sufficiently low-noise or power capability were simply not obtainable. Further, even when such transistors first became available it was still several years before the requisite design tools were sufficiently refined for regular use. Now these obstacles have largely been overcome and extensive CAEE and computer simulation techniques can be used.

As a result microwave transistor amplifiers, oscillators and many other types of circuits have been designed in both MIC and MMIC form. Active diode devices, such as Gunn (TED) and IMPATT, are still used in some instances, particularly at the high millimeter-wave frequencies (beyond 80 GHz as a ball-park) where transistors cannot yet be employed.

Amplifier circuits are generally designed for either narrowband or broadband applications and microwave oscillators are also fully designable around the types of transistors listed above. Operating frequencies can range from below 1 GHz up to frequencies well beyond 40 GHz.

Hybrid microwave integrated circuits, generically termed MICs, are frequently used for many amplifier and oscillator configurations but monolithic microwave integrated circuits (MMICs), usually fabricated with gallium arsenide but sometimes with silicon, are becoming increasingly common for many applications. The present situation, in the early 1990s, allows the co-existence of MICs and MMICs and it is felt that, to some extent, this co-existence will probably continue for many years.

There are a number of existing texts on active MIC/MMIC design that provide excellent bases for the various design requirements.[1-3]

Computer-aided microwave design and engineering is now used extensively in this industry for both MIC and MMIC design purposes. An overview of several packages available for this purpose is provided in Appendix D. The most popular types of package tend to be available through Compact Software (SuperCompact™) and Eesof, with their Touchstone™ series. Another popular package is provided by Hewlett-Packard and is termed MDS (microwave design software). Amongst further computer-aided design packages for microwave purposes we include, with brief overview discussions in Appendix D, TeMCAD, Toast, ESACAP and Esope.

TeMCAD is available from British Telecommunications Research Laboratories in England and is designed primarily to run on advanced Hewlett-Packard workstations. It is intended mainly for microwave and RF circuit layout and manufacturing applications, with inbuilt advantages in terms of repeatability and design—anticipating various tolerance effects during specific manufacturing processes. Toast is available from CaroLine Habigand in France, ESACAP is available through the European Space Agency and Esope can be obtained from Thomson of France.

Yield is one of the most important parameters in the integrated circuits industries, including the production of both MICs and MMICs. There are problems associated with this yield parameter for both types of technology and, in an attempt to improve the situation, Kazuo Eda et al.[4] have developed a novel thin-film technology to create miniature hybrid microwave integrated circuits (they term these MHMICs). The processes used in this technology allow for the full integration of thin-film resistors and capacitors together with conventional microstrip lines. The team that developed MHMICs show their techniques for grounding the thin-film capacitors to the substrate and they demonstrate a 14 GHz power amplifier using this technology. The specific advantages of MHMIC technology include size reduction (to two-thirds that of a conventional MIC), a reduction in the number of discrete parts by a factor of three, a halving of the MIC mounting and wiring processes and a cost reduction due to the use of relatively inexpensive substrates. There is also, and this is particularly important, the high yield obtained and additionally a relatively high performance due to a significant decrease in the values of parasitic elements. The circuit diagram of the 14 GHz amplifier built on an MHMIC basis is shown in Figure 9.1.

Fig. 9.1. A solid-state power amplifier (SSPA) module constructed with miniature hybrid microwave ICs (MHMICs) fabricated by novel thin-film technology. (© 1990 IEEE. Reprinted, with permission, from Eda *et al.*[4])

In the next section of this chapter we consider in detail the specific topic of amplifiers in both MIC and nominally MMIC technologies.

9.2 AMPLIFIERS

Amplifiers can be optimized for low-noise or moderate-to-high gain, or substantial power output. Using GaAsFETs, HEMTs or HBTs, amplification can be obtained at frequencies extending well into the millimetre-wave regions. Oscillators can also be designed around the transistor devices.

There are many circuit configurations and we shall only concentrate on a restricted number here—focusing on relatively simple single-ended designs and then describing distributed and balanced amplifiers. Another approach, namely feedback amplifiers, is also used quite often. But we shall consider this topic outside the scope of the text and refer to several excellent references for this design approach.[1-3]

Any microwave transistor amplifier requires the circuit arrangements indicated in the general block form of Figure 9.2. The d.c. biasing circuit is fairly standard, it does not involve any microwave constraints, and it will not be discussed here. The low-pass filters ('bias circuits') can have one of several forms, some details of possibilities being given in Chapter 8. It is the realization of the input and output networks, and occasionally feedback networks, that is of principal significance and this will occupy our attention here.

A slightly more detailed presentation of the device with its input and output networks is given in Figure 9.3. Here the intervening networks are labelled 'optimizing' rather than 'matching', since Γ_s in particular will be anything but the value for matching.

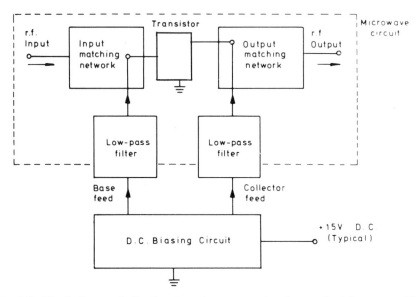

Fig. 9.2. Block diagram indicating the microwave networks required for an amplifier

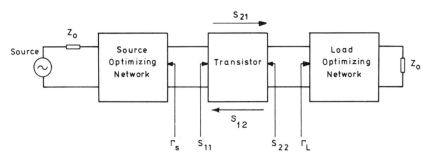

Fig. 9.3. The scattering parameter (S_{nm}) and reflection coefficients (Γ) associated with a microwave transistor amplifier

9.2.1 A basic low-noise MIC amplifier

In this section we consider the design of a relatively simple low-noise amplifier. A suitable overall specification of initial design goals is first listed:

 (a) Amplifier noise figure $F < 3$ dB
 (b) Associated gain > 8 dB
 (c) Output VSWR $r < 1.5$
 (d) Frequency range 4 ± 0.2 GHz
 (This represents a 'fairly modest' 10 per cent band.)

The next requirement is to list the scattering parameters for the selected transistor. We shall base the design on a transistor which has the following scattering parameters, all measured at 4 GHz with the bias conditions (specially the drain current) set for minimum noise figure:

$$S_{11} = 0.522\underline{/169°} \quad S_{21} = 1.681\underline{/26°}$$
$$S_{12} = 0.049\underline{/23°} \quad S_{22} = 0.839\underline{/-67°}$$

Before any further design calculations are attempted, it is essential to check the stability and maximum available gain. This can be done using a standard computer program (for example 'GMAX', written in BASIC[5] or any other suitable package). In this case the indications are that the circuit is likely to be stable and the maximum available gain exceeds 13 dB.

As with any amplifier, an optimum source impedance ($Z_{s,opt}$) exists for a minimum noise figure. With $Z_{s,opt}$ an associated optimum reflection coefficient Γ_o can be determined based upon the system characteristic impedance Z_0 which is usually 50 Ω. These quantities are best determined by measurement and in this case the values are

$$\Gamma_o = 0.476\underline{/166°}$$
$$F_{min} = 2.5 \text{ dB}$$

It should be carefully verified that this transistor, with a source whose reflection coefficient is $0.476\underline{/166°}$, will still yield *some* power gain (a very low noise 'amplifier' which does not provide gain is not a very useful system element!). Computer routines yielding automatically plotted graphs can assist in this verification. Alternatively, especially over this narrow band, a family of constant-gain contribution circles may be overlaid upon a family of constant noise figure circles on a Smith chart. The details will not be dealt with here because we are interested in the development of the microstrip network. It is found that, in the region where the noise figure is about 2.6 dB, the transducer power gain lies between 8.2 and 9.2 dB— which is considered adequate and meets the specification.

We are therefore satisfied to realize an input network yielding the value of $\Gamma_o = 0.476\underline{/166°}$.

The input optimizing network is designed first

The primary purpose of the input optimizing network is to realize the optimum source impedance for minimum noise. The main design steps are:

(1) Convert Γ_o to an equivalent normalized admittance Y_{NF}.
(2) Realize the susceptive component of Y_{NF}, that is B_{NF} with a short-circuited $\lambda_g/8$ stub.
 (The $\lambda_g/8$ length yields an inductive element, jZ_0).

(3) Realize G_{NF} by means of a $\lambda_g/4$ impedance transformer, transforming back to the original 50 Ω generator.

With care, the Smith chart can be used for all these operations.

We shall assume that this part of the design exercise has been completed here and the characteristic impedances of the $\lambda_g/8$ stub and $\lambda_g/4$ transformer are now calculated. (These may also be checked on a Smith chart.)

The values are

$$Y_{NF} = G_{NF} + jB_{NF} = 2.5 - j0.75$$

or, upon 'denormalizing' (i.e. dividing through by Z_0),

$$Y_F = G_F + jB_F = 0.051 - j0.015 \ S$$

with $-jB_F = 1/jZ_0$, we therefore get, for the $\lambda_g/8$ stub, a characteristic impedance of

$$Z_0 = \frac{1}{0.015} = 65.8 \ \Omega$$

For the $\lambda_g/4$ transformer, having characteristic impedance Z_{01},

$$Z_{in} = \frac{Z_{01}^2}{50} = \frac{1}{G_F} = \frac{1}{0.051}$$

Therefore,

$$Z_{01} = \sqrt{50 \times 19.6} = 31.3 \ \Omega$$

In a computer-orientated routine it would be particularly helpful for a Smith chart display to become generated, but the actual computations would be carried out using the transmission line expressions. These are, writing $\Gamma_o = |\Gamma_o| \exp(-j\phi_o)$ and using denormalized quantities,

$$Z_F = \frac{(1-|\Gamma_o|^2)Z_0}{1+|\Gamma_o|^2 - 2|\Gamma_o|\cos\phi_0} + j\frac{2Z_0|\Gamma_o|\sin\phi_o}{1+|\Gamma_o|^2 - 2|\Gamma_o|\cos\phi_o} \qquad (9.1)$$

Also, the usual impedance-to-admittance conversion is

$$Y_F = \frac{1}{Z_F} \qquad (9.2)$$

with the value of Z_F substituted after evaluation using eqn (9.1). The network topology is dictated within the computer program. This approach allows some flexibility in the design values and rapid indication of the sensitivities to various parameter changes. The final input network can now be drawn and is shown in Figure 9.4.

The characteristic impedances are indicated for each microstrip line and we note that these are capable of being realized without difficulty in microstrip. (Values below 10 Ω or above 110 Ω become difficult to

(a) Block diagram

(b) Circuit requirements (schematic plan view)

Fig. 9.4. Input noise-matching network for the low-noise transistor amplifier whose design details are described in the text. (Reproduced by permission of Hewlett-Packard Company, Palo Alto, Calif.)

manufacture.) Discontinuity effects associated with the right-angled bend will be very small at only 4 GHz and may be neglected. The area of microstrip around this bend will also have quite enough physical area to allow a small bonding pad or wire to go directly to the base of the transistor.

Again, at a frequency of 4 GHz, the short-circuit termination presents little difficulty. It can be formed using a wire carefully bonded to the microstrip and taken through a substrate via to the ground plane.

Finally, the output matching network is designed

One of the design specification requirements was an output VSWR of better than (i.e. less than) 1.5. Ideally, the VSWR would be that of a perfect match, i.e. unity, and that value is obtained when the load is conjugately matched. It will be assumed that true conjugate matching is desired, at least at the centre frequency.

Now the output reflection coefficient is also dependent upon the input network reflection coefficient and the expression required, which is derived from S-parameter theory, is

$$S'_{22} = S_{22} + \frac{S_{12}S_{21}\Gamma_s}{1 - S_{11}\Gamma_s} \tag{9.3}$$

In this case $\Gamma_s = \Gamma_o = 0.476\underline{/166°}$, so S'_{22} must first be calculated. This calculation yields

$$S'_{22} = 0.846\underline{/-72°}$$

It is important to note that, while $|S'_{22}|$ does not differ very much from $|S_{22}|$, the associated phase angles *do* differ, to the extent of 5°, which is a significant change in electrical length.

A practical approach is to start with the output impedance Z'_{22} of the transistor, which it is desired to match. In this case the difference between the effective output impedance Z'_{22} and the load impedance Z_0 is much greater than exists with the input circuit. It is consequently found that a two-element network will not provide sufficient bandwidth to meet the specification and a five-element network is therefore developed. Again, the design may proceed using a Smith chart—or a computer technique may be used. The steps are as follows:

(1) Mainly for collector bonding convenience, a length of 50 Ω microstrip line is extended from the collector location to the next element of the output matching network. A convenient section amounts to approximately 60° of electrical length.

(2) A short-circuited stub of characteristic impedance Z_{0s} and $\lambda_g/8$ in length (hence having an input reactance jZ_{0s} Ω) is connected to the network end of the 60° line. This is designed to tune out *most*, but not all, of the susceptance at this plane in the network. The remaining sections of the network will finally tune out all the susceptance. It is found that $Z_{0s} = 29$ Ω, and the normalized impedance at this plane in the network is then

$$Z_{Np} = 1.25 - j1.0$$

(3) Now, we choose to employ just a length of microstrip line between the load and the final shunt matching element. (This length of line will be the 9.1° line shown on the complete output-matching network of Figure 9.5.) Thus the load is matched (real part of normalized admittance equal to 1.0) at all planes everywhere along this length of line and it remains for us to properly connect a matching element somewhere along this line.

(4) With the network designed as far as step (2) the normalized admittance Y_N looking towards the transistor must be forced to equal

$$Y_N = 1 - j0.535$$

which is obtained from the Smith chart or by calculation. The equivalent impedance is

$$Z_N = 0.782 + j0.422$$

and this will be the input impedance to a section of transmission line having a characteristic impedance Z_{0l} and distance l from the short-

Fig. 9.5. Output power-matching network for the low-noise transistor amplifier whose design details are described in the text. (Reproduced with permission of Hewlett-Packard Company, Palo Alto, Calif.)

circuit stub. This input impedance has the form as given by eqn (1.13). That is, after denormalizing Z_N,

$$Z = Z_{0l}\left[\frac{Z_p + jZ_{0l}\tan\beta l}{Z_{0l} + jZ_p\tan\beta l}\right] \tag{9.4}$$

where Z_p is the denormalized value of Z_{Np} as already determined. Equation (9.4) shows that we have some freedom of choice regarding l and Z_{0l}. If we rather arbitrarily choose $Z_{0l} = 80\ \Omega$ then the electrical length βl must equal 64.8°.

(5) Following this length of line l we must overtune the susceptive part of the admittance (Y) there. It is necessary to overtune so that the final length of transmission line can tune out the remaining susceptance of appropriate sign. An open-circuited stub of characteristic impedance Z_{0st} having a length $l_{os} < \lambda_g/4$ provides a shunt susceptance equal to

$$jY_{0st}\tan\beta l_{os}$$

If $Z_{0st} = 25.5\ \Omega$ a length $\beta l_{os} = 19.2°$ ensures that this susceptance is sufficient.

(6) As explained at the beginning of step (3), a final length of microstrip line completes the match. This length has to be 9.1°.

These last two elements, (5) and (6), also provide some flexibility to the output tuning of the amplifier. The complete output-matching network is shown in Figure 9.5 and it should be observed that all the elements can quite easily be produced using microstrip lines.

9.2.2 High-gain narrowband amplifier design

Many systems require relatively high-gain amplifiers in addition to low-noise or high-power and in this section we indicate the design technique applicable

to microstrip realisation of such designs. By 'narrowband' we generally mean circuits having bandwidths less than 10 per cent of their centre frequencies.

The overall configuration will be identical to that presented in Figures 9.2 and 9.3 and now the aim is to design the input and output microstrip networks so that high-gain is achieved. From transmission-line theory, the source reflection coefficient yielding maximum power gain is given by

$$\Gamma_{s,m} = \frac{Z_{s,m} - Z_0}{Z_{s,m} + Z_0} \tag{9.5}$$

where the quantities have been previously defined, except that now $\Gamma_{s,m}$ and $Z_{s,m}$ refer to maximum-providing values. The source equivalent impedance for maximum power gain is therefore expressed as

$$Z_{s,in} = Z_0 \frac{1 + \Gamma_{s,m}}{1 - \Gamma_{s,m}} \tag{9.6}$$

where

$$\Gamma_{s,m} = |\Gamma_{s,m}| \cos \underline{/\Gamma_{s,m}} + j \sin \underline{/\Gamma_{s,m}} \tag{9.7}$$

is the source reflection coefficient yielding maximum available power gain.

When eqn (9.7) is substituted into eqn (9.6) and the resulting expression is rearranged an expression of the form of the earlier eqn (9.1) is obtained, but of course with maximal quantities replacing the original low-noise quantities:

$$Z_{s,in} = Z_0 \frac{1 - |\Gamma_{s,m}|^2 + j2|\Gamma_{s,m}| \sin \underline{/\Gamma_{s,m}}}{1 + |\Gamma_{s,m}|^2 - 2|\Gamma_{s,m}| \cos \underline{/\Gamma_{s,m}}} \tag{9.9}$$

The output matching network must also be designed for maximum power gain and the load equivalent impedance for this network takes the following similar form:

$$Z_{lo} = Z_0 \frac{1 - |\Gamma_{l,m}|^2 + j2|\Gamma_{l,m}| \sin \underline{/\Gamma_{l,m}}}{1 + |\Gamma_{l,m}|^2 - 2|\Gamma_{l,m}| \cos \underline{/\Gamma_{l,m}}} \tag{9.10}$$

We can now list the design procedure for this type of amplifier:

(1) Convert the source and load reflection coefficients to the source and load equivalent impedances.
(2) Find the source and load equivalent admittances from these impedances.
(3) Realize the susceptance components with either a short or an open-circuited three-eighths-wavelength stub.
(4) Realize the conductance components with quarter-wavelength transformers.

As an example of high-gain narrowband amplifier design we consider a 6 GHz amplifier originally given by Liao.[1]

Design example

A GaAsFET amplifier is required to provide maximum power gain over a 400 MHz bandwidth centered on 6 GHz. The device's measured parameters (all at 6 GHz) are

$$\Gamma_{s,m} = 0.761\underline{/177°} \qquad \Gamma_{l,m} = 0.719\underline{/104°} \qquad G_{a,max} = 12 \text{ dB}$$

The design starts as usual with the input matching network.

From eqn (9.7) the source equivalent impedance is:

$$Z_{s,in} = \frac{50(1-|0.761|^2)+j2\times50\times\sin(177°)}{1+|0.761|^2-2|0.761|\cos(177°)}$$
$$= 6.77 + j1.28 \text{ } \Omega$$

Convert $Z_{s,in}$ to $Y_{s,in}$

$$Y_{s,in} = \frac{1}{Z_{s,in}} = \frac{1}{6.77+j1.28} = 0.15 - j0.02 \text{ S}$$

Use a three-eighths-wavelength open-circuited stub to represent this shunt admittance such that $Y = jY_0 \tan(\beta l)$. Such a stub behaves as a shunt inductor of susceptance $-jY_0$. We therefore have the characteristic impedance for this stub

$$Z_{01} = \frac{1}{Y_0} = \frac{1}{\text{Im}[Y_{s,in}]} = \frac{1}{0.02} = 50.0 \text{ } \Omega$$

Next, a quarter-wave transformer will transform the source resistance of 50 Ω to the source equivalent conductance of 0.15 S. This transformer has the characteristic impedance

$$Z_{02} = \sqrt{\frac{50}{0.15}} = 18.26 \text{ } \Omega$$

This completes the input matching network.

We now require the output matching network. For this we first find the load equivalent impedance from eqn (9.10)

$$Z_{l0} = \frac{50(1-|0.719|^2)+j2\times50\times0.719\sin(104°)}{1+|0.719|^2-2\times0.719\cos(104°)}$$
$$= 12.83 + j37.30 \text{ } \Omega$$

Now convert Z_{l0} to Y_{l0}:

$$Y_{l0} = \frac{1}{Z_{l0}} = \frac{1}{12.83+j37.30} = 0.008-j0.024 \text{ S}$$

Next, we use precisely the same method of employing a three-eighths-wavelength stub that was used for the input network, to realize an equivalent shunt inductor of susceptance $-jY_0$. Its characteristic impedance must be

$$Z_{03} = \frac{1}{Y_0} = \frac{1}{\text{Im}[Y_{10}]} = \frac{1}{0.024} = 41.67 \ \Omega$$

Finally, we use a quarter-wave transformer to transform between this network and the 50 Ω load impedance. The characteristic impedance of this transformer line section is

$$Z_{04} = \sqrt{\frac{50}{0.008}} = 79 \ \Omega$$

The final circuit layout is shown in Figure 9.6. Actual physical designs of the microstrip sections for this circuit are left as an exercise for the reader.

A further important development that one must consider is the effects of significantly increasing the centre frequency. For example, the European satellite communications bands include 11.5 GHz and there is intense activity around bands including 22 GHz and even higher—into the millimeter-wave regions in fact. Again, at this stage it is left to the reader to carefully enter into these considerations (issues such as choice of substrate and adjustments for discontinuities should be addressed on the basis of information provided earlier in this text).

We now have design steps for low-noise and high-gain amplifier stages. Such stages represent, for example, the first two cascaded requirements for 'low-noise block' (LNB) or 'down-converter' modules in satellite receivers and they are followed by a microwave mixer from which the intermediate-

Fig. 9.6. Circuit layout for the narrowband amplifier design example (after Liao[1] and reprinted, with permission)

frequency (IF) is extracted. After this comes the IF amplifier which drives the signal, now centered on about 1.35 GHz, via the down-lead cable to the TV set. In many designs of production LNBs most of the circuitry is monolithically integrated within one or more MMICs.

The early 1990s state-of-the-art is well represented by a 22 GHz LNB designed by Imai and Nakakita.[6] This circuit uses three HEMTs in the pre-amplifier, to achieve the required low-noise and gain at 22 GHz, and a local oscillator based upon a 21.4 GHz GaAsFET oscillator stabilized by a dielectric resonator. The overall noise figure is less than 2.8 dB over the 22.5–23.0 GHz frequency range and the local oscillator achieves a frequency variation of less than 600 kHz over a −20°C to +60°C temperature excursion. The overall conversion gain achieved is 47–48 dB over the full frequency range. Imai and Nakakita used a simple E-field probe transition from the waveguide input to the microstrip circuit. This performed adequately, yielding a maximum insertion loss of only 0.2 dB over this narrow bandwidth.

The IF amplifier was realized with three cascaded packaged MMICs such that it provided a noise figure below 2.5 dB and a flat gain of 32 dB over the entire IF band. A block diagram of the complete LNB, a structural diagram of the mixer and also of the local oscillator are indicated in Figures 9.7, 9.8 and 9.9. It is the pre-amplifier that is of primary interest here and this is designed on a 0.38 mm thick alumina substrate. Input and output matching networks are designed around each HEMT, using the principles described above, and the resulting stages are then directly cascaded. Finally the matching networks of the noise-critical first and second stages are trimmed to minimise the total noise figure.

9.2.3 Design of a decade-bandwidth distributed amplifier

For certain applications, notably spread-spectrum, electronic warfare (EW) and instrumentation, solid-state microwave amplifiers capable of broadband operation are required. There are several circuit approaches leading to broadband MIC and MMIC design, dependent on the desired combinations of bandwidth, centre frequency, noise etc. One such design approach is the

Fig. 9.7. Block diagram of 22 GHz band low noise down converter. (Reprinted, with permission, from Imai and Nakakita.[6])

Fig. 9.8. Construction of mixer. (Reprinted, with permission, from Imai and Nakakita.[6])

Fig. 9.9. Circuit layout of local oscillator for 22 GHz LNB (Reproduced, with permission, from Imai and Nakakita.[6])

distributed amplifier and a specific example, originally due to Sweet[3] (pp 152–162) is outlined here.

Sweet initially calculates FET parameters leading to the individual device g_m but we shall assume this has been done and state the outcome—which amounts to a circuit comprising four FET devices to be designed within a four-section distributed amplifier (for further details Sweet's text is recommended). For distributed amplifiers the inductance per section of the 'artificial' lines is the square of the gate/drain characteristic impedance multiplied by the gate-source capacitance and, in this example, the result is 0.55 nH.

The most economical method of realizing inductors having this value is with narrow microstrip transmission lines and an expression for this is given in Chapter 8 (eqn 8.15). This expression is repeated here:

$$l = \frac{\lambda_g}{2\pi} \sin^{-1}\left(\frac{2\pi f L}{Z_0}\right) \qquad (9.11)$$

Now it is fundamental for the distributed amplifier that the upper cut-off frequency determines the gate-line and drain-line inductances. So the upper cut-off frequency (20 GHz) must be used here and, assuming that 100 Ω microstrip lines with an effective microstrip permittivity of 8.5 (at 20 GHz) are employed, substitution into the above expression gives:

$$l = \frac{\{3\times10^{10}/(\sqrt{8.5}\times20\times10^9)\}}{2\pi} \sin^{-1}\left\{\frac{2\pi(20\times10^9)(0.55\times10^{-9})}{100}\right\}$$

So $l = 0.62$ mm.

The strip width is calculated from the narrow-strip synthesis expressions, eqns (3.17) and (3.18) and hence all the basic design criteria are known for the inductive microstrip lines.

The drain line loading capacitance value is the difference between the gate-source and drain-source capacitances and Sweet calculates this to be 0.15 pF. In practice, however, separate physical capacitors are not required because small drain line microstrip parasitic capacitances intrinsically serve the purpose to advantage.

The overall topology of this amplifier is shown in Figure 9.10 and detail of the gate line sections is indicated in Figure 9.11. The drain-line section layout is broadly similar except that small strip fingers are provided for final connection to the drains instead of the four connections shown for each gate.

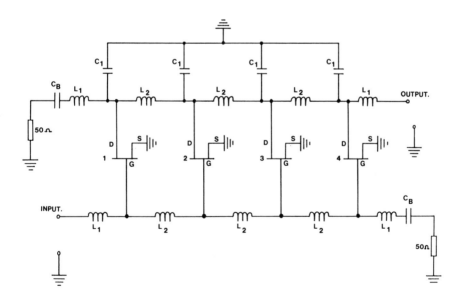

Fig. 9.10. Overall topology of the 2–20 GHz MMIC distributed amplifier (after Sweet[3], reprinted with permission)

Fig. 9.11. Detail of the gate line sections for the amplifier of Figure 9.10 (after Sweet[3], reprinted with permission)

Having completed the preceding design procedures the next steps are: select an appropriate transistor (GaAsFET, HEMT, etc), satisfy output power or noise figure specifications and choose a suitable bias circuit.

In general the chosen device may have C_{gs} and C_{ds} values that differ from those used above. In which case iterations are necessary to finalize the design.

In his design example Sweet assumes that the actual device chosen has values for these capacitances that are identical to those calculated.

The next step is to check that the total DC current required by the amplifier is below the specified limit and to determine the width of each gate finger of the chosen four-fingered FETs. From this data, combined with the equivalent circuit capacitances, resistances and g_m, a full model for the transistor is created using a computer modelling and simulation tool such as Compact Software's SuperCompact™, Eesof's Touchstone™, Hewlett-Packard's MDS™ or one of several other software tools outlined in Appendix D.

Although the distributed amplifier described here is designed for relatively high gain (a 'gain block'), power considerations are always important and noise aspects may be significant. Therefore appropriate power levels are calculated for the amplifier. The saturated drain current, maximum output power, d.c. supplied power and actual power output are all determined in the reference given.[3]

The choice of bias circuit is application-specific to any amplifier design and, in this instance with such a broadband amplifier, Sweet shows that an inductor–capacitor bias circuit is ineffective because the inductor cannot be designed free of resonances over the full 2–20 GHz band. An active load bias technique is recommended, where an additional FET is used with its

gate and source directly interconnected and its gate width chosen such that I_{dss} is equal to the total current drawn by the entire amplifier. Such an active load provides an RF resistance in the region of several hundred ohms and is much smaller than an 'equivalent' inductor.

The computer-generated layout of one section for a distributed amplifier is shown in Figure 9.12. Examples of industry products featuring state-of-the-art distributed amplifiers are illustrated in Figures 9.13 and 9.14. The Texas Instruments TGA8300 chip provides 6.5 dB gain with 18 dBm typical output power over the 2–18 GHz bandwidth. In addition, this amplifier features input and output VSWRs both below 2.0 and a noise figure of less than 7 dB over the band. The chip has four identical FETs, each with gates having 189 μm width and 0.5 μm (recessed) length. Two power supply rails are required: +8 V and 0 to −5 V, and the chip dissipates 1 Watt average power. Eesof's Touchstone™ and also Compact's SuperCompact™ software tools were used for the design of this chip.

The M 221D amplifier manufactured by Avantek's foundry is depicted in Figure 9.14. This operates over the same frequency range as Texas Instruments' chip, but the circuit is implemented differently. There are two stages in this design, each with six sections and each section comprising two

Fig. 9.12. Computer-generated layout of just one section ('cell') for a distributed amplifier (photo courtesy of Eesof, Inc. 1990)

Fig. 9.13. Texas Instruments' TGA 8300 distributed amplifier (photo courtesy of Texas Instruments, Inc. 1990)

cascade-connected FETs. This approach results in a much higher gain amplifier (16 dB) than the Texas Instruments chip, but with a somewhat lower output power at 16 dBm. The M221D amplifier operates from a single-rail + 6 V supply and draws 200 mA. It has a noise figure of 6 dB over the band.

9.2.4 Balanced amplifiers

In 1964, several years before the advent of microwave transistors, Engelbrecht and Kurokawa invented the balanced transistor amplifier and during the following year the design theory of such amplifiers was developed.[7] More than a quarter of a century later balanced amplifiers are available commercially from several vendors. This circuit configuration is now described since it possesses certain advantages over the single, cascaded, or distributed stage approaches.

In this type of design, transistors are selected in pairs for operating in each amplifier stage and both the input and signal circuits are connected by means of 3 dB directional couplers. The general arrangement is shown in Figure 9.15 (the transistors, used in the active circuits T_a and T_b, may in principle be of any appropriate type). The complete practical circuit will of

Fig. 9.14. The Avantek M221D distributed amplifier (photo courtesy of Avantek, Inc. 1990)

course require d.c. bias arrangements, and source and load optimization networks.

Balanced amplifiers have been manufactured which operate with only a few dB noise figure with substantial accompanying gain. These capabilities apply to operation over at least octave bandwidths, up to a maximum frequency around 40 GHz (early 1990s state-of-the-art).

Lange couplers are usually employed and GaAsFETs or HEMTs form the favoured active devices.

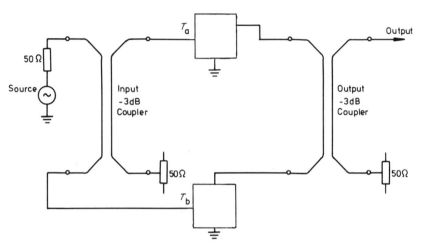

Fig. 9.15. Topology for a balanced amplifier

The 50 Ω matched loads shown in Figure 9.15 are essential to preserve proper operation of the directional couplers and it can also be shown that they assist in maintaining a low noise figure.[7] These loads may be formed in thin or thick-film resistive material and compensated if necessary to achieve good broad-band performance.

It is useful to compare the balanced against the single-device amplifier stage; this is presented in the following lists:

Advantages for a single-device amplifier:

 (a) Only one device is required, thus reducing component and assembly cost.

 (b) Only about one-half of the d.c. supply power is consumed.

 (c) Less circuitry and therefore less substrate area is required.

 (d) There is no requirement for rather 'exotic' circuit elements such as the Lange coupler.

Advantages for the balanced amplifier:

 (a) Input and output impedances are automatically matched in an amplifier optimized for noise figure or output power.

 (b) Good phase linearity is obtained even when using 'conventional' 3 dB microstrip couplers. With Lange couplers the phase linearity is excellent and hence the group-delay distortion is very small (unless the load is significantly mismatched).

 (c) Gain compression and intermodulation characteristics are good.

 (d) Short-circuit and open-circuit stability is guaranteed.

(e) Overall sensitivity to variations in device parameters is low, *provided that selected pairs of devices are used.*

(f) The amplifier gain can be controlled over wide ranges by the *d.c. bias* with little effect on gain flatness or impedance matching.

(g) In some systems the reliability may be enhanced because of the two parallel circuits afforded by this arrangement. If one device fails a little finite gain is usually still available, although over a much reduced band, due to the second device.

(h) Power amplifiers can readily be designed in this topology.

For all these circuits, single-device as well as balanced, it is entirely likely that 'flip-chips' will increasingly be incorporated instead of the more familiar wire-bonded chips—for performance and production reasons. In such situations there is no facility for realizing inductances with bond wires, and it becomes quite essential to evolve circuits that avoid such inductances altogether (bearing in mind that microstrip characteristic impedances greater than about 110 Ω should also be avoided).

Computer-aided design, simulation (particularly important) and test are regular requirements in microwave laboratories and within the industry as a whole. Currently, in most situations, combinations of proprietary packages and in-house routines are run on sufficiently powerful desk-top computers or 'workstations'. Typically over 2 MB of on-board RAM is required and a fast processor is an advantage. Suitable software languages include Pascal and C—although BASIC and FORTRAN are still sometimes used. With these facilities the designer can, for example, study the effects on overall performance of tolerances in the Lange couplers used in a balanced amplifier.

Work is in progress, in several microwave laboratories, on the development of millimeter-wave balanced amplifiers using modified Lange couplers. These typically cover the 26–40 GHz frequency range.

Ladbrooke[2] provides a scattering parameter analysis of the balanced amplifier, although this will not be repeated here. This analysis shows that, in the ideal case of perfect balance, both input and output reflection coefficients are zero, i.e. there are exact matches at both input and output. Also, the magnitude of the forward parameter S_{21} is identical to that of each individual amplifying device, i.e. there is no increase in gain despite the use of two transistors instead of just one. In his text Ladbrooke also points out that this type of circuit provides the capability of delivering almost twice the power available from each transistor as a consequence of additive amplification.

A design example of a balanced amplifier is given by Sweet[3] (pp 162–168). The specification calls for a bandwidth of 4–10 GHz, which is particularly interesting because it is well over one octave. The minimum small-signal gain must be 5 dB, input and output return loss 10 dB minimum, saturated power output 1 W and d.c. bias 6 V with a maximum supply current of

450 mA. The final circuit is to be a hybrid MIC fabricated on an alumina substrate (relative permittivity 9.9).

A balanced reactively-matched amplifier is selected because it provides flexibility in terms of tailored output impedance for broadband power matching and good input/output matches as described above. Each of the two elementary amplifying circuits, for insertion into the balanced topology, are single-ended GaAsFET circuits and the layout for each is shown in Figure 9.16. The L-type matching networks in which the shorted stubs are connected close to the gate and drain are designed with this configuration to allow for matching the lower impedance levels here. The power FETs used in this design require relatively large gate widths and hence low input and output impedances. These circuit 'cells' (Figure 9.16) connect into the active circuit blocks, T_a and T_b, shown in Figure 9.15.

The first step in the design is to consider mid-band, in this case 7 GHz, and to initially make the shorted stubs $\lambda_g/4$ long with the series lines $\lambda_g/8$ in length. For the alumina substrate, operated at 7 GHz, the effective microstrip permittivity is then 6.3 and the lengths are:

$$\lambda_g/4 = \frac{3\times10^{10}}{4\times\sqrt{6.3}\times7\times10^9} = 4.267 \text{ mm}$$

and

$$\lambda_g/8 = 2.134 \text{ mm}$$

The stub widths are first chosen to be the minimum allowable by limitations of the thin-film on alumina MIC fabrication process, which is around 80 μm, and the series line widths are initially selected to be a midway compromise between the 50 Ω system impedance and the input or output impedance (as appropriate) of the transistor.

Detailed considerations for the active device are provided in Sweet's text and are essentially outside the scope of this book. Suffice it to say here that

Fig. 9.16. Single-ended GaAsFET amplifying circuit designed for later insertion into a balanced amplifier (after Sweet[3], reprinted with permission) (two of these identical designs will be required)

a Mitsubishi MGF2116 device was chosen since it met the gain, power and frequency requirements. Each FET is self-biased in this circuit, so the source contacts are grounded by employing the usual parallel capacitor/resistor combination. The capacitor is effectively a short circuit at 4 GHz and therefore yields the RF ground whilst the 6 Ω resistor leaves the gate-source biased at 1.2 V. A 40 pF capacitor has a 1 Ω reactance at 4 GHz and is therefore probably large enough in this design. In practice the resistor and capacitor could be discrete chips, or they may comprise film layer and overlay structures respectively.

The design of the Lange couplers follows the general method presented in Chapter 6.

A typical computer work-station layout for a balanced amplifier design is shown in Figure 9.17. The larger gaps are for GaAsFET chip insertion and the smaller gaps accommodate capacitors.

A two-stage balanced amplifier module with SMA coaxial RF ports, manufactured by Miteq, is shown in Figure 9.18. This amplifier covers the frequency range 4–8 GHz. Further examples of commercially available balanced amplifiers are illustrated in Figures 9.19 and 9.20. The first module (Figure 9.19) embodies three balanced amplifiers and a bandpass filter whilst

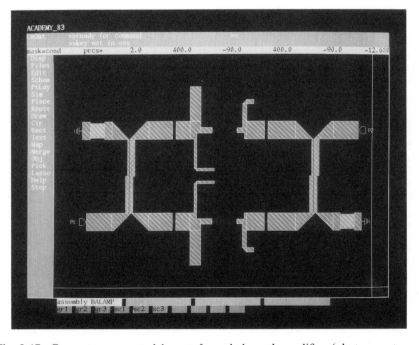

Fig. 9.17. Computer-generated layout for a balanced amplifier (photo courtesy of Eesof, Inc. 1990)

Fig. 9.18. A typical two-stage balanced amplifier (photo courtesy of Miteq, Inc. 1990)

Fig. 9.19. A solid state microwave module involving three balanced amplifier stages and a parallel-coupled bandpass filter (photo courtesy of Avantek, Inc. 1990)

Fig. 9.20. A relatively complex module comprising a total of 17 balanced amplifiers (photo courtesy of Avantek, Inc. 1990)

the next example (Figure 9.20) is more complex, having seventeen balanced amplifiers in its configuration.

The photograph of Figure 9.21 shows a high-efficiency 8.5–10.5 GHz power module from Avantek. This amplifier provides +36.5 dBm (almost 4.5 Watts) minimum output power, nearly 30 per cent typical power-added efficiency and a large-signal gain of typically 16 dB. This open-carrier module has dimensions of 9.1×25.0×2.8 mm, and operates from a dual-voltage d.c. supply of +9 V and −5 V. It is designed to operate in either CW or pulsed mode, with drain voltage modulation. The module comprises a two-stage power GaAs MMIC driver and a pair of 7.2 mm gate length GaAsFET power devices combined using a balanced configuration. All active devices are silicon nitride passivated. The input and output MMIC matching circuits, the couplers and the input and output discrete transistor matching circuits are fabricated variously on alumina or barium titanate substrates using thin-film technology. All the module components are eutectically bonded onto a machined, multilevel gold-plated metal carrier.

The final photograph of a commercially available balanced amplifier is shown in Figure 9.22. This is Texas Instruments' product TGM8014 and, adopting a rather similar design philosophy to that of Avantek, Texas employ two MMIC internal amplifiers with external hybrid MIC Lange couplers, matching and bias circuitry. The TGM8014 is intended primarily for military markets and it provides 0.6 W of average output power, at 1 dB gain compression, over the band 6–18 GHz (this threefold bandwidth was

Fig. 9.21. A high efficiency 4.5 W power module operating over the 8.5–10.5 GHz band (photo courtesy of Avantek, Inc. 1990)

probably achieved by reducing the calculated coupling gap dimension 's' as described in Chapter 6 for Lange couplers). As a result of the broad bandwidth the average small-signal gain (6 dB) of this circuit is quite modest and a maximum VSWR of 3 is specified at both input and output. Once again a dual-voltage DC supply is used with +8 V and −1.5 V and the maximum current drawn from the positive DC supply rail is 650 mA. A novel feature of the TGM8014 amplifier is adjustable voltage dividers provided on the thin-film network to optimize the gate bias for each MMIC.

As mentioned above, when describing the distributed amplifier, both Eesof's Touchstone™ and Compact's SuperCompact™ software tools were used for the design of this balanced amplifier.

9.2.5 Some state-of-the-art amplifier developments

The output power capabilities available from solid-state amplifiers, at least partly based on MICs, are continually being increased and some particularly high-power amplifiers are under development in several laboratories. In most cases these high power designs use waveguide power combiners at their outputs rather than Lange couplers because the latter cannot handle the power levels involved. Recent developments of this type (principally for satellite communications transmitters) have demonstrated an average power

Fig. 9.22. Example of a broadband (6–18 GHz) balanced amplifier (photo courtesy of Texas Instruments, Inc. 1990)

capability of 120 W over a 7.9–8.4 GHz bandwidth (using, for example, Fujitsu power GaAsFETs). A good example of the type of power-combining technology necessary in this kind of amplifier is provided by Ezzeddine and Hung.[8] These workers disclose a new power-combining technique using a dual-ridged structure in a rectangular waveguide. They demonstrate a 1–2 Watt amplifier with an average gain of about 15 dB over the 14–14.5 GHz communications band.

Amongst other developments, broadband, transimpedance and ultra-broadband low-noise amplifiers are described by Perennec et al.[9] The first two classes of amplifiers are important in high-speed (Gbit/s) optical fibre communications. Perennec et al. also describe an MMIC amplifier operating from 1 MHz to 10 GHz for a 9.8 Gbit/s optical receiver and a low-noise transimpedance amplifier, using NEC 32000 GaAsFETs, providing about 23 dB of gain over the 0.3–2.3 GHz band. Noise figure is below 3.4 dB and both input and output VSWRs are less than 2.0 over this bandwidth.

Another amplifier described by Perennec and his colleagues covers the very broad band extending from 0.1 to 30 GHz. This amplifier provides approximately 10.5 dB gain, which remains quite constant over the entire

band, and a maximum noise figure of 6 dB (about 3.2 dB at the higher frequencies). Input and output VSWRs are generally below 2.0, although the input VSWR slightly exceeds 2.0 in mid-band. The Perennec group use their own in-house optimization software based upon a double-matching real frequency method.

An alternative approach to CAD for MMICs is reported by Sharma et al.[10] This group claims a first-pass design success for their technique— the concomitant minimization of design risk results in a factor-of-three reduction in chip cost. Sharma et al. point out that presently available design tools tend to lose accuracy as the frequencies extend above about 15 GHz due to the inability to account for effects such as coupling, spurious radiation, excitation of surface waves and interaction with package modes.

Sharma's group considered a 75 mm diameter wafer containing over 600 elements for experimental characterization. These elements included: 207 lumped structures (inductors and MIM capacitors), 275 distributed structures included the various microstrip discontinuities and special structures such as via holes, air-bridges and constant-R networks. They verified the models by comparing the simulated and measured performance of 3–6 GHz and 6–10 GHz amplifiers. Interestingly, although the overall gain discrepancies are small it must be said that the reflection coefficients (S_{nn}), and therefore the prediction of the degree of match, are in considerable divergence in terms of the comparison between simulated and measured values. This is particularly noticeable in the case of the 6–10 GHz amplifier.

9.3 OSCILLATORS

As with amplifiers there also exists a wide range of possible types for oscillators. Various systems require swept oscillators, using varactors for a narrow-band sweep or YIG for broadband. Voltage-controlled oscillators (VCOs) are required for microwave FM, instrumentation and other purposes and new developments are regularly reported on VCOs.[11, 12] For example, Dupois et al.[11] describe a hybrid MIC Ku-Band VCO, designed around an NE71000 GaAsFET, with a tuning range extending from 12.6 to 15.5 GHz, an average power output of 12 dBm and a phase noise of −70 dBc/Hz at an offset of 10 kHz from the carrier. These workers consider that the available relationships for designing microstrip (i.e. width and length synthesis) placed limitations on their design and that improvements should make their design approach viable for oscillators at frequencies up to 20 GHz.

It has long been appreciated that surface-acoustic-wave (SAW) technology also provides a basis for stable oscillator design. Fleischmann et al.[12] have taken this technology a stage further by developing a low-noise 2.65 GHz VCO based on this type of stabilizing technology (they use the term acoustic

surface transverse wave, or STW). Fleishmann's team use STW delay lines operating at the fifth harmonic to achieve phase noise levels of -93 dBc/Hz at 1 kHz and -63 dBc/Hz at 100 Hz from the carrier. Their oscillator design, which has an output power of 3 dBm and a tuning range of 750 kHz, is centred on a low-noise silicon MMIC amplifier with a matched gain of 15.2 dB at 2.65 GHz.

Alternatively, fixed but highly frequency-stable oscillators are often required—for example, a circuit with low FM noise that is also stabilized using a dielectric resonator to form a 'dielectric resonator oscillator' (DRO). Most types of oscillator have been realised in MIC form but we shall limit our considerations to the DRO and recent developments here.

9.3.1 Example of a DRO

In many existing microwave and millimeter-wave systems highly-stable oscillators use one of the following two techniques:

(i) Locking a relatively unstable microwave oscillator to a highly-stable but lower frequency quartz oscillator.

(ii) Using a mechanically tuned microwave resonator embodying a low thermal expansion material such as Kovar. Again, a relatively unstable microwave oscillator is used.

The use of stable dielectric resonators represents a more recent technique— one that is much more compatible with MIC/MMIC realisation. Dielectric resonators are available to operate at specific frequencies ranging from around 1 GHz to almost 40 GHz. At the lower limit the resonator becomes rather large and tends to dominate the real estate whilst at the higher frequencies the resonator 'puck' becomes somewhat 'uncontrollably small' and sensitive to production variations. These types of resonators have Q-factors in the 1000 to over 40 000 range and thus oscillators controlled by such resonators are characterized by very low FM noise and an output frequency that is very temperature-stable.

It is important to couple the dielectric resonator to the appropriate microstrip line in some optimum way—either for maximum degree of coupling or for maximum output power from the oscillator. The general layout applicable to this situation is shown in Figure 9.23, where the dielectric resonator puck may have to slightly overlap the microstrip line for maximum coupling (in which case it would be necessary to include the layer of low-permittivity material).

Design example

The example given here follows that originally provided by Sweet[3] (pp 211–216).

Fig. 9.23. Structural diagram of dielectric resonator/microstrip coupling scenario

The specifications include: an operating frequency of 8.4 GHz, a minimum power output of +10 dBm, FM noise not to exceed −110 dBc at 1 MHz from the carrier in a 1 kHz bandwidth and temperature stability to be +5 MHz over the temperature range 0 to +50°C.

This circuit is based on a GaAsFET. An effective method of providing the required negative resistance in a fixed-frequency oscillator is to incorporate an inductor between gate and ground, i.e. common-gate inductive feedback (as Sweet points out, this technique is not suitable for broadband oscillators). It is not necessary for the negative resistance to be calculated and designed in detail since it is only required that this resistance be significant for oscillation, which is easily obtained at these frequencies.

The fundamental circuit topology for this DRO is shown in Figure 9.24. Here the resonator itself is modelled as a parallel tuned circuit and its coupling to the GaAsFET's source is represented as an ideal transformer with an adjustable turns ratio. The drain-to-ground inductance and capacitance form a matching circuit for the output.

A specially selected dielectric resonator puck is required for the specified frequency of 8.4 GHz. In this case a Murata Erie type P/N DRD077SC034 is selected which has a diameter of 7.7 mm, a thickness of 3.4 mm and a relative permittivity of 28.6. The elements of this resonator's equivalent circuit are calculated from its Q-factor (5000) and its resonant frequency (8.4 GHz), and an initially assumed value for either L or C.

The choice of active device, calculation of noise and assessment of stability, and choice of bias circuit are beyond the scope of this text but are

Fig. 9.24. Circuit topology for the 8.4 GHz low-noise narrowband DRO design example

all considered by Sweet. The gain/frequency requirements are well met by an NEC 71000 transistor and the calculated noise-to-carrier ratio is 33 dB better than the specification demands. It is also shown in the same reference that the temperature drift with this type of oscillator is of the order of 4 ppm per degree Celsius.

The circuit is self-biased through the 50 Ω microstrip line to which the resonator is coupled, terminated in a resistor of approximately 50 Ω value which also acts as a source-bias resistor. A 5 V bias supply, fed through a choke inductor, is provided for the drain.

In the practical layout of this oscillator two 0.025 mm thick alumina substrates are required, attached by a metal grounding rib on which the GaAsFET is die-attached. The resonator puck, coupled to the 50 Ω microstrip line, occupies one substrate whereas the output capacitance/inductance matching circuit, together with the drain bias feed and RF output line, are located on the second substrate. The common-gate feedback, required to yield the negative resistance as described above, is provided simply by the bond wire from the gate to the rib ground.

This circuit can be optimized, including all parasitic elements associated with the practical layout, and simulated using one of the packages available from organizations such as Compact Software or Eesof.

9.3.2 DRO oscillator developments

Seawright *et al.*,[13] in their paper on the optimum synthesis of (in their example) a 13.8 GHz microstrip DRO, show that oscillator output power is a highly non-linear function of DRO position relative to the microstrip. They also show that this position depends upon whether the circuit is unshielded, shielded (but untuned) or shielded and tuned. Assuming that such a circuit is likely to be both shielded and tuned in practice, then the

results due to Seawright *et al.* indicate that the centre of the dielectric resonator should be less than 3.2 mm from the edge of the microstrip. At distances greater than 3.2 mm the output power decreases rapidly.

It should be noted that this is a different situation from the maximum coupling condition considered above.

One of the most significant attributes of a DRO is the temperature stability of its output frequency and Massias *et al.*[14] have reported a novel analysis and design technique for highly temperature-stable DROs. They define a parameter involving the unloaded-Q of the dielectric resonator, the coupling and a circuit-related quantity and develop a fast method, based on measurements of the temperature stability of a given DRO, for analysing and designing new DROs.

It is sometimes necessary to synchronize several DROs simultaneously as well as to further improve the oscillator phase noise (note that DROs inherently reduce FM noise but not necessarily phase noise and this can become a serious problem). One method for achieving both objectives is to injection-lock a DRO and to synchronize it with a multitone signal, and such a technique has been described by Golubicic.[15] In this work an NE71083 GaAsFET has its gate microstrip circuit (50 Ω) coupled to the dielectric resonator and the multitone signal is injected to the source, matched by an L-topology microstrip circuit.

There are several alternative topologies and design approaches that have been described in the literature and significant new developments continue to be reported.[16, 17] A design method for DROs based upon parallel feedback is reported by Filicori *et al.*[16] In this approach the dielectric resonator is coupled to two microstrip lines located in a parallel feedback path between source and drain of a GaAsFET and the schematic layout is

Fig. 9.25. Microstrip implementation of the parallel feedback DRO. (Reprinted, with permission, from Fillicori, Monaco and Vannini.[16])

shown in Figure 9.25. This oscillator runs at a stable frequency of 3 GHz and is simulated using a 'SPICE' circuit simulator.

Although DROs have achieved a strong position in this sector of solid-state microwave circuit realization other approaches may be 'on the horizon'. For example, given the improvements in microstrip Q-factor achievable with superconductivity some oscillator (and indeed complete subsystem) designs using superconductive microstrip may emerge from current work. It would also be interesting to see whether SAW oscillators, perhaps combined with microstrip, could be improved using superconductivity.

The hybrid mode coupling of dielectric resonators to microstrip lines has been considered in detail by Liang and Zaki.[17] Their results apply to frequencies in the range 11.71–14.91 GHz and, in all cases, the external-Q increases smoothly with the separation between the resonator and the microstrip line. However, the design implications reported by Seawright et al.[13] (described above) may be of more significance for oscillators.

9.4 ACTIVE MICROWAVE FILTERS

These represent relatively recent developments and some discussion on them is included here for completeness. The filters described in Chapter 8 all embody passive structures—microstrip L/C derived sections, coupled microstrip resonators, etc. By the use of active elements in filters and related circuits inherently high Q-factors, voltage-tuning and internal gain may be obtained. Active filters free the designer from the restrictions of significant in-band insertion loss, corner rounding and poor skirt selectivity commonly encountered with the more familiar passive circuits. The active device is usually a GaAsFET or a HEMT, although it can in principle be any microwave amplifying device.

Two examples are now briefly presented describing active microwave filters. Chang and Itoh[18] describe a varactor-tuned, multipole active microwave bandpass filter based upon an NE71083 GaAsFET. The basic structure is indicated in Figure 9.26 and the complete layout is shown in Figure 9.27. In order to obtain an improved aspect ratio for the coupled line section, the filters are designed with a characteristic impedance of 82 Ω. Transformation of this impedance to the 50 Ω external environment is achieved with tapered microstrips. The two-pole version of this filter exhibits a tuning range of 430 MHz with an 80 MHz 3 dB bandwidth centred on 10 GHz (i.e. a 0.8 per cent bandwidth). In each case the passband insertion loss is typically 0±1 dB.

Jiao et al. report microwave frequency-agile active filters for MIC and MMIC applications.[19] Their designs use modified microstrip hairpin-shaped split-ring resonators parallel-coupled to an active loop and they have demonstrated both bandpass and bandstop filters. For the bandstop filters

Fig. 9.26. The basic structure of a one-pole active tunable filter. (© 1990 IEEE. Reprinted, with permission, from Chang and Itoh.[18])

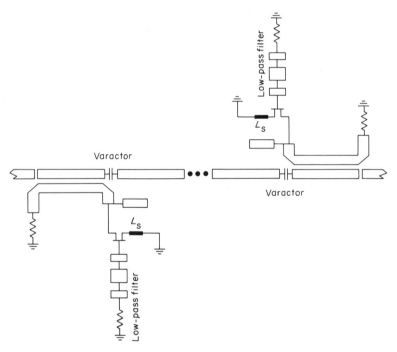

Fig. 9.27. The multipole active tunable filter. The negative resistance is realized by a GaAsFET circuit. (© 1990 IEEE. Reprinted, with permission, from Chang and Itoh.[18])

the entire split hairpin lines and their active loops are themselves parallel-coupled to a main through-connecting microstrip line. At the centre frequency of 3.1 GHz a rejection of 45.3 dB is obtained. The bandpass filters exhibit typically 0.03 dB insertion losses together with a return loss

of 35.9 dB in-band. The frequency tuning range is approximately 47 MHz and the 3 dB bandwidth is 3.17 MHz (i.e. 0.1 per cent).

These types of developments offer the potential for effective high-Q resonators using entirely planar technologies (unlike dielectric resonators). Therefore, this direction of technology may hold a key for future all-MMIC filters and oscillators.

9.5 MICROSTRIP CIRCUITS FOR HIGH-SPEED PULSE SYSTEMS

There now exists, in high-speed digital electronic systems, a requirement for properly designing microstrip interconnections for emitter-coupled logic (ECL) or GaAs ICs. At present most implementations demand standard DIL packages (or 'flat-packs' for surface-mounting) on printed-circuit boards. The microstrip transmission lines must then be formed on these printed-circuit boards. Where very high density integration is essential, in spacecraft or guided missiles for example, ECL or other high-speed logic ICs have to be mounted as chips on to a microstrip circuit on alumina or a similar substrate.

We have briefly considered high-speed pulse propagation along microstrip, towards the end of Chapter 3, and crosstalk between such lines running parallel, in Section 6.14. Here we are concerned with circuit interconnections.

It is most important that the lines interconnecting the ICs are matched/terminated closely so that reflections can be reduced to a minimum. Coekin[20] provides a detailed account of reflection effects and their minimization; we will not reproduce details of the general problems here. If we consider the problem of matching just one line interconnecting between two ICs, the methods illustrated in Figure 9.28 show what can be done. In both cases, Figure 9.29(a) and (b), the necessary d.c. bias is supplied to the input of ECL circuit (2) through the matching resistor/s themselves. If a specially tapped supply is available a single 50 Ω resistor is suitable as shown in Figure 9.28(a). Where only the standard -5.2 V ECL supply is available two resistors act as the required potential divider. The parallel combination of these resistors must be made equal to the characteristic impedance of the microstrip lines, that is 50 Ω. On a printed-circuit board discrete metal-film resistors with greatly shortened wires are usually quite suitable, although flat-chip resistors would be preferable electrically and from the viewpoint of reliability. When the system is hybrid-integrated on to, say, an alumina substrate, the resistors may be manufactured in either thick-film or thin-film technology as mentioned earlier (Chapters 2 and 8).

Some difficulties arise when a number of ICs are required to be fed via a single IC with its output transmission line. As we can see from Figure 9.29, we may consider two approaches. In Figure 9.29(a) the fed circuits

(a) Tapped supply

(b) Standard supply

Fig. 9.28. Requirements for transmission line interconnections between two ECL high-speed digital circuits: alternative techniques

are partly connected in parallel along the line and partly placed at the end of the line. A single shunt resistance Z_0' at the far output of the line provides the matched termination; this is therefore called *parallel output matching*. Alternatively, we could connect all the fed circuits in shunt at the far output of the line and match just the feeding circuit (1). The first arrangement is by far the most satisfactory: the response is faster and intermediate circuits (2, 3, and 4) do not suffer from reflected pulses. Coekin[20] shows that it is important to choose Z_0' correctly. The value should be

$$Z_0' = \frac{Z_0}{\sqrt{(1+C_d/C)}} \tag{9.12}$$

where Z_0 = microstrip characteristic impedance (often 50 Ω)
 C = microstrip capacitance per unit length
 C_d = effective extra distributed capacitance, equal to the sum of the

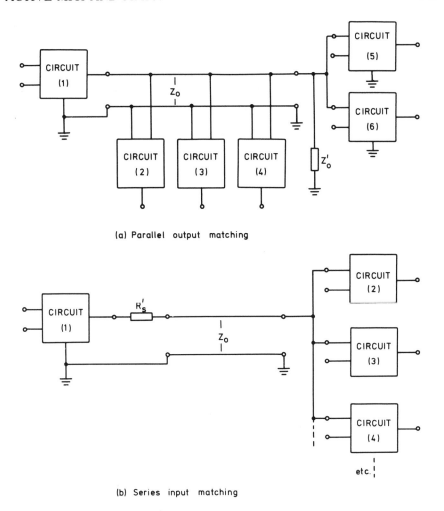

(a) Parallel output matching

(b) Series input matching

Fig. 9.29. Alternative matching arrangements for substantial fan-out in transmission-line-interconnected high-speed digital circuits

input capacitances due to circuits (2), (3), and (4), divided by the line length

This sytem is the version which is generally used in practice, and Z_0' may be realized as a single resistance or a pair of resistances, as described for Figure 9.28.

As pulse systems become even faster, into the gigabits per second ranges, microstrip interconnections are often used but the effects of discontinuities now become very significant. Menzel[21] has investigated the behaviour of

microstrip structures under these very high speed pulse conditions. A practical trapezoidal input pulse shape is considered, with a rise-time of the order of tens of picoseconds and pulse durations of the order of hundreds of picoseconds (down to 100 ps). A Fourier series is written for these pulses and the effects of the microstrip circuit elements are analysed in terms of scattering parameters. Both alumina and plastic substrates are studied theoretically and experimentally. Microstrip pulse-transforming filters are examined and a matched filter is also studied.

GaAs digital circuits, with systems implementation, are described in the text by Haigh and Everard.[22] Various advanced examples of developments in this field have been reported in the literature.[23-26]

REFERENCES

1. Liao, S. Y., *Microwave Circuit Analysis and Amplifier Design*, Prentice-Hall, Inc., Reading, Massachusetts, 1987
2. Ladbrooke, P. H., *MMIC design: GaAsFETs and HEMTs*, Artech House, Inc., 1989
3. Sweet, A., *MIC and MMIC Amplifier and Oscillator Circuit Design*, Artech House, Inc., 1990
4. Eda, K. *et al.*, 'Miniature hybrid microwave ICs using a novel thin-film technology', *IEEE MTT-S International Microwave Symposium* 1990, Digest of Papers, 419–422
5. Tipping, R. G., 'GMAX', in 1975 *Microwave Journal Engineers' Handbook and Buyers' Guide*, 1975, 30–31 (BASIC program analysing stability and max available gain for a two-port network, in *S*-parameters)
6. Imai, K. and Nakakita, H. 'A 22GHz-band low noise down converter for satellite broadcast receivers' *Proceedings of the 19th European Microwave Conference*, London, Sept. 1989, 549–554
7. Kurokawa, K., 'Design theory of balanced transistor amplifiers', *Bell System Tech. J.*, October 1975, 1675–1698
8. Ezzeddine, A. K. and Hung H-L. A., 'A novel power combiner for MIC and MMIC amplifiers' *IEEE MTT-S International Microwave Symposium* 1990, Digest of Papers, 479–482
9. Perennec, A. *et al.*, 'Optimization of gain, VSWR and noise of the broadband multistage microwave MMIC amplifier by the real frequency methods. Synthesis in lumped and distributed elements', *IEEE MTT-S International Microwave Symposium* 1990, Digest of Papers, 363–366
10. Sharma, A. K. *et al.*, 'Computer-aided design techniques for microwave monolithic integrated circuits' *IEEE MTT-S International Microwave Symposium* 1990, Digest of Papers, 555–558
11. Dupuis, A., Hausner, J., and Russer, P., 'Hybrid integrated Ku-Band VCO', *Proceedings of the 19th European Microwave Conference*, London, Sept. 1989, 1009–1014
12. Fleischmann, B. *et al.*, 'A high frequency low noise acoustic STW VCO', *Proceedings of the 19th European Microwave Conference*, London, Sept. 1989, 1015–1020
13. Seawright, S. W. J. *et al.*, 'Optimum DRO Synthesis' *Proceedings of the 19th European Microwave Conference*, London, Sept. 1989, 406–411

14. Massias, M., Howes, M., and Postoyalko, V., 'A novel analysis and design technique for temperature stable DROs' *Proceedings of the 19th European Microwave Conference*, London, Sept. 1989, 418–423
15. Golubicic, Z., 'Injection locked DRO synchronized by multitone signal', *Proceedings of the 19th European Microwave Conference*, London, Sept. 1989, 395–400
16. Filicori, F., Monaco, V. A., and Vannini, G., 'A design method for parallel feedback dielectric resonator oscillators', *Proceedings of the 19th European Microwave Conference*, London, Sept. 1989, 412–417
17. Liang, X-P., and Zaki, K. A., 'Hybrid mode coupling of dielectric resonators to microstrip lines', *IEEE **MTT-S** International Microwave Symposium* 1990, Digest of Papers, 395–398
18. Chi-Yang, C., and Tatsuo, Itoh, 'A varactor-tuned, active microwave band-pass filter', *IEEE **MTT-S** International Microwave Symposium* 1990, Digest of Papers, 499–502
19. Jiao, X. H. *et al.*, 'Microwave frequency agile active filters for MIC and MMIC applications', *IEEE **MTT-S** International Microwave Symposium* 1990, Digest of Papers, 503–506
20. Coekin, J. A., *High-speed pulse techniques*, Pergamon Press Ltd, Oxford, 1975, 196–205
21. Menzel, W., 'Analysis and design of microstrip circuits for Gbit/s-pulse applications', *AEU,* **33**, No. 7/8, 1979, 309–313
22. Haigh, D. and Everard, J. (Editors) *GaAs Technology and its Impact on Circuits and Systems'*, Peter Peregrinus Ltd, 1989 (Chapter 6, by Flynn, B. W., Reekie, H. M., and Mavor, J.)
23. Abe, M. *et al.*, 'HEMT LSI technology for high-speed computers', *IEEE GaAs IC symposium* Tech. Digest, 1983, 158–161
24. Nansed, B. A. and Gilbert, B. K., 'A 32-bit 200 MHz GaAs RISC for high-throughput signal processing environment', *IEEE Micro*, December 1987, 8–20
25. Kiefer, D. and Heightly, J., 'A GaAs implemented supercomputer system', *IEEE GaAs IC Symposium* Tech. Digest, 1987, 158–161
26. Singh, H. P. *et al.*, 'GaAs low-power integrated circuits for a high speed digital signal processor', *IEEE Trans. Electron Devices*, **ED-36**, 1989, No. 2, pp 240–249

10 Microstrip Antennas

10.1 FUNDAMENTAL ASPECTS

The bulk of this text is aimed at designers of microstrip circuits, as the title of the book implies. Yet it is a fact that open (i.e. unshielded) microstrip inherently radiates a significant fraction of the electromagnetic energy originally transferred to it.

For circuits this phenomenon is a problem and the radiation must be minimised or at least contained, but the radiation capability is also a useful attribute which has been exploited in many types of microstrip antennas.

This chapter provides only a brief overview of the subject of microstrip antennas and is certainly not intended as a design basis. There exist several well researched texts that provide detailed information on microstrip antenna theory and design and the reader who needs to design such an antenna is referred to these sources.[1-4] It is noteworthy to reflect on the fact that most microwave engineers tend not to think naturally in terms of microstrip antennas (or, for that matter, other planar antennas) but instead they usually first consider 'dish' reflector designs. At present, reflector antennas maintain a strong advantage where gain is of great importance—although recent advances in quasi-planar zoned structures have demonstrated strong competition for 'dish' reflectors on a production basis. Indeed this could have considerable implications for many microwave systems, including satellite communications.

However, where high antenna gain is not of paramount importance and where low cost (an important attribute of microstrip antennas) is of prime significance, as in domestic satellite television receivers, these types of antennas are likely to find favour. Their relative compactness and ruggedness are also significant, in aerospace applications for example. Given the continuing trend towards higher powered satellites there are also several applications in satellite communications where microstrip antennas represent a good choice in spite of their relatively low gain.

Some basic microstrip radiating elements are shown in Figure 10.1. The narrow-line resonant radiating element shown in Figure 10.1(a) is very

(a)

(b)

(c)

Fig. 10.1. A range of basic microstrip radiating elements. (Reprinted, with permission, from James, Hall and Wood.[1])

elementary but it also obviously requires some form of feed or illumination and a form of direct-feed connection to such a resonator is indicated in Figure 10.1(b). A number of microstrip radiating elements are usually arranged to form an array and a simple 'array' of proximity-fed resonant radiators is shown in Figure 10.1(c).

The choice of radiating elements and their array configurations (see later in this chapter) is dependent upon the radiation characteristics required, the most significant of which are:

bandwidth
polarisation
efficiency
pattern shape and scan coverage.

James *et al.*[1] show that the antenna Q-factor of a microstrip resonant radiating element is given by

$$Q_r = \frac{2\pi f_r \mathscr{E}_s(1/h)}{P_r} \tag{10.1}$$

where $\mathscr{E}_s(1/h)$ is the stored energy, P_r is the power radiated and f_r is the frequency. To find the total Q-factor both dielectric and conductor losses (P_d and P_c) must also be taken into account, giving the result

$$Q_r = \frac{2\pi f_r \mathscr{E}_s(1/h)}{P_r + P_d + P_c} \tag{10.2}$$

This expression shows that increasing losses yield an increase in the bandwidth but, as James *et al.*[1] point out, it is important to consider the effect on antenna efficiency, given by

$$\eta = \frac{P_r}{P_r + P_d + P_c} \times 100\% \tag{10.3}$$

which indicates that any increase in bandwidth as a result of an increase in losses is accompanied by a proportional reduction in efficiency. For most types of substrates (except when extremely thin) the radiated power dominates the resonant behaviour of the antenna and Q-factors in the range 20–200 are common.

James *et al.* also describe the problem of the feed structures which, if designed in a completely microstrip form, would interfere electromagnetically with the desired properties of the antenna. It is therefore necessary to ensure that the feeders are realised in some form of shielded structure such as triplate.

Whilst any microstrip structure, of whatever complexity, will radiate to some extent it is necessary to consider many fundamental constraints in order to effectively design an efficient microstrip antenna. As previously argued in the case of microstrip circuits, but with differing reasons now, the initial decision must be the choice of the main substrate. We talk in terms of main substrate because some microstrip antenna structures use multiple substrates in a layered arrangement. The 'main' substrate referred to here is that upon which the antenna metallization is deposited.

Further initial design criteria will include other features also related closely to the intended application and both the type of array and the feed method are of fundamental importance. Design details centre upon the radiation mechanism of open-circuit microstrip terminations, particularly the determination of the end admittance as a function of frequency. It should be emphasized here that the end admittance involves several effects (previously mentioned in Chapter 5) and these include equivalent conductances associated with surface-wave losses as well as radiation losses. It is therefore important to separate the component applicable to the radiation losses.

This radiation conductance can be given as a family of curves applicable to a specific substrate, with microstrip width as parameter. Measured results have been published for the radiation conductance G_r and, for a polymer substrate with a permittivity of 2.32, the value rises with frequency. It is about 0.05 mS at 3.5 GHz, for a microstrip width of 1.08 mm, and approximately 9 mS at 15 GHz for a microstrip width of 9.63 mm.[1] Polymer-based materials are often used as the main substrate in microstrip antennas so this type of data is useful in such design.

For CAD purposes it is necessary to have a closed-form expression for the radiation conductance and the following expression has been given by James *et al.*[1]

$$G_r \simeq \frac{4\pi}{3Z_0} \left(\frac{h}{\lambda_0^2} w_{\text{eff}} \right) \frac{1}{\sqrt{\epsilon_{\text{eff}}}} \tag{10.4}$$

This expression holds for both h/λ_0 and $w_{\text{eff}}/\lambda_0 \ll 1$.

So far, we have not said anything about the important topic of the distribution of antenna radiation in space. A fundamental determination of this follows from a study of the rectangular patch antenna. This is defined as a rectangular microstrip resonator designed to be an integral number of half-wavelengths long. A useful first approximation is to consider the total radiation to be due to the combined effect of radiation from each of the resonator ends (i.e. neglecting mutual coupling between these ends). For details on the development of radiation pattern expressions the interested reader is referred to the work of James and Wilson.[5]

The design of antenna elements then makes use of the expressions for Q-factors given earlier in this chapter.

10.2 MICROSTRIP PATCH ANTENNAS AND LINEAR ARRAYS

The basic concept of a microstrip patch antenna, driven against a ground plane, is shown in Figure 10.2. In practice the actual antenna may take on any one of a wide variety of shapes, often rectangular but also frequently some form of circular structure. These antenna configurations can be folded into three-dimensional shapes to suit missile or aerospace applications and fed via coaxial, stripline or waveguide transmission arrangements. Patches that are short-circuited along an edge are advantageous for generating sum and difference radiation patterns and this approach has been used with both semicircular and rectangular patches (in combination).

A recent development of a blade antenna, for use on the skin of an aircraft and based on microstrip elements, has been described by Jairam.[6] This antenna was originally required to operate at 550 MHz with a 5 per cent bandwidth and a VSWR of 2. For developmental convenience a scaled

Fig. 10.2. Simple conceptual diagram of a microstrip patch antenna driven against a ground plane. (Reprinted, with permission, from James, Hall and Wood.[1])

model was constructed operating at 2.8 GHz. Some details of the antenna and its feed network are shown in Figure 10.3.

Another basic design constraint with antennas is the bandwidth and, for a microstrip patch behaving in the manner of an *RLC* tuned circuit, James *et al.*[1] show that the usable bandwidth B_u is given by

$$B_u = \frac{100(S-1)}{\sqrt{S}}\left(\frac{8}{3\epsilon_r}\right)\left(\frac{h}{\lambda_0}\right) \quad \text{per cent} \tag{10.5}$$

where S is the VSWR (this expression holds for $S > 1$).

The Q-factor of a rectangular patch may be calculated in several ways, one being to follow the Fabry–Perot approach given in Appendix B. The resulting expressions are given by James *et al.* as:

$$Q_R = \frac{3\epsilon_r}{8}\left(\frac{\lambda_0}{h}\right) \quad \text{for } \epsilon_r > 2 \tag{10.6}$$

$$Q_R = \frac{\pi^2\sqrt{\epsilon_r}}{4\pi^2 - 16\sqrt{\epsilon_r}}\left(\frac{\lambda_0}{h}\right) \quad \text{for } \epsilon_r < 2 \tag{10.7}$$

(Actually, it is also indicated that these formulae both estimate to within 10 per cent, so eqn (10.6) could be used exclusively for approximate calculations.)

Most practical microstrip antennas combine several patch (or similar) elements to form arrays. Feed methods represent a vital consideration for antenna arrays and, in the microstrip case, corporate feed structures in which branched microstrip lines (taken in pairs and themselves often fed by triplate lines) separate out to deliver the power to the elements.

A simple edge (or parallel) coupled linear array of microstrip resonators is shown in Figure 10.1(c). In this structure the radiating elements are half-wavelength resonators, themselves spaced symmetrically half a wavelength apart along the feeding microstrip line. Disadvantages with this approach

Fig. 10.3. Blade antenna using microstrip patch elements. (a) Blade antenna. (b) Feed network. (Reprinted, with permission, from Jairam.[6])

include losses due to the resonator coupling and the requirement for a matched load which loses precious power and adds an undesirable constraint to the design and performance.

Therefore a more common approach is that of the comb line array in which the resonant microstrip elements are formed again by half-wavelength lines, but this time stubs are directly connected to the microstrip feed line and spaced a full wavelength apart. Although a matched load is still required at the final end of such an array, much less power is absorbed in this than in the case of the parallel-coupled structure and overall antenna performance is acceptable for some applications. It is clear that stub-length-tapering provides a way of overcoming the inherent difficulty of achieving sufficient

coupling between stubs over the bandwidth (-16 dB is the maximum otherwise achievable with quarter-wavelength stub spacings).

James et al.[1] describe several variations on this approach, notably:

Linear travelling-wave comblines (with slitted stubs to force the wave to travel up and down each stub efficiently).
Swept-frequency scanned comblines.
A quasi-snake line array.

Other types of microstrip array antennas include:

Discontinuity arrays (in which radiation arises from the discontinuities associated with interleaved fingers, or bends).
Serpent lines (sinusoidal, trapezoidal and zig-zag).
Chain and Franklin antennas.
Rampart line arrays.
Leaky wave and slot arrays (these use higher-order modes—they convert a large proportion of the incident quasi-TEM energy into modes such as TE_{01} by incorporating a chain of fundamental mode suppressing slots).

The chain, Franklin and rampart arrays are all capable of generating (or receiving) radiation with circular, parallel or perpendicular polarization, achieved simply by adjusting the geometry of the structures in terms of the fractional wavelengths in the line lengths and separations.

10.3 REAR-FED AND MULTIPLE-LAYER ARRAYS

Rear-fed arrays, usually with stacked (or multiple-layer) structures, are becoming increasingly important. With these circular discs or rings, or more conventional rectangular microstrip elements, can be incorporated into the design and a typical elementary feed structure is shown in Figure 10.4. It should be noted that this is coaxial fed, with the coaxial line extended to join the actual antenna element situated on the top substrate. The feed structure is positioned on the lower substrate which has the higher permittivity, hence allowing much greater freedom over the design of the feed network at the same time as ensuring that the antenna elements operate close to their maximum efficiency. This antenna, with the upper disc driven, operates in a dual frequency mode.

If, instead of the coaxial and lower disc feed, a directly-fed rectangular microstrip element forms the feed structure then wide bandwidth operation is obtained, typically 18 per cent bandwidth at 10 GHz.

Multiple-beam operation is also possible with microstrip patch arrays. Hall and Vetterlin[7] describe such an array that provides a one-dimensional beam set by combining the function of a beamformer within the antenna array itself. This antenna was designed to provide five beams and to operate

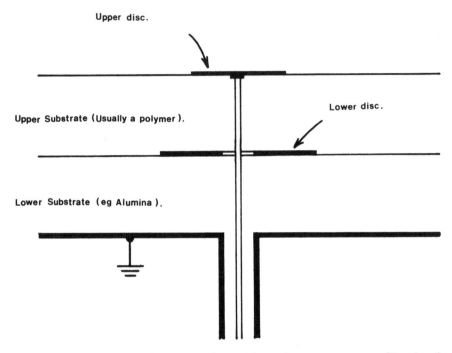

Upper disc.

Upper Substrate (Usually a polymer).

Lower disc.

Lower Substrate (eg Alumina).

Fig. 10.4. Elementary feed structure for a microstrip antenna array. (Reprinted, with permission, from James, Hall and Wood.[1])

at 8.2 GHz. A conceptual diagram is shown in Figure 10.5. In this figure the upper and lower substrates are clearly visible with their resonant arrays and meander feed lines respectively. In this antenna the beam directions are determined by relatively narrow meander line phase-shifting sections in each feed line. For the experimental design both the patch arrays and the feed lines were etched on a 0.79 mm thick glass-loaded PTFE substrate. The antenna gain was 19.5 dB at centre frequency (minimum 15 dB over a 10 per cent bandwidth) and the sidelobe levels were generally better than 12 dB down on the main beam level.

Another method for feeding stacked microstrip antennas has been described by Heberling *et al.*[8] This antenna was developed for use with a GPS (Global Positioning satellite System) receiver and was required to operate at 1.227 GHz and 1.575 GHz, with right-handed circular polarization. A further requirement was for a single feed line and no impedance-matching network. This design was conceptually based upon the two-disk coaxial structure described above, but realised in patch structures. Both microstrip and what is referred to as a notch feed are described.

Ostermann[9] has reported a logarithmic zig-zag antenna providing a maximum gain of 4.5 dB over the 2.9–3.5 GHz frequency range and Splitt[10]

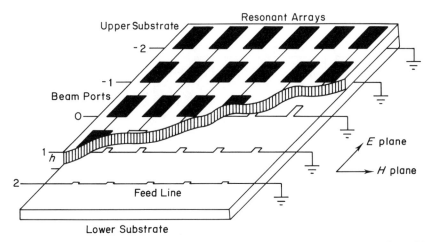

Fig. 10.5. Multiple beam microstrip array. (Reprinted, with permission, from Hall and Vetterlein.[7])

has described a moment method for predicting the performance of electromagnetic and edge-fed coupled antennas.

Other interesting developments include a dual polarised antenna for a polarization shift-keying transponder,[11] an integrated multichannel multifrequency array[12] and an intelligent electronically spinning TACAN antenna.[13]

Smith and Jefferson's polarisation shift-keying design concept is intended to provide a novel method of data transmission.[11] It operates at 2.45 GHz and embodies a dual-fed microstrip disc antenna element. Negev and Samson's[12] multichannel multifrequency antenna achieves high isolation between channels, low sidelobes and an absence of 'grating' lobes. Each major element of this array comprises a rectangular microstrip patch driven at a relatively low frequency in the band of interest. Into these patches are cut two rectangular non-coated 'windows' to allow emergence of the radiation from two inner patches, driven at a relatively high frequency. There are separate feed networks for the low and the higher frequencies and the resulting arrangement is shown in Figure 10.6. Sidelobe levels are typically better than 30 dB down on that of the main beam. Foti et al.[13] report a TACAN (TACtical Air Navigation) antenna based upon a microstrip dipole vertical line array. Their design requires relatively sophisticated feed networks embodying 'thinned' Butler matrices, 'see-saw' phase shifters and a 16-way radial waveguide power combiner. The elevation radiation pattern for their antenna, designed for a centre frequency of 1.087 GHz, indicates a sidelobe level 21.7 dB below that of the main beam.

Apart from strictly microstrip antennas, triplate-type structures have also been reported.[14] The design due to Haneshi et al.[14] overcomes a problem with microstrip antennas intended for use in satellite TV reception—namely

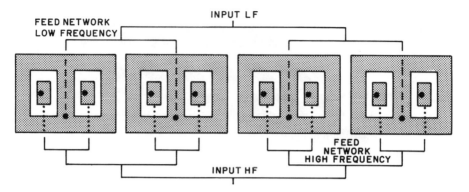

Fig. 10.6. The integrated multichannel microstrip antenna array. (Reprinted, with permission, from Negev and Samson.[12])

the radiation of an undesirable mode. This design is based upon ring-type elements and a typical set of four of these rings, with their triplate feeds, is shown in Figure 10.7. The final antenna exhibits a bandwidth of about 3 per cent centred on 11.85 GHz.

Although all the examples outlined here are directed at frequencies generally below 12 GHz it must be emphasized that there is no fundamental

$$d_1 = d_2 = 0.68 \, \lambda_0$$

F_1, F_2: feed point

$$F_1 = F_2 \exp(-90°)$$

T_a, T_m: quarter transformer

Fig. 10.7. Basic subarray-unit composed of two pairs-elements. (Reprinted, with permission, from Haneishi et al.[14])

frequency limitation and microstrip antennas have been designed and operated at millimeter-wave frequencies.

It should finally be mentioned that microstrip antenna-type structures are not only used in far-field situations (e.g. for communications or radar) but are also employed in applications exploiting their near-field characteristics. Hyperthermia applicators capable of coupling microwave energy into the human body in a highly localised manner represent a good example of a non-communications application.[15] Frequencies ranging from 200 MHz to 1.2 GHz are commonly used and the radiating elements tend to be much simpler than those applicable to the communications or radar fields.

REFERENCES

1. James, J. R., Hall, P. S. and Wood, C., 'Microstrip antenna theory and design', *IEE Electromagnetic Waves Series* No 12, Peter Peregrinus Ltd., London, 1981
2. Bahl, I. J. and Bhartia, P., *'Microstrip Antennas'* Artech House, Inc., 1980
3. Stutzman, L. and Thiele, G., *'Antenna theory and design'*, John Wiley & Son, New York, 1981
4. James, J. R. and Hall, P. S. (Editors), *'Handbook of microstrip antennas'*, Peter Peregrinus Ltd., London, 1989
5. James, J. R. and Wilson, G. J., 'Microstrip antennas and arrays Pt 1—fundamental action and limitations', *IEE J. Microwave, Optics and Acoustics*, **1**, 165–174
6. Jairam, H. 'Novel blade antenna using microstrip patch elements' *Proceedings of the 19th European Microwave Conference*, London, Sept. 1989, 1118–1122
7. Hall, P. S. and Vetterlein, S. J. 'Microstrip patch array with multiple beams' *Proceedings of the 19th European Microwave Conference*, London, Sept. 1989, 343–348
8. Heberling, D., Schmitt, H. J., and Kirchlinde, T. 'Simple feeding technology for stacked microstrip antennas' *Proceedings of the 19th European Microwave Conference*, London, Sept. 1989, 155–160
9. Ostermann-S, R. H., 'A new logarithmic zig-zag microstrip antenna', *SMBO International Microwave Symposium*, Brazil, 24–27 July 1990, 139–144
10. Splitt, G., 'Moment method for electromagnetically and edge-feed coupled microstrip antennas', *SMBO International Microwave Symposium*, Brazil, 24–27 July 1990, 131–138
11. Smith, D. and Jefferson, R. L., 'Dual polarised microstrip antenna design for a polarisation shift keying microwave transponder' *Proceedings of the 19th European Microwave Conference*, London, Sept. 1989, 149–154
12. Negev, M. and Samson, C., 'Integrated multi-channel multi-frequency microstrip antenna array', *Proceedings of the 19th European Microwave Conference*, London, Sept. 1989, 355–360
13. Foti, S. J. *et al.*, 'An intelligent electronically spinning tacan antenna' *Proceedings of the 19th European Microwave Conference*, London, Sept. 1989, 959–965
14. Haneishi, M. *et al.*, 'Radiation properties of circularly polarised triplate-type planar antenna' *Proceedings of the 19th European Microwave Conference*, London, Sept. 1989, 161–166
15. Kobayash, D. *et al.*, 'Multi-microstrip applicator for heating and temperature measurement' *IEEE MTT-S International Microwave Symposium* 1990, Digest of Papers, 533–536

Appendix A Analysis of Parallel-coupled TEM-mode Transmission Lines

The material of Chapter 6 requires some concepts and expressions which are applicable to parallel-coupled TEM-mode transmission lines. Appropriate basic results are developed in this appendix.

The four-port configuration which was introduced as Figure 1.12 of Chapter 1 is required again here and therefore it is repeated as Figure A.1. This configuration can readily be analysed by connecting matched terminations Z_0 at ports 2, 3, and 4, and driving the circuit from a 2-V source having an impedance which is also equal to Z_0. The resulting 'correctly terminated' configuration is shown in Figure A.2. Later on in the analysis it is shown that Z_0 is explicitly related to the even-mode and odd-mode characteristic impedances associated with this coupled-line arrangement.

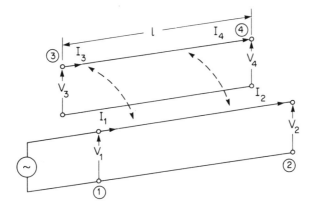

Fig. A.1. Two parallel-coupled transmission lines

Fig. A.2. 'Correctly terminated' parallel-coupled transmission lines

Note that for this analysis, which applies exactly for TEM mode only, we assume the lines to be totally embedded in a uniform dielectric. As a consequence of this, the odd- and even-mode characteristic impedances appear, but the phase velocity has a single value. This means that we can derive useful basic impedance relationships and also the principal frequency dependence of voltage ratios which still apply quite accurately to cases where the propagation is not strictly TEM (especially microstrip). However, the even-mode and odd-mode phase velocities are of different values in non-TEM situations and this leads to some significant refinements of the theory. In particular, non-TEM couplers possess finite (i.e. very imperfect) directivity. These features are discussed in more detail in Chapter 6.

We now proceed with a detailed analysis of the TEM-mode parallel-coupled lines. The arrangement of Figure A.2 may be analysed by first considering excitation under even-mode conditions alone, then odd-mode conditions alone, and finally combining the results. The notation is given in Figure A.3. In this arrangement it is assumed that losses may be neglected so that the propagation coefficient, $\gamma = \alpha + j\beta$, reduces to the phase coefficient $j\beta$ only.

Since the structure is physically symmetrical, it can be seen (Figure A.3) that only *two* circuits actually have to be solved. We will take Figure A.3(a) (ii) and (b) (ii). Then, later, we can use the fact that the total voltages and currents on the original structure are a superposition of the even- and odd-mode solutions as follows:

$$\left.\begin{aligned} V_1 &= V_{1e} + V_{1o} & I_1 &= I_{1e} + I_{1o} \\ V_2 &= V_{2e} + V_{2o} & I_2 &= I_{2e} + I_{2o} \\ V_3 &= V_{1e} - V_{1o} & I_3 &= I_{1e} - I_{1o} \\ V_4 &= V_{2e} - V_{2o} & I_4 &= I_{2e} - I_{2o} \end{aligned}\right\} \tag{A.1}$$

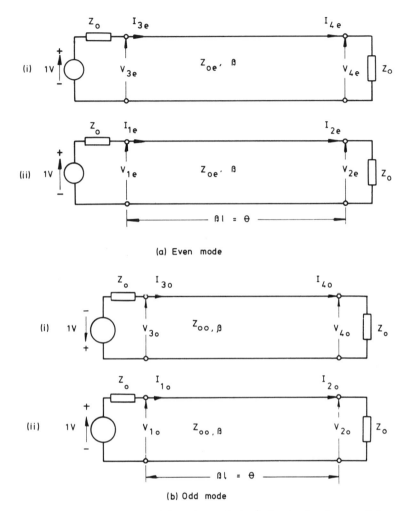

Fig. A.3. Separated even- and odd-mode equivalent circuits for the parallel-coupled lines

Now, the *ABCD* matrix for any transmission line was given as eqn (1.29). This can be used to express the voltage-current relations for lines (a) (ii) and (b) (ii) in Figure A.3 as follows:

$$\begin{bmatrix} V_{1e} \\ I_{1e} \end{bmatrix} = \begin{bmatrix} \cos\theta & jZ_{0e}\sin\theta \\ jY_{0e}\sin\theta & \cos\theta \end{bmatrix} \begin{bmatrix} V_{2e} \\ I_{2e} \end{bmatrix} \tag{A.2}$$

and

$$\begin{bmatrix} V_{1o} \\ I_{1o} \end{bmatrix} = \begin{bmatrix} \cos\theta & jZ_{0o}\sin\theta \\ jY_{0o}\sin\theta & \cos\theta \end{bmatrix} \begin{bmatrix} V_{2o} \\ I_{2o} \end{bmatrix} \tag{A.3}$$

Applying the terminal conditions $V_{2e} = Z_0 I_{2e}$, $V_{2o} = Z_0 I_{2o}$, $V_{1e} + I_{1e} Z_0 = 1$, and $V_{1o} + I_{1o} Z_0 = 1$ results in the voltage and current expressions:

$$V_{1e} = \frac{Z_{0e} Z_0 \cos\theta + j Z_{0e}^2 \sin\theta}{2 Z_{0e} Z_0 \cos\theta + j(Z_{0e}^2 + Z_0^2)\sin\theta} \qquad (A.4)$$

$$I_{1e} = \frac{Z_{0e} \cos\theta + j Z_0 \sin\theta}{2 Z_{0e} Z_0 \cos\theta + j(Z_{0e}^2 + Z_0^2)\sin\theta} \qquad (A.5)$$

$$V_{2e} = \frac{Z_0 Z_{0e}}{2 Z_{0e} Z_0 \cos\theta + j(Z_{0e}^2 + Z_0^2)\sin\theta} \qquad (A.6)$$

$$I_{2e} = \frac{Z_{0e}}{2 Z_{0e} Z_0 \cos\theta + j(Z_{0e}^2 + Z_0^2)\sin\theta)} \qquad (A.7)$$

$$V_{1o} = \frac{Z_{0o} Z_0 \cos\theta + j Z_0^2 \sin\theta}{2 Z_{0o} Z_0 \cos\theta + j(Z_{0o}^2 + Z_0^2)\sin\theta} \qquad (A.8)$$

$$I_{1o} = \frac{Z_{0o} \cos\theta + j Z_0 \sin\theta}{2 Z_{0o} \cos\theta + j(Z_{0o}^2 + Z_0^2)\sin\theta} \qquad (A.9)$$

$$V_{2o} = \frac{Z_0 Z_{0o}}{2 Z_{0o} Z_0 \cos\theta + j(Z_{0o}^2 + Z_0^2)\sin\theta} \qquad (A.10)$$

$$I_{2o} = \frac{Z_{0o}}{2 Z_{0o} Z_0 \cos\theta + j(Z_{0o}^2 + Z_0^2)\sin\theta} \qquad (A.11)$$

The next step consists of an important simplification and amounts to forcing the composite original circuit to become *matched* to the feed lines of characteristic impedance Z_0. Therefore, the total input impedance Z_{in} has to equal Z_0:

$$Z_{in} = Z_0 = \frac{V_1}{I_1} = \frac{V_{1e} + V_{1o}}{I_{1e} + I_{1o}} \qquad (A.12)$$

Substituting eqns (A.4), (A.5), (A.8), and (A.9) into eqn (A.12) gives

$$Z_0 = \frac{Z_0(Z_{0o} D_e + Z_{0e} D_o)\cos\theta + j(Z_{0o}^2 D_e + Z_{0e}^2 D_o)\sin\theta}{(Z_{0o} D_e + Z_{0e} D_o)\cos\theta + j Z_0(D_e + D_o)\sin\theta} \qquad (A.13)$$

where

$$D_e = 2 Z_{0e} Z_0 \cos\theta + j(Z_{0e}^2 + Z_0^2)\sin\theta \qquad (A.14)$$

and

$$D_o = 2 Z_{0o} Z_0 \cos\theta + j(Z_{0o}^2 + Z_0^2)\sin\theta \qquad (A.15)$$

The equality (eqn A.13) will be established if

$$Z_0^2(D_e + D_o) = Z_{0o}^2 D_e + Z_{0e}^2 D_o$$

Substituting, using eqns (A.14) and (A.15) for D_e and D_o, results in

$$2Z_0^3(Z_{0e}+Z_{0o})\cos\theta+j(Z_{0e}^2Z_0^2+Z_{0o}^2Z_0^2+2Z_0^4)\sin\theta$$
$$= 2Z_{0e}Z_{0o}Z_0(Z_{0e}+Z_{0o})\cos\theta+j(Z_{0e}^2Z_0^2+Z_{0o}^2Z_0^2+2Z_{0e}^2Z_{0o}^2)\sin\theta$$

This equality will be satisfied if

$$\boxed{Z_0^2 = Z_{0e}Z_{0o}}$$ (A.16)

Inserting this relationship into eqns (A.4) to (A.11) gives

$$V_{1e} = \frac{Z_0\cos\theta+jZ_{0e}\sin\theta}{2Z_0\cos\theta+j(Z_{0e}+Z_{0o})\sin\theta}$$ (A.17)

$$I_{1e} = \frac{\cos\theta+jZ_{0o}Y_0\sin\theta}{2Z_0\cos\theta+j(Z_{0e}+Z_{0o})\sin\theta}$$ (A.18)

$$V_{2e} = \frac{Z_0}{2Z_0\cos\theta+j(Z_{0e}+Z_{0o})\sin\theta}$$ (A.19)

$$I_{2e} = \frac{1}{2Z_0\cos\theta+j(Z_{0e}+Z_{0o})\sin\theta}$$ (A.20)

$$V_{1o} = \frac{Z_0\cos\theta+jZ_{0o}\sin\theta}{2Z_0\cos\theta+j(Z_{0e}+Z_{0o})\sin\theta}$$ (A.21)

$$I_{1o} = \frac{\cos\theta+jZ_{0e}Y_0\sin\theta}{2Z_0\cos\theta+j(Z_{0e}+Z_{0o})\sin\theta}$$ (A.22)

$$V_{2o} = \frac{Z_0}{2Z_0\cos\theta+j(Z_{0e}+Z_{0o})\sin\theta}$$ (A.23)

$$I_{2o} = \frac{1}{2Z_0\cos\theta+j(Z_{0e}+Z_{0o})\sin\theta}$$ (A.24)

Final expressions for the total (complete structure—Figure A.1) voltages and currents can now be obtained by substituting eqns (A.17) to (A.24) into the group of eqns (A.1):

$$V_1 = V_{1e}+V_{1o} = \frac{2Z_0\cos\theta+j(Z_{0e}+Z_{0o})\sin\theta}{2Z_0\cos\theta+j(Z_{0e}+Z_{0o})\sin\theta} = 1$$ (A.25)

$$I_1 = I_{1e}+I_{1o} = \frac{2\cos\theta+jY_0(Z_{0e}+Z_{0o})\sin\theta}{2Z_0\cos\theta+j(Z_{0e}+Z_{0o})\sin\theta} = \frac{1}{Z_0}$$ (A.26)

$$V_2 = V_{2e}+V_{2o} = \frac{2Z_0}{2Z_0\cos\theta+j(Z_{0e}+Z_{0o})\sin\theta}$$ (A.27)

$$I_2 = I_{2e}+I_{2o} = \frac{2}{2Z_0\cos\theta+j(Z_{0e}+Z_{0o})\sin\theta}$$ (A.28)

$$V_3 = V_{1e} - V_{1o} = \frac{j(Z_{0e} - Z_{0o})\sin\theta}{2Z_0\cos\theta + j(Z_{0e} + Z_{0o})\sin\theta} \tag{A.29}$$

$$I_3 = I_{1e} - I_{1o} = \frac{jY_0(Z_{0o} - Z_{0e})}{2Z_0\cos\theta + j(Z_{0e} + Z_{0o})\sin\theta} \tag{A.30}$$

$$V_4 = V_{2e} - V_{2o} = \frac{0}{2Z_0\cos\theta + j(Z_{0e} + Z_{0o})\sin\theta} = 0 \tag{A.31}$$

$$I_4 = I_{2e} - I_{2o} = \frac{0}{2Z_0\cos\theta + j(Z_{0e} + Z_{0o})\sin\theta} = 0 \tag{A.32}$$

These voltages and currents can be further simplified by defining a coupling parameter C such that

$$\boxed{C = \frac{Z_{0e} - Z_{0o}}{Z_{0e} + Z_{0o}}} \tag{A.33}$$

This definition plus the relationship $Z_0^2 = Z_{0e}Z_{0o}$ (eqn A.16) allows us to write:

From (A.25): $V_1 = 1$ (just the input voltage, as defined) (A.34)

From (A.27): $V_2 = \dfrac{\sqrt{1-C^2}}{\sqrt{1-C^2}\cos\theta + j\,\sin\theta}$ \hfill (A.35)

From (A.29): $V_3 = \dfrac{jC\sin\theta}{\sqrt{1-C^2}\cos\theta + j\,\sin\theta}$ \hfill (A.36)

From (A.31): $V_4 = 0$ \hfill (A.37)

The maximum degree of coupling occurs when the length of the coupled structure is $\lambda_g/4$, i.e. when $\theta = \pi/2$. This condition, set into eqns (A.34) to (A.37) yields:

$$V_1 = 1, \text{ as eqn (A.34)}$$

$$V_2 = -j\sqrt{1-C^2} \tag{A.38}$$

$$V_3 = C \tag{A.39}$$

$$V_4 = 0, \text{ as eqn (A.37)} \tag{A.40}$$

Equation (A.39) gives the mid-band (maximum) extent of coupling C, expressed in terms of impedances by eqn (A.33). Equation (A.38) shows that the *transmitted* voltage V_2 (*not* electromagnetically coupled) depends upon C and is 90° out-of-phase with V_1, and hence also with V_3. This feature gives rise to the names often used for couplers based upon this parallel-coupled line principle:

90° hybrid, or quadrature coupler

(There is zero output on port 4, for a theoretically perfect structure.)

The principal frequency response characteristics, $|V_2(\theta)|$ and $|V_3(\theta)|$ may be obtained from eqns (A.35) and (A.36) respectively. Where we note that the electrical length θ is also given by

$$\theta = \frac{2\pi l}{\lambda_g} \qquad\qquad (A.41)$$

$$= \frac{2\pi f l}{v}$$

Since the physical length l of the structure is fixed and we assume the velocity of propagation v to be constant, eqn (A.41) shows that θ is directly dependent on the frequency f.

In design procedures we shall usually be confronted with the problem of determining the physical dimensions of some proposed parallel-coupled structure. The cross-sectional dimensions, line separation and line diameters or widths, can be determined knowing the even-mode and odd-mode characteristic impedances Z_{0e} and Z_{0o}, where the precise relationships depend upon the proposed structure. Thus we require Z_{0e} and Z_{0o} as functions of Z_0 and C. These relationships may be obtained from eqns (A.33) and (A.16). The coupling factor C is often expressed in decibels. If we call this value C' and remember that we have a voltage ratio, then from eqn (A.33):

$$C' = 20 \log \left| \frac{Z_{0e} - Z_{0o}}{Z_{0e} + Z_{0o}} \right| \, \text{dB} \qquad\qquad (A.42)$$

Combining this with eqn (A.16), $Z_0^2 = Z_{0e} Z_{0o}$, yields the initial design equations:

$$Z_{0e} = Z_0 \sqrt{\frac{1 + 10^{C'/20}}{1 - 10^{C'/20}}} \qquad\qquad (A.43a)$$

and

$$Z_{0o} = Z_0 \sqrt{\frac{1 - 10^{C'/20}}{1 + 10^{C'/20}}} \qquad\qquad (A.43b)$$

The application of these equations is discussed in Chapter 6.

BIBLIOGRAPHY

The book cited below is especially interesting for its treatment of lines, waves, and antennas. Most of the material given in this appendix follows that developed in that book, with some further elaboration regarding the use of the results.

Brown, R. G.: *The Transmission of Electric Energy*, Ronald Press, New York, 1973.

Appendix B External *Q*-factor of a Straight-edged, Open-circuited Microstrip Resonator

The arrangement is shown schematically in Figure B.1.

The reflection coefficient Γ due to either end is given by

$$\Gamma = \frac{Y_0 - Y}{Y_0 + Y} = \frac{(Y_0 - G) - jB}{(Y_0 + G) + jB} \tag{B.1}$$

Rationalizing we obtain

$$\Gamma = \frac{(Y_0 - G)(Y_0 + G) - B^2 - j2BY_0}{(Y_0 + G)^2 + B^2} \tag{B.2}$$

or

$$\Gamma = |\Gamma|\exp(-j\theta) = \rho\exp(-j\theta) \tag{B.3}$$

where ρ and θ are the magnitude and phase angle of the reflection coefficient respectively, given by

$$\rho = \frac{\{(Y_0^2 - G^2 - B^2)^2 + 4B^2 Y_0^2\}^{1/2}}{(Y_0 + G)^2 + B^2} \tag{B.4}$$

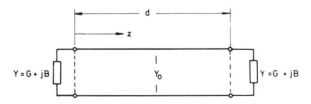

Fig. B.1. Equivalent equal loads $(G+jB)$ at both ends of a straight, symmetrical, open-ended resonator

and

$$\theta = \tan^{-1}\left(\frac{2BY_0}{Y_0^2 - G^2 - B^2}\right) \tag{B.5}$$

These expressions will be required shortly, but first we must find the conditions for the maximum electric field E_{max} in the cavity, since this corresponds to the resonance condition. We now proceed to obtain the conditions for E_{max} in terms of the reflection coefficient Γ, the phase coefficient β, and the resonator length. It is initially necessary to consider the multiple reflections occurring within the resonator (see Figure B.2). The field amplitude E_0 is merely a reference field, so that other relative field strengths may be developed. The phase coefficient $\beta = \omega_0\sqrt{\epsilon_{eff}(f)}/c$. From the fields shown on the diagram (Figure B.2) we may add all the components at distance z to obtain the total field at that location:

$$E_x(z) = E_0\exp(-j\beta z) + \Gamma E_0\exp\{-j\beta(2d-z)\} + \cdots$$
$$+ \Gamma^2 E_0\exp\{-j\beta(2d+z)\} + \cdots \tag{B.6}$$

or

$$E_x = E_0\exp(-j\beta z)\sum_{n=0}^{\infty}\Gamma^{2n}\exp(-j2\beta nd) + \cdots$$

$$+ \Gamma E_0\exp\{-j\beta(2d-z)\}\sum_{n=0}^{\infty}\Gamma^{2n}\exp(-j2\beta nd) \tag{B.7}$$

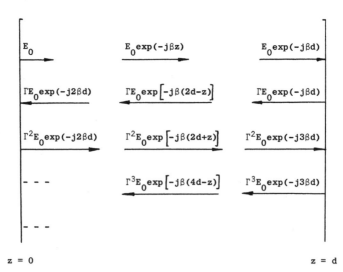

Fig. B.2. Multiple-reflection field components in the straight open-ended resonator of Figure B.1

whence

$$E_x = \frac{E_0[\exp(-j\beta z) + \Gamma\exp\{-j\beta(2d-z)\}]}{1 - \Gamma^2\exp(-j2\beta d)} \tag{B.8}$$

The resonance of the microstrip resonator, which corresponds to a maximum of this electric field E_x, occurs when the denominator (D, say) of eqn (B.8) is a minimum.

Writing magnitudes we require a minimum for

$$|D| = |1 - \Gamma^2\exp(-j2\beta d)|$$

or, inserting the polar form of Γ given earlier,

$$|D| = |1 - \rho^2\exp\{-j2(\beta d + \theta)\}| \tag{B.9}$$

and this is a minimum when

$$2(\beta d + \theta) = 2\pi l \tag{B.10}$$

where $l = 1, 2, 3, ...,$ i.e. an integer.

In the 'ideal' case where Y is either a true open circuit ($G = 0$, $B = 0$) or a perfect short circuit ($G = \infty$, $B = 0$, or $B \neq 0$) we obtain $\theta = 0$ and the almost trivial solution to eqn (B.10), that is $\beta d = \pi l$—whence $d = l\lambda_0/2$.

In our more practical case with finite Y the full resonance condition of eqn (B.10) is required.

From eqn (B.9), the minimum value of $|D|$ is clearly

$$|D|_{min} = 1 - \rho^2 \tag{B.11}$$

Now, as D increases, E_x will exhibit the usual resonant response curve as shown in Figure B.3. It is well known that the Q-factor is then given by

$$Q_e = \frac{\omega_0}{2\Delta\omega} \tag{B.12}$$

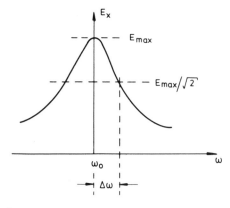

Fig. B.3. Resonator frequency response

where E_{max}, at ω_0, is reduced to $E_{max}/\sqrt{2}$ at $(\omega \pm \Delta\omega)$. Equation (B.12) can also be written in terms of the phase coefficient

$$Q_e = \frac{\beta}{2\Delta\beta} \tag{B.13}$$

where β is the phase coefficient at frequency ω_0. We therefore now require an expression for $\Delta\beta$, in order to finally obtain a formula for Q_e as a function of ρ.

The field $E_{max}/\sqrt{2}$ occurs when $|D|$ is increased to the value $\sqrt{2}(1-\rho^2)$ and the corresponding β and θ values are $(\beta \pm \Delta\beta)$ and $(\theta \pm \Delta\theta)$ respectively. Therefore we can write

$$\Delta(\beta d + \theta) = \Delta\beta d + \Delta\theta$$

$$= \Delta\beta\left(d + \frac{\Delta\theta}{\Delta\beta}\right) \tag{B.14}$$

To find the required quantity $\Delta\beta$, we replace $(\beta d + \theta)$ by $\{(\beta d + \theta) + \Delta(\beta d + \theta)\}$ in the expression for $|D|^2$ and equate the result to $\{\sqrt{2}(1-\rho^2)\}^2 = 2(1-\rho^2)^2$. First, expand, the expression in β, d, θ, using eqns (B.10) and (B.14): $(\beta d + \theta) + \Delta(\beta d + \theta) = \pi l + \Delta\beta(d + \Delta\theta/\Delta\beta)$. Hence,

$$|D| = |1 - \rho^2\exp\{-2j\pi l + \Delta\beta(d + \Delta\theta/\Delta\beta)\}|$$

$$= |1 - \rho^2\exp\{-j2\Delta\beta(d + \Delta\theta/\Delta\beta)\}|$$

$$= |1 - \rho^2\cos2\Delta\beta(d + \Delta\theta/\Delta\beta) + \rho^2 j\sin2\Delta\beta(d + \Delta\theta/\Delta\beta)| \tag{B.15}$$

We may also expand $|D|^2$ in complex product:

$$|D|^2 = DD^*$$

or

$$|D|^2 = \{1 - \rho^2\cos2\Delta\beta(d + \Delta\theta/\Delta\beta)\}^2 + \rho^4\sin^2 2\Delta\beta(d + \Delta\theta/\Delta\beta)$$

$$= (1 - \rho^2)^2 + 2\rho^2\{1 - \cos2\Delta\beta(d + \Delta\theta/\Delta\beta)\} \tag{B.16}$$

This is now equated to $2(1-\rho^2)^2$ to give

$$(1 - \rho^2)^2 = 2\rho^2\{1 - \cos2\Delta\beta(d + \Delta\theta/\Delta\beta)\} \tag{B.17}$$

Since, for moderate or high Q-factor resonators, $\Delta\beta$ must be small ($\Delta\beta \ll \beta$), we may use the small-angle (δy) approximation

$$1 - \cos\delta y \simeq \tfrac{1}{2}(\delta y)^2$$

giving

$$(1 - \rho^2)^2 = 4\rho^2\Delta\beta^2(d + \Delta\theta/\Delta\beta)^2$$

whence

$$\Delta\beta = \frac{1-\rho^2}{2\rho(d+\Delta\theta/\Delta\beta)} \qquad (B.18)$$

Substituting this into eqn (B.13) gives

$$Q_e = \frac{\beta\rho(d+\Delta\theta/\Delta\beta)}{1-\rho^2} \qquad (B.19)$$

The quantity $\Delta\theta/\Delta\beta$ is still required, and we obtain this by differentiating eqn (B.5):

$$\frac{\Delta\theta}{\Delta\beta} = \frac{\Delta}{\Delta\beta}\left[\tan^{-1}\left\{\frac{2BY_0}{(Y_0^2-G^2-B^2)}\right\}\right] \qquad (B.20)$$

The result of this can be written down if we first note that

$$\frac{\mathrm{d}}{\mathrm{d}x}\left(\tan^{-1}\frac{x}{a}\right) = \frac{a}{x^2+a^2}\frac{\mathrm{d}}{\mathrm{d}x}\left(\frac{x}{a}\right)$$

Then

$$\frac{\Delta\theta}{\Delta\beta} = \frac{Y_0^2-G^2-B^2}{(Y_0^2-G^2-B^2)^2+4B^2Y_0^2}\frac{\Delta}{\Delta\beta}\left(\frac{2BY_0}{Y_0^2-G^2-B^2}\right) \qquad (B.21)$$

In order to fully evaluate this we really need to know the functional dependence of B and G on β.

However, because the susceptance is predominantly of a capacitive nature ($B \simeq \omega C$), B is approximately a linear function of β and then

$$\frac{\Delta B}{\Delta\beta} \simeq \frac{\theta}{\beta} \qquad (B.22)$$

Taking this result and substituting, first in eqn (B.21) and finally in eqn (B.19) we find

$$Q_e = \frac{\rho(\beta d+\theta)}{1-\rho^2}$$

that is

$$Q_e = \frac{\pi l\rho}{1-\rho^2} \qquad (B.23)$$

This is the 'classic' result for the external Q-factor where a resonator is equally end-loaded and it is easily evaluated.

The calculation of B and G follow from ρ and θ as expressed by eqns (B.4) and (B.5). G is quite easily determined by combining (B.4) and (B.5):

$$\rho = \frac{2BY_0\sqrt{1+\cot^2\theta}}{(Y_0+G)^2+B^2} \tag{B.24}$$

or

$$G = \left(\frac{2BY_0\sqrt{1+\cot^2\theta}}{\rho} - B^2\right)^{1/2} \tag{B.25}$$

To obtain B we assume that $G \ll Y_0$ and determine the approximate $\tan\theta$ expression from eqn (B.5), yielding

$$\tan\theta \simeq \frac{2BY_0}{Y_0^2-B^2} \tag{B.26}$$

or

$$B = \frac{Y_0(-1\pm\sqrt{1+\tan^2\theta})}{\tan\theta} \tag{B.27}$$

We have to choose the correct sign on the root in this equation, and to decide this we study the effects of the extreme condition $B \ll Y_0$. This gives, from eqn (B.26),

$$\tan\theta \simeq \frac{2B}{Y_0} \simeq \theta$$

whence

$$B \simeq \frac{Y_0}{2}\tan\theta \simeq \frac{Y_0\theta}{2} \tag{B.28}$$

If we set the condition $\theta \ll 1$, $\tan\theta \approx \theta$, in eqn (B.27) we obtain

$$B \simeq \frac{Y_0(-1\pm\sqrt{1+\theta^2})}{\theta} \simeq \frac{Y_0\{-1\pm(1+\theta^2/2)\}}{\theta}$$

This can only equal $Y_0\theta/2$ (as eqn B.28) if the positive root is taken; hence the appropriate expression for B is

$$B = \frac{Y_0(-1+\sqrt{1+\tan^2\theta})}{\tan\theta} \tag{B.29}$$

In a situation where the initial assumption, $G \ll Y_0$, is not valid, an iterative procedure must be carried out in order to find a corrected expression for $\tan\theta$. At each stage up-dated values of B and G are obtained and the procedure is only halted when a satisfactory convergence to limiting values has been obtained.

In the above expressions Y_0, the characteristic admittance of the resonator, is equal to $1/Z_0$ ($Y_0 = 0.02$ S when $Z_0 = 50$ Ω, $Y_0 = 0.1$ S when $Z_0 = 10$ Ω, etc.).

The phase angle θ of the reflection coefficient can be readily obtained from eqn (B.10):

$$\beta d + \theta = \pi l$$

where

$$\beta = \frac{c}{f \sqrt{\epsilon_{\text{eff}}(f)}}$$

$d =$ known length (physically measured)

$l =$ (integer) order of resonance

Appendix C An Outline of Scattering Parameter Theory

C.1 INTRODUCTION

Circuit analysis and design for frequencies ranging from audio to some hundreds of megahertz proceeds on the basis of admittances, impedances and dimensionless factors. Transistors, for example, have input impedance, output impedance, mutual conductance and current gain. At microwave frequencies it is neither practical nor meaningful to measure voltage, current or impedance; at least, not in any direct manner. It is important to appreciate that the circuits are distributed—and so are the voltages and currents (even across a given plane in a transmission line). On the other hand useful quantities such as 'voltage reflection coefficient' and microwave power can be quite readily measured. A very significant set of parameters known as 'scattering parameters' or more briefly S-parameters embody the effect of reflection or through-transmission of microwave power for any network. It matters not whether the network is passive or active, nor whether it comprises one port or one hundred ports; scattering parameters remain very useful.

In order to see why a new parameter set is highly desirable we first review the more 'conventional' sets of parameters.

C.2 PARAMETERS FOR LOWER FREQUENCIES

It is well known that several parameter sets are available for the characterization of a two-port network (i.e. a 'four-terminal' network) as illustrated in Figure C.1.

Although each parameter, in any set, should strictly be written as a function of complex frequency s (or at least as a function of operator $j\omega$)

Fig. C.1.

this fact is generally left understood. For example, in admittance parameters, y_{11} really means $y_{11}(s)$, or $y_{11}(j\omega)$.

A few significant examples are now given of the linear network equations appropriate to some parameter sets.

Impedance parameters, or z-parameters:

$$V_1 = z_{11}I_1 + z_{12}I_2 \tag{C.1a}$$

$$V_2 = z_{21}I_1 + z_{22}I_2 \tag{C.1b}$$

Admittance parameters, or y-parameters:

$$I_1 = y_{11}V_1 + y_{12}V_2 \tag{C.2a}$$

$$I_2 = y_{21}V_1 + y_{22}V_2 \tag{C.2b}$$

Hybrid parameters, or h-parameters:

$$V_1 = h_{11}I_1 + h_{12}V_2 \tag{C.3a}$$

$$I_2 = h_{21}I_1 + h_{22}V_2 \tag{C.3b}$$

Chain parameters, or $ABCD$ parameters:

$$V_1 = AV_2 - BI_2 \tag{C.4a}$$

$$I_1 = CV_2 - DI_2 \tag{C.4b}$$

For active circuits, particularly discrete transistors, the h-parameters have been used quite extensively at much lower frequencies and the y-parameters at frequencies ranging from mf through vhf. Letter subscripts rather than numeric subscripts have been used. For example, considering a bipolar transistor connected in common-emitter (e) configuration the y-parameters become:

$$y_{ie} = y_{11} \qquad y_{re} = y_{12} \qquad y_{fe} = y_{21} \qquad y_{oe} = y_{22}$$

The first letter subscripts give a direct indication of the meaning of each parameter: i means 'input', r means 'reverse' (transfer), f means 'forward' (transfer) and o means 'output'. Thus h_{oe} is the output conductance in common-emitter configuration, y_{re} is the reverse transfer admittance for the common-emitter configuration, etc. The reasons for choosing h-parameters

at lower frequencies and y-parameters at higher frequencies are largely to do with measurement convenience. It can be seen from inspection of eqns (C.3) that each h-parameter may be measured by alternately open-circuiting the input port and short-circuiting the output port, hence producing signal current zero and signal voltage zero respectively. For lower frequency devices, at least, this has proved a satisfactory procedure in view of bias requirements and device resistances. Under radio-frequency conditions, perhaps some hundreds of kilohertz upwards, open circuits become harder to produce and less reliable. The y-parameters have become attractive at such frequencies because, as can be seen from eqns (C.2), each y-parameter is obtainable by alternately short-circuiting output and input ports.

All the parameters are interrelated (any set of N-port parameters can be exactly related to any other set) and the relationships are given in matrix notation in Table C.1.

Table C.1. Matrix parameter interrelationships

	$[z_{mn}]$	$[y_{mn}]$	$[h_{mn}]$	$\begin{bmatrix} A & B \\ C & D \end{bmatrix}$
$[z_{mn}]$	$\begin{matrix} z_{11} & z_{12} \\ z_{21} & z_{22} \end{matrix}$	$\begin{matrix} \dfrac{y_{22}}{\Delta_y} & -\dfrac{y_{12}}{\Delta_y} \\ -\dfrac{y_{21}}{\Delta_y} & \dfrac{y_{11}}{\Delta_y} \end{matrix}$	$\begin{matrix} \dfrac{\Delta_h}{h_{22}} & \dfrac{h_{12}}{h_{22}} \\ -\dfrac{h_{21}}{h_{22}} & \dfrac{1}{h_{22}} \end{matrix}$	$\begin{matrix} \dfrac{A}{C} & \dfrac{\Delta_A}{C} \\ \dfrac{1}{C} & \dfrac{D}{C} \end{matrix}$
$[y_{mn}]$	$\begin{matrix} \dfrac{z_{22}}{\Delta_z} & -\dfrac{z_{12}}{\Delta_z} \\ -\dfrac{z_{21}}{\Delta_z} & \dfrac{z_{11}}{\Delta_z} \end{matrix}$	$\begin{matrix} y_{11} & y_{12} \\ y_{21} & y_{22} \end{matrix}$	$\begin{matrix} \dfrac{1}{h_{11}} & -\dfrac{h_{12}}{h_{11}} \\ \dfrac{h_{21}}{h_{11}} & \dfrac{\Delta h}{h_{11}} \end{matrix}$	$\begin{matrix} \dfrac{D}{B} & -\dfrac{\Delta_A}{B} \\ -\dfrac{1}{B} & \dfrac{A}{B} \end{matrix}$
$[h_{mn}]$	$\begin{matrix} \dfrac{\Delta_z}{z_{22}} & \dfrac{z_{12}}{z_{22}} \\ -\dfrac{z_{21}}{z_{22}} & \dfrac{1}{z_{22}} \end{matrix}$	$\begin{matrix} \dfrac{1}{y_{11}} & -\dfrac{y_{12}}{y_{11}} \\ \dfrac{y_{21}}{y_{11}} & \dfrac{\Delta_y}{y_{11}} \end{matrix}$	$\begin{matrix} h_{11} & h_{12} \\ h_{21} & h_{22} \end{matrix}$	$\begin{matrix} \dfrac{B}{D} & \dfrac{\Delta_A}{D} \\ -\dfrac{1}{D} & \dfrac{C}{D} \end{matrix}$
$\begin{bmatrix} A & B \\ C & D \end{bmatrix}$	$\begin{matrix} \dfrac{z_{11}}{z_{21}} & \dfrac{\Delta_z}{z_{21}} \\ \dfrac{1}{z_{21}} & \dfrac{z_{22}}{z_{21}} \end{matrix}$	$\begin{matrix} -\dfrac{y_{22}}{y_{21}} & -\dfrac{1}{y_{21}} \\ -\dfrac{\Delta_y}{y_{21}} & -\dfrac{y_{11}}{y_{21}} \end{matrix}$	$\begin{matrix} -\dfrac{\Delta_h}{h_{21}} & -\dfrac{h_{11}}{h_{21}} \\ -\dfrac{h_{22}}{h_{21}} & -\dfrac{1}{h_{21}} \end{matrix}$	$\begin{matrix} A & B \\ C & D \end{matrix}$

The quantity Δ is the determinant in each case. For example:

$$\Delta_y = y_{11}y_{22} - y_{12}y_{21}$$

Every parameter, in every set of parameters, is sensitive to change in frequency, bias conditions, or temperature *and this important fact must always be remembered*.

For design at higher frequencies, especially beyond uhf, y-parameters become less useful and a new concept is required. There are two principal reasons for this:

(1) The difficulty in realizing accurate, well-defined, broadband, short-circuits as required for y-parameters.
(2) The considerable chance that the device or amplifier may become unstable and oscillate under short-circuit test.

C.3 SCATTERING (S) PARAMETERS

It will already be well known that the reflection coefficient of a 'one-port network' can quite easily be measured. This quantity (Γ) is simply related to the impedance (Z_L) of such a 'one-port':

$$\Gamma = \frac{Z_L - Z_0}{Z_L + Z_0} \tag{C.5}$$

where Z_0 is the characteristic impedance of the connecting transmission line.

When a network is more complicated, that is one consisting of two or more ports, then quantities beyond reflection coefficients are required to fully characterize its behaviour. We shall be able to make considerable use of two-port networks and these may well include active (i.e. amplifying) elements. One basic way of identifying four parameters is suggested, rather fancifully, in Figure C.2. Here Γ_{in} and Γ_{out} are respectively the input and output port reflection coefficients, μ_F is a forward transmission parameter and μ_R is a reverse transmission parameter. These four 'parameters' have

Fig. C.2.

only been mentioned in passing at this stage. They have not been properly defined and are not in the notation used for S-parameters.

It was previously mentioned that power is another quantity which can be directly measured at microwave frequencies. The S-parameters will lead on explicitly to power relationships (such as amplifier power gain) and, conceptually, these S-parameters are based upon the incident and reflected proportions of power around the network. More precisely, quantities having the dimensions 'square-root of power' are considered and are generally referred to as 'waves'. A transmission parameter (or 'transfer' parameter) through a circuit or a junction is a relative measure of the amplitude and phase of the transmitted wave as compared with the amplitude and phase of the incident wave. In other words, the directly measurable quantities are the amplitudes and phase angles of the waves reflected, or scattered, from a junction or port relative to the incident-wave amplitudes and phase angles. In view of the linearity of the field equations for a variety of important microwave devices, the scattered-wave amplitudes are linearly related to the incident-wave amplitudes. The matrix describing this linear relationship is called the scattering matrix.

Scattering matrices are therefore particularly significant and useful in a microwave transmission environment. Impedance as such cannot, however, be discarded since the characteristic impedance Z_0 of the transmission lines will be vitally and inextricably involved with the elements of these scattering matrices. Other circuit impedances will be shown to be interrelated with certain elements of the matrices or 'scattering parameters' (transistor equivalent circuit impedances and mutual conductance are a good example).

A basic treatment, applicable to circuits where the characteristic impedance is real and positive, is given here.

C.3.1 Scattering parameters for a 2-port network

Filters can be synthesized and microwave transistors can be characterized in terms of the scattering or S-parameters applicable to 2-port networks. These restricted versions of S-parameters are therefore very important and worthy of detailed examination. Figure C.3 shows a 2-port network connected to a load Z_L by means of a length of transmission line and fed from a source, with impedance Z_s, also by means of a length of transmission line.

The transmission lines will be assumed to have identical characteristic impedances Z_0. The equivalent voltages $V_{i(n)}$ are incident voltage waves travelling along the transmission line towards port n, and the equivalent voltages $V_{r(n)}$ are reflected voltage waves travelling along the transmission line away from port n ($n = 1, 2$).

Since each reflected wave must be a linear combination of both a port 1 incident term and a port 2 incident term we have:

$$V_{r1} = x_{11}V_{i1} + x_{12}V_{i2} \qquad \text{(C.6a)}$$

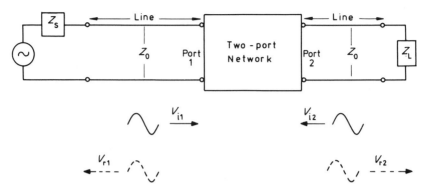

Fig. C.3.

and

$$V_{r2} = x_{21}V_{i1} + x_{22}V_{i2} \tag{C.6b}$$

where x_{nm} depends upon the precise characteristics of the network. We shall not need to dwell upon the precise nature of these x_{nm} coefficients.

Now the incident power wave travelling along the transmission line towards port 1 has a value given by

$$P_{i1} = \frac{V_{i1}^2}{Z_0} \tag{C.7}$$

The square root of this power, conventionally given the symbol a, is simply related to the incident voltage V_{i1} by

$$a_1 = \sqrt{P_{i1}} = \frac{V_{i1}}{\sqrt{Z_0}} \tag{C.8}$$

Similar arguments apply to the power waves reflected from port 1, and incident upon and reflected from port 2 to yield:

$$b_1 = \frac{V_{r1}}{\sqrt{Z_0}} \tag{C.9}$$

$$a_2 = \frac{V_{i2}}{\sqrt{Z_0}} \tag{C.10}$$

and

$$b_2 = \frac{V_{r2}}{\sqrt{Z_0}} \tag{C.11}$$

When both sides of both eqns (C.6a) and (C.6b) are divided through by $\sqrt{Z_0}$ the variables automatically become a_n and b_n and the coefficients are then called the scattering parameters.

The expressions can then be written:

$$b_1 = S_{11}a_1 + S_{12}a_2 \tag{C.12a}$$

$$b_2 = S_{21}a_1 + S_{22}a_2 \tag{C.12b}$$

In matrix form:

$$[b] = [S][a]$$

As a reminder, it has to be appreciated that:

$$a = \sqrt{\text{Incident power}}$$

and

$$b = \sqrt{\text{Reflected power}}$$

The network diagram may now be re-sketched, with these 'root power' variables and scattering parameters included, and the result is Figure C.4.

Using this diagram, and equations (C.12a) and (C.12b), the following definitions can readily be established.

C.3.2 Definitions of 2-port S-parameters

1. S_{11}: If the output line is matched, $Z_L = Z_0$, the load cannot reflect power, $a_2 = 0$ and eqn (C.12a) then yields

$$S_{11} = \left.\frac{b_1}{a_1}\right|_{a_2 = 0} \tag{C.13}$$

By alternately matching the input and output the remaining three parameters are defined as follows:

2. S_{22}:

$$S_{22} = \left.\frac{b_2}{a_2}\right|_{a_1 = 0} \tag{C.14}$$

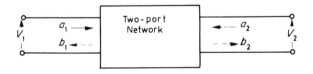

Fig. C.4.

3. S_{21}:

$$S_{21} = \left.\frac{b_2}{a_1}\right|_{a_2=0} \tag{C.15}$$

4. S_{12}:

$$S_{12} = \left.\frac{b_1}{a_2}\right|_{a_1=0} \tag{C.16}$$

Under these definitions, specific physical meanings can be ascribed to the scattering or S-parameters as follows:

1. S_{11} is the input reflection coefficient (i.e. Γ_{11}) when the output is matched.
2. S_{22} is the output reflection coefficient (i.e. Γ_{22}) when the input is matched.
3. S_{21} is the forward transfer (or 'transmission') coefficient with the output matched. (S_{21} *strongly influences the characteristics of a filter; and the gain of an amplifier.*)
4. S_{12} is the reverse transfer (or 'transmission') coefficient with the input matched. (S_{12} becomes particularly important in amplifier stability considerations.)

Whereas S_{21} and S_{12} may be complicated functions of network characteristics, the reflection coefficient parameters S_{11} and S_{22} are readily expressed in terms of driving-point impedances. It is already known that the reflection coefficient Γ for an impedance Z loading a transmission line is:

$$\Gamma = \frac{Z-Z_0}{Z+Z_0} \tag{C.17}$$

Therefore

$$S_{11} = \frac{Z_{11}-Z_0}{Z_{11}+Z_0} \tag{C.18}$$

and

$$S_{22} = \frac{Z_{22}-Z_0}{Z_{22}+Z_0} \tag{C.19}$$

where Z_{11} and Z_{22} are, respectively, the input and output driving-point impedances for the 2-port network when the appropriate opposite port is matched.

The scattering parameters are *small-signal* quantities and therefore to remain consistent with the other small-signal parameters, a lower-case s should strictly be used. However, we have retained the conventional upper-case S here.

Also, it is understood that each parameter is (like any small-signal parameter) a function of frequency, bias conditions, and temperature. The strict frequency function $S(j\omega)$ is left understood and just S is used.

C.3.3 Evaluation of scattering parameters

A simple example of a shunt susceptive element is used to illustrate this (see Figure C.5).

To find S_{11} assume the output side to be matched so that $a_2 = 0$. The reflection coefficient on the input side is then:

$$S_{11} = \frac{Y_0 - Y_{in}}{Y_0 + Y_{in}} \qquad (C.20)$$

But in this case $Y_{in} = Y_0 + jB$, because the (matched) output line shunts the susceptance jB, therefore:

$$S_{11} = \frac{Y_0 - Y_0 - jB}{2Y_0 + jB} = \frac{-jB}{2Y_0 + jB} \qquad (C.21)$$

From symmetry considerations it is clear that:

$$S_{22} = S_{11} \qquad (C.22)$$

Next, S_{12} can be evaluated by finding the transmitted 'voltage' b_2 (i.e. \sqrt{power}) with the output line matched. This is the $a_2 = 0$ condition. For a pure shunt 1-port element we must have:

$$b_2 = S_{21}a_1 \qquad \text{(see also eq (C15) here)}$$

We can combine expressions to obtain

$$\begin{aligned} b_2 &= a_1 + b_1 \\ &= a_1(1 + S_{11}) \end{aligned} \qquad (C.23)$$

Since, with this matched load $a_2 = 0$ situation, we also have

$$S_{21} = 1 + S_{11} = S_{12} \qquad (C.24)$$

and we have already found S_{11} as eqn (C.21). Therefore, after substituting and clearing we obtain:

$$S_{12} = \frac{2Y_0}{2Y_0 + jB} \qquad (C.25)$$

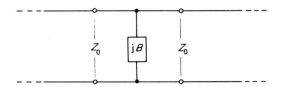

Fig. C.5.

C.3.4 Measurement of scattering parameters

A network analyser system is usually employed for this purpose. This comprises principally:

1. A swept-frequency generator (various bands within ~200 MHz to 18 GHz; higher using waveguide or a special small diameter coaxial system).

2. An oscilloscope-type display capable of facilitating a Smith chart or polar plot.

3. A magnitude/phase angle display unit to provide Bode plots of test circuit response to a base of frequency.

4. A harmonic frequency down-converter. This unit converts the microwave test signal frequencies to lower (usually vhf) frequencies which can be processed by the various display units. It is important to note that the precise frequency interrelationships, the relative phases and the relative amplitudes are all retained at the output of this harmonic frequency converter.

5. The S-parameter test set. This device has two measuring ports so that, when required, S_{11}, S_{12}, S_{21} and S_{22} can all be determined 'at the flick of a switch'.

6. (Optional) Frequency counter. This is actually quite important in filter design and development work, where frequencies are required to be known to accuracies of at least a few megahertz.

An outline block diagram of a suitable system is shown in Figure C.6.

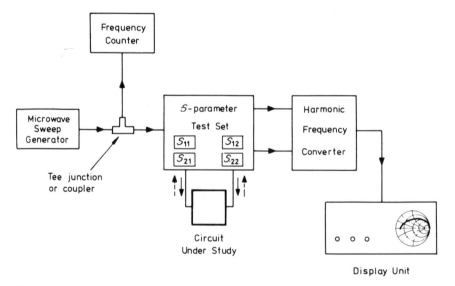

Fig. C.6.

The initial calibration of the system is very important and it is necessary to calibrate the display on the basis of:

(a) Short circuits: to calibrate $S_{11} = S_{22} = 1.0\underline{/180°}$
and
(b) 50 Ω through-line: to interconnect ports 1 and 2 and calibrate $S_{12} = S_{21} = 1.0\underline{/0°}$.

The calibration must be completed over the full bandwidth required and step (b) is normally performed using a 50-Ω standard microstrip line which is inserted into a special jig or 'fixture'. Several manufacturers supply such jig arrangements. The purpose of this special jig is to accept the package shape(s) of the microwave transistor, or other device, whose S-parameters are desired. It is also necessary to provide dc bias, under proper control, for the inserted device under study.

For the measurement of complete circuits, whether active or passive, separate coaxial feed lines (and re-calibration) are often necessary.

C.3.5 Some S-parameter relationships which are particularly useful in filter design

(In this subsection only lossless, reciprocal junctions are treated. Therefore the following results are neither valid for transistors, nor for circuits with ferrite elements.)

Filters to be studied here are passive circuits with lossless and reciprocal properties. For any such networks:

$$S_{12} = S_{21} \qquad (C.26)$$

(This is really a statement directly defining the 'reversibility' or reciprocity of the network.)

Now, if unit power flows into the network, a fraction $|S_{11}|^2$ is reflected and a further fraction $|S_{12}|^2$ is transmitted through the network. This is illustrated by the following power flow breakdown for this case as indicated in Figure C.7.

Thus, we may write:

$$|S_{11}|^2 + |S_{12}|^2 = 1 \qquad (C.27)$$

It is customary to express each quantity in terms of complex conjugate products. Thus eqn (C.27) becomes:

$$S_{11}S_{11}^* + S_{12}S_{12}^* = 1 \qquad (C.28)$$

By similar arguments the following relationships also apply:

$$S_{22}S_{22}^* + S_{12}S_{12}^* = 1 \qquad (C.29)$$

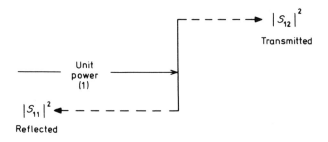

Fig. C.7.

and

$$S_{11}S_{12}^* + S_{12}S_{22}^* = 0 \tag{C.30}$$

These equations can all be derived in a more rigorous mathematical manner. They are concisely expressed in an important general relationship for lossless, reciprocal networks known as the unitary condition:

$$[S][S]^* = 1 \tag{C.31}$$

From eqns (C.28) and (C.29):

$$S_{11} = S_{22}$$

and substituting this in eqn (C.28), again:

$$[S_{12}]^2 = [S_{21}]^2 = 1 - [S_{11}]^2 \tag{C.32}$$

It will be convenient to use this relationship in filter synthesis.

C.3.6 Multiport *S*-parameters

The simple 2-port scattering matrix extends naturally to cover multiport situations. An '*N*-port' network is shown in Figure C.8.

The scattering matrix for this becomes:

$$
\begin{bmatrix} b_1 \\ b_2 \\ \dots \\ b_N \end{bmatrix}
\begin{bmatrix} S_{11} & S_{12} & S_{13} & \dots & S_{1N} \\ S_{21} & S_{22} & S_{23} & \dots & S_{2N} \\ \dots & \dots & \dots & \dots & \dots \\ S_{N1} & S_{N2} & S_{N3} & \dots & S_{NN} \end{bmatrix}
\begin{bmatrix} a_1 \\ a_2 \\ \dots \\ a_N \end{bmatrix}
\tag{C.33}
$$

or, more briefly and concisely,

$$[b] = [S][a] \tag{C.34}$$

3-port scattering matrices are required in microwave circuits involving circulators. It is, however, comparatively rare to have to operate with

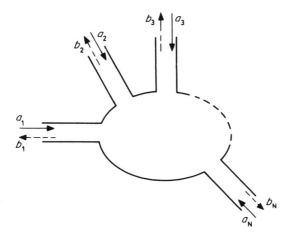

Fig. C.8.

complex networks having four or more ports. The *3-port junction* must be analysed here because the results are essential for reflection amplifier work. Specifically, we must determine the effect of attempting to simultaneously match all 3-ports of a lossless, reciprocal 3-port junction of transmission lines. We start by noting the complete 3-port scattering matrix:

$$[S] = \begin{bmatrix} S_{11} & S_{12} & S_{13} \\ S_{21} & S_{22} & S_{23} \\ S_{31} & S_{32} & S_{33} \end{bmatrix} \tag{C.34}$$

If the junction can be matched then we have

$$S_{11} = S_{22} = S_{33} = 0 \tag{C.35}$$

Also, by reciprocity,

$$S_{21} = S_{12} \qquad S_{31} = S_{13} \qquad S_{32} = S_{23} \tag{C.36}$$

Substituting the conditions described by eqns (C.35) and (C.36) into eqn (C.34) we obtain:

$$[S] = \begin{bmatrix} 0 & S_{12} & S_{13} \\ S_{12} & 0 & S_{23} \\ S_{13} & S_{23} & 0 \end{bmatrix} \tag{C.37}$$

The conditions described under Section C.3.5 are also demanded for this 3-port junction. They may be expressed thus:

$$\sum_{m,n=1}^{3} |S_{mn}|^2 = 1 \tag{C.38}$$

and

$$\sum_{m,n=1}^{3} S_{1n}S_{mn}^* = 0 \qquad (C.39)$$

where both eqns (C.38) and (C.39) can be conveniently combined into the unitary scattering matrix:

$$S\tilde{S}^* = \begin{bmatrix} 1 & 0 & 0 \\ 0 & 1 & 0 \\ 0 & 0 & 1 \end{bmatrix} \qquad (C.40)$$

(another form of this is eqn (C.31) given earlier). We note from eqn (C.37) that only six terms are involved here, and expand the unitary matrix to yield:

$$S_{12}S_{12}^* + S_{13}S_{13}^* = 1 \qquad (C.41)$$

$$S_{12}S_{12}^* + S_{23}S_{23}^* = 1 \qquad (C.42)$$

$$S_{13}S_{13}^* + S_{23}S_{23}^* = 1 \qquad (C.43)$$

We also have

$$S_{13}S_{23}^* = 0 \qquad (C.44a)$$

$$S_{12}S_{13}^* = 0 \qquad (C.44b)$$

$$S_{12}S_{23}^* = 0 \qquad (C.44c)$$

To satisfy these eqns (C.44) we must have:

$$S_{13} = 0 \quad \text{or} \quad S_{23} = 0 \quad \text{or} \quad S_{12} = 0$$

If we initially suppose that $S_{12} = 0$, then eqn (C.41) gives:

$$S_{13}S_{13}^* = 1$$

and eqn (C.42) gives

$$S_{23}S_{23}^* = 1$$

hence

$$S_{13}S_{13}^* + S_{23}S_{23}^* = 2$$

This directly contradicts the necessary result expressed as eqn (C.43), being double the required quantity, and thus an intolerable inconsistency

exists. We therefore conclude that a lossless, *reciprocal* 3-port junction *cannot* be simultaneously matched at all 3-ports. In fact, a special non-reciprocal arrangement is required and this realized in practice by employing a microwave ferrite component called a *circulator*.

C.3.7 Signal-flow-graph techniques and *S*-parameters (a brief introduction)

So-called signal-flow graphs form a logical and convenient means for analysing microwave circuits of virtually any form and complexity, provided only that they can be treated as linear and *S*-parameters are applicable. Typical problems requiring this kind of analysis include measurement error assessment and amplifier gain calculations.

To set up equivalent *S*-parameter signal-flow graphs, each port of the network is given two 'nodes'—one for the incident wave (a_k) and the other for the reflected wave (b_k). Branches, represented as lines between these nodes, carry the appropriate scattering parameters for each particular section of the network. Thus, the input reflection and reverse transfer sections appear as shown in Figure C.9. When the forward transfer and output reflection parameters are included the full flow-graph for the 2-port network appears as shown in Figure C.10.

Fig. C.9.

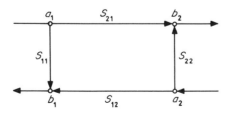

Fig. C.10.

C.4 SCATTERING TRANSFER (or 'T') PARAMETERS

Because the basic S-parameters relate reflected waves to incident waves (both at the output and the input) they are not immediately useful when dealing, as often in communication circuits and systems, with cascaded networks. Instead, there exists a related set of parameters that are based upon the output waves as the independent variables and the input waves as the dependent variables. These are called the scattering transfer (or 'T') parameters and their matrix is, for a two-port network:

$$\begin{bmatrix} b_1 \\ a_1 \end{bmatrix} = \begin{bmatrix} T_{11} & T_{12} \\ T_{21} & T_{22} \end{bmatrix} \begin{bmatrix} a_2 \\ b_2 \end{bmatrix} \tag{C.35}$$

or

$$[\text{i}] = [T][\text{o}] \tag{C.36}$$

where i means 'input' and o means 'output'. Since the same network, the same conditions and the same 'root power' waves are involved, these T-parameters are exactly and fairly simply related to the original S-parameters. This relationship is:

$$\begin{bmatrix} T_{11} & T_{12} \\ T_{21} & T_{22} \end{bmatrix} = \begin{bmatrix} \left(-\dfrac{S_{11}S_{22}-S_{12}S_{21}}{S_{21}} \right) & \left(\dfrac{S_{11}}{S_{21}} \right) \\ -\left(\dfrac{S_{22}}{S_{21}} \right) & \left(\dfrac{1}{S_{21}} \right) \end{bmatrix} \tag{C.37}$$

C.4.1 Cascaded 2-port networks: the utility of T-parameters

Amplifier units are invariably multi-stage, with the individual stages being connected in cascade. Also, each stage will necessarily feature input and output matching networks. If each stage and matching network can be characterized in T-parameters then the overall response can be evaluated readily. This is indicated by considering two networks in cascade as shown in Fig C.11.

Here, the individual T-matrices are:

$$\begin{bmatrix} b_1 \\ a_1 \end{bmatrix} = \begin{bmatrix} T_{11} & T_{12} \\ T_{21} & T_{22} \end{bmatrix} \begin{bmatrix} a_2 \\ b_2 \end{bmatrix} \tag{C.38}$$

and

$$\begin{bmatrix} b_1' \\ a_2' \end{bmatrix} = \begin{bmatrix} T_{11}' & T_{12}' \\ T_{21}' & T_{22}' \end{bmatrix} \begin{bmatrix} a_2' \\ b_2' \end{bmatrix} \tag{C.39}$$

Fig. C.11.

But, because of the uninterrupted flow between the networks,

$$\begin{bmatrix} a_2 \\ b_2 \end{bmatrix} = \begin{bmatrix} b_1' \\ a_1' \end{bmatrix} \tag{C.40}$$

Therefore, eqns (C.38) and (C.39) can be combined to yield the overall result:

$$\begin{bmatrix} b_1 \\ a_1 \end{bmatrix} = \begin{bmatrix} T_{11} & T_{21} \\ T_{21} & T_{22} \end{bmatrix} \begin{bmatrix} T_{11}' & T_{12}' \\ T_{21}' & T_{22}' \end{bmatrix} \begin{bmatrix} a_2' \\ b_2' \end{bmatrix} \tag{C.41}$$

Two aspects are important here:

(a) The matrices must be multiplied in the proper order (matrix multiplication is not commutative).

(b) The proper order to use is that in which the networks are actually connected.

This formulation is well suited to computer-aided design. For a transistor, the S-parameters would have to be measured over the frequency band of interest. This data would then be stored in the computer, after automatic conversion to T-parameters. For a passive network such as a matching network a combination of theoretical and measured data would be exploited (to allow for flexible and interactively introduced design adjustments). Again, this data would usually start as S-parameter information and would be automatically converted to T-parameters.

After matrix multiplication the program would allow reverse conversion to overall S-parameters and computation of overall gain, etc, with, very often, automatic plotting of the response.

Appendix D Overview of Proprietary CAD/CAE Packages for RF and Microwave Applications

D.1 BRIEF HISTORICAL BACKGROUND

Microwave CAD has a history dating back about a quarter of a century. As described in the main text of this book, throughout the 1960s and into the 1970s the operating frequency capabilities of transistors steadily increased and these devices were increasingly connected into thin-film or thick-film circuits to create the early hybrid microwave integrated circuits (MICs).

Over those two decades many organisations developed their own computer-based circuit simulators and optimization routines. The software was usually fairly straightforward—it ran on the (usually timeshared) large mainframe computers of the day and 'design' was anything but interactive. One of the first commercially available microwave simulators was developed by John Trudell in the USA. This was known as 'Magic' and it ran with relatively simple input/output formats and very basic graphics by today's standards. Although Magic played a significant role in several important microwave designs of the early 1970s it ultimately failed commercially due to lack of support for the product. The seeds had been sown, however, to stimulate the microwave industry into an increasing awareness of the competitive necessity for CAD and CAE. Later in the 1970s Les Besser, as President of what was then termed Compact Engineering (USA) introduced a well supported set of microwave CAD tools which were used for many successful module designs resulting in highly saleable products.

Meanwhile for computers the ratio of computer processing power to physical size continued to increase dramatically—interactive 'screen-keyboard' computing took hold naturally and strongly almost everywhere (e.g. with Digital's VAX systems). Industry analysts began to forecast the arrival of the 'one per desk' (OPD) concept in the fairly near future, and even the dawn of the now-powerful OPD CAD/CAE workstation concept.

Later on, Compact Engineering, renamed Compact Software, was acquired by the Communications Satellite Corporation who developed the SuperCompact™ package for use on interactive machines such as Digital's VAXs. SuperCompact™ continues today as a strong product in the market, with its circuit optimization, MIC simulation and production yield prediction facilities. Using stored output data for designs created under this software tool, Autoart™ outputs the mask layouts required for manufacture of MICs and Gas Station™ for MMICs.

In 1984 a start-up North American company, EEsof of California, introduced to the market its Touchstone™ family of microwave and RF CAD/CAE tools for use on personal computer workstations. This product soon proved to be highly competitive with SuperCompact and in 1985 EEsof offered its answer to Autoart™ for artwork preparation, namely MiCAD™.

During the past few years several other companies have entered the market with different types of CAD/CAE packages. In the UK for example British Telecom now offer their TeMCAD™ package. This is currently believed to represent the only UK originated and supported microwave CAD/CAE package available on the open market.

By fall 1987 Hewlett Packard had announced their microwave design system (MDS™—the HP 85150A CAE design system). Now Thomson of France also offer their package known as Esope™, and Caro-Line Habigand (also of France) offer packages they term TOAST™ and ESACAP™.

Most recently EM simulators have become available and these are discussed in Section D.3.

D.2 CIRCUIT-BASED CAE TOOLS

Amongst the most important circuit-based design tools available today we have: Touchstone™ (EEsof), SuperCompact™ (now offered by Communications Consulting Corp.), MDS™ (Hewlett-Packard), Esope™ (Thomson), and TeMCAD™ (British Telecom). Table D.1 provides a comparison of various selected features associated with these packages.

Whilst there is clearly overlap in several areas, each package also offers a specific range of benefits which would appeal to various types of users. For example, Touchstone™, SuperCompact™ and MDS™ may appeal more to product designers working with fundamental design data inputs and

Table D.1. Summary of selected features offered by a range of CAD/CAE microwave engineering software tools

Feature	Super-Compact	Touch-stone	TeMCAD	MDS	Esope
Menu driven input	X	X	X	X	X
Rubilyth/plotter output	X	X	X	X	X
Internal cell libraries	X	X	X	X	X
User-expandable cells				X	X
Suite of device package outlines (1)				X	X
Suite of 'standard' microwave design tools (1)	X	X	X	X	X
Automatic compensation for manufacturing process variations (2)			X	X	
Microstrip design to at least 60 GHz			X		
Components called up by engineering variables			X		
Automated filter programs	X	X	X(3)	X	X
Electronic circuit design (1)			X		
Electronic PCB layout (1)			X		
Microwave active device characterization (4)	X	X	X	X	X
Circuit analysis	X	X			X
Circuit optimization (4)	X	X		X	X
Routines for refining and correcting designs	X	X		X	X
Non-linear simulation	X(5)	X(6)			

Gaps in the table do not necessarily mean current non-availability since each product is under virtually continuous development. This is the best assessment at the time of writing.
(1) Available within libraries.
(2) An important advantage for high-accuracy artwork (e.g. for filters).
(3) Now also includes full synthesis and design of microstrip bandpass filters.
(4) Important for active circuit design (e.g. amplifiers).
(5) Through Microwave Harmonica, first introduced in 1990.
(6) Through Libra, first introduced in 1988.

requiring simulation and optimization, etc. TeMCAD™, on the other hand, would appear to be particularly attractive to the product manufacturing side with its inherent facilities for compensating the effects of process variations, ease of accessing components and availability of design tools for electronics.

All the packages are demonstrably capable of outputting mask layers for MMICs as well as for MICs. Both SuperCompact™ and Hewlett-Packard's MDS™ incorporate TriQuint Semiconductors' MMIC library of foundry design modules within their packages.

Regarding prices, it is noted that at the time of writing EEsof's Libra package is offered at $24 000 and Compact Software's Microwave Harmonica is priced at $28 700 to $34 500 per seat. It is obviously up to the intending user to carefully determine the best value for money to suit his or her particular purpose. EEsof also offers a package it terms Academy, priced at $14 000, which is designed as a framework for analog circuit and system simulation.[1]

In 1990 the first examples of new, relatively low-priced design packages became available. MMICAD, provided by Optotek at a unit price of $3500, is orientated toward the development of designs at MMIC foundries (both analog and digital). SANA, available from its originators Webb Laboratories and priced at $3450, is intended for simulation, analysis, and optimization of high-speed digital circuits and also structures such as antennas. There also exist proprietary packages for simpler, more restrictive design requirements that are useful in their own right. For example, the TeCAD package for designing parallel-coupled bandpass filters in either microstrip or triplate form. This is available from Defence Technology Enterprises in the UK (contactable through the present author). There is also WaveMaker 3.0 supplied by Barnard Microsystems (UK) and Jansen Microwave (Germany).

The majority of these microwave design tools will operate on most popular types of desk-top personal computer workstations, including IBM PCs with sufficient speed and on-board RAM and the Hewlett-Packard 200 and 300 series machines favoured by many microwave companies. Esope, however, is not specified to run on Hewlett Packard computers (again, it is stressed that this was the situation at the time of writing).

Workstations suitable for running these types of packages should, in general, have at least 1 MB on-board RAM, mouse input operation and have at least a 16-inch (preferably 19-inch) colour display.

The competitive advantages of using these CAD/CAE software tools stem from a number of factors, including speed and flexibility of the design operations and repeated access to cell libraries. One very important advantage overall is the potential for reducing the cost of production. The use of these tools on computer workstations avoids the relatively primitive and laborious 'cut 'n' strip' methods and facilitates automatic correction for manufacturing process variations and, potentially at least, raw material (e.g. substrate) parameter variations. With these advantages production yield can also be substantially improved.

These CAD/CAE packages are all steadily progressing to the stage of 'full portation', i.e. the ability to be run on any (or almost any) types of appropriate workstation. Important types of workstation will continue to include, amongst several others: Hewlett-Packard (particularly the 200 or 300 series), IBM PC/XT and PS/2, Sun, HP-Apollo, and MAC2.

All packages incorporate microstrip models and some provide facilities for several alternative types of transmission structures. For example,

TOAST™ (from Caro-Line Habigand of France) offers slotline, coplanar waveguide, inverted microstrip and suspended stripline (i.e. 'triplate').

D.3 EM SIMULATORS

A relatively recent development in CAE concerns electromagnetic (EM) simulators. Conventionally, the accurate and reliable analysis and synthesis of microwave structures and circuits requires a full electromagnetic approach. With regard to metallic waveguides, for example, it has long been understood that only a complete three-dimensional 'picture' of the internal fields and wall currents can provide an engineer with the full information required to use such transmission media in microwave design. Complex circuits using waveguides have, however, been amenable to systematic design by modelling the transmission structures in terms of 'TEM-like' lines. Such models have considerable limitations and do not cope accurately and comprehensively with more complex and combined structures such as ridged waveguide, coaxial-to-waveguide transitions, or feeders to antennas. EM simulators provide the advantage of full electromagnetic analysis leading to a deeper understanding and hence more accurate and reliable design.

Field simulation enables an engineer to analyse the electromagnetic behaviour of a structure whilst avoiding physical construction and measurements. As described in general terms at the beginning of this Appendix, the availability of sufficiently powerful and amenable computers now allows finite-element or similar analyses to be performed in a cost-effective manner. The various transmission media suited to MIC or MMIC realisation are capable of being more completely analysed using EM simulators.

The output obtained from EM simulators is typically available as a full three-dimensional projection display on the workstation screen. This display shows details of all the fields, electric and magnetic, associated with the structure under design. The input requirements include mechanical details regarding the structure as well as driving signals. As an example, consider a coaxial-to-ridged-waveguide transition operated in 'X-band' i.e. 8.2 to 12.4 GHz. The analysis divides the space into 10 439 second-order, tangential-vector finite elements and the matrix has 52 000 by 52 000 elements. The solution requires 48 MB of RAM and takes two hours to run on a Hewlett-Packard 834 computer.

A summary of some EM simulator packages is provided in Table D.2. Sonnett Software's EM package, for example, uses the method of moments and is primarily intended for the design of planar circuits including all electromagnetic effects. The package may be used for microwave or very high-speed digital circuit design. The structures can be of considerable complexity and, where the level of complexity demands, the package may be run on a Cray supercomputer.

Table D.2. Aspects of some commercially available EM simulators for microwave applications

Company	Package Name	Year (1)	Platforms	Price in $ US
EEsof	EM-sim	1989	Apollo, HP, IBM, Sun	not quoted
Hewlett-Packard	High-frequency structure simulator	1990	Apollo, HP, Sun	not quoted
Infolytica Corp	Magnet 2D, 3Dx	1988	Apollo, DEC, HP, IBM, IBM PC, Sun	$14 999+
MacNeal-Schwendler Corp	MSC\EMAS	1989	Apollo, Convex, DEC, HP, IBM, Sun	$1000–$2500 per month (2)
Sonnet Software	EM	1989	Apollo, Cray, HP, IBM, Sun	$30 000

1. The year that this general type of software was first available from the vendor.
2. Available only on a rental basis at these rates.
(Most of the information for this table was obtained from ref. 2.)

New developments in this field are frequently reported in the literature. For example, Rautio[3] describes the implementation of triangular cells in an electromagnetic analysis of arbitrary microstrip circuits. This worker describes a relatively simple solution to this potentially complex problem, using Galerkin analysis and 'roof-top' basis functions.

REFERENCES

1. 'User-friendly CAE', (ed IEEE Spectrum Staff), *IEEE Spectrum*, November 1990, 90–93
2. Cendes, Z. C., 'EM simulators = CAE tools', *IEEE Spectrum*, November 1990, 73–77 (concluded on p 93)
3. Rautio, J. C., 'Triangular cells in an electromagnetic analysis of arbitrary microstrip circuits', 1990 *MTT-S Digest*, 701–704

Index